The colobine monkeys of Africa and Asia have a digestive system unique among primates. This system, analogous to that of ruminants, allows them to exploit foliage and seeds as food, and opens niches that are closed to other mammals. From a Miocene origin, the colobines have radiated to inhabit a wide range of forest and woodland habitats in the Old World tropics, where they are often the most abundant arboreal mammal.

Many long-term studies of colobine ecology and social behaviour have been undertaken, but until now no synthesis of this work has been available. This book provides this synthesis, setting it within the context of colobine evolutionary history, anatomy and physiology. It portrays the adaptive radiation that has resulted from the interaction between the special features of colobines and a set of forest ecosystems that vary in their flora, fauna, chemistry and seasonality.

This book compares and contrasts the colobines with other mammalian groups, particularly ruminants and other primates. It will be of relevance to workers in evolutionary ecology, primatology, physical anthropology and tropical biology, and will remain a significant and useful reference for many years to come.

COLOBINE MONKEYS

COLOBINE MONKEYS:

their ecology, behaviour and evolution

Edited by

A. GLYN DAVIES

Department of Anthropology, University College London

JOHN F. OATES

Department of Anthropology, Hunter College, and the CUNY Graduate School, New York

Published by the Press Syndicate of the University of Cambridge
The Pitt Building, Trumpington Street, Cambridge CB2 1RP
40 West 20th Street, New York, NY 10011–4211, USA
10 Stamford Road, Oakleigh, Melbourne 3166, Australia

First published 1994

Printed in Great Britain at the University Press, Cambridge

A catalogue record for this book is available from the British Library

Library of Congress cataloguing in publication data

Colobine monkeys : their ecology, behaviour, and evolution / edited by
A. Glyn Davies, John F. Oates.
p. cm.
Includes index.
ISBN 0 521 33153 6 (hardback)
1. Cercopithecidae. 2. Cercopithecidae – Adaptation. I. Davies,
A. Glyn. II. Oates, John F., 1944– .
QL737.P93C57 1995
599.8′2 – dc20 94–14232 CIP

ISBN 0 521 33153 6 hardback

WV

Contents

List of contributors *page* viii
Preface xi
1 What are the colobines? *John F. Oates and A. Glyn Davies* 1
2 Evolutionary history of the colobine monkeys in paleoenvironmental perspective *Eric Delson* 11
3 The diversity of living colobines *John F. Oates, A. Glyn Davies and Eric Delson* 45
4 The natural history of African colobines *John F. Oates* 75
5 The ecology of Asian colobines *Elizabeth L. Bennett and A. Glyn Davies* 129
6 Functional morphology of colobine teeth *Peter W. Lucas and Mark F. Teaford* 173
7 Functional anatomy of the gastrointestinal tract *David J. Chivers* 205
8 Digestive physiology *Robin N. B. Kay and A. Glyn Davies* 229
9 Colobine food selection and plant chemistry *Peter G. Waterman and Karen M. Kool* 251
10 Colobine populations *A. Glyn Davies* 285
11 Colobine monkey society *Paul N. Newton and Robin I. M. Dunbar* 311
12 Conclusions: the past, present and future of the colobines *John F. Oates and A. Glyn Davies* 347
References 359
Index 402

Contributors

Elizabeth L. Bennett
World Wide Fund for Nature Malaysia and the Wildlife Conservation Society, 7 Jalan Ridgeway, 93200 Kuching, Sarawak, Malaysia

David J. Chivers
Wildlife Research Group, Department of Anatomy, University of Cambridge, Tennis Court Road, Cambridge CB2 1QS, UK.

A. Glyn Davies
Department of Anthropology, University College London, Gower Street, London WC1E 6BT, UK.

Eric Delson
Department of Anthropology, Lehman College and the Graduate School of the City University of New York, and Department of Vertebrate Paleontology, American Museum of Natural History, Central Park West at 79 Street, New York, NY 10024, USA.

Robin I. M. Dunbar
Department of Anthropology, University College London, Gower Street, London WC1E 6BT, UK.

Robin N. B. Kay
Rowett Research Institute, Bucksburn, Aberdeen AB2 9SB, UK.

Karen M. Kool
School of Biological Science, University of New South Wales, P.O. Box 1, Kensington, New South Wales 2033, Australia.

Peter W. Lucas
Department of Anatomy, University of Hong Kong, Li Shun Fan Building, 5 Sasson Road, Hong Kong.

Paul N. Newton
Wolfson College, Oxford University, Oxford OX2 6UD, UK.

John F. Oates
Department of Anthropology, Hunter College and the Graduate School of the City University of New York, 695 Park Avenue, New York NY 10021, USA.

Mark F. Teaford
Department of Cell Biology and Anatomy, The Johns Hopkins University School of Medicine, 725 North Wolfe Street, Baltimore, Maryland 21205, USA.

Peter G. Waterman
Department of Pharmaceutical Sciences, University of Strathclyde, Royal College, 204 George Street, Glasgow G1 1XW, UK.

Preface

We responded with enthusiasm when Cambridge University Press suggested that we produce a book on the ecology of colobine monkeys. Until now there has been no compilation and synthesis of the extensive but scattered knowledge about the natural history of this major group of African and Asian primates, and we therefore hope that this book will help to fill a large gap in the primatological literature.

Colobines featured in some of the earliest scientific writings on the behaviour and ecology of wild primates, writings such as the perceptive reports by Charles McCann (1928, 1933) on the Indian langurs and Angus Booth (1957) on the olive colobus. But in the great surge of primate field studies that began in the early 1960s, colobines received much less attention than most other primates, especially their close cercopithecine relatives and the apes. Their neglect relative to cercopithecines is probably in part a consequence of some of their ecological differences. Many cercopithecines (such as macaques, baboons and vervet monkeys) flourish in relatively open woodland and savanna habitats, in which they can often be readily observed. In contrast, the great majority of colobine populations live in moist forests, in which observational studies are more difficult, while many populations occur in countries that have presented serious political or logistical obstacles to field research. Further bias in knowledge of the two groups has arisen because of the widespread use of various baboon species as 'models' for the behaviour of early humans.

Gradually, the balance has been redressed, as field workers have expanded their efforts into ever more challenging locales. Even so, only a few species of colobine have been the subject of more than one or two careful field studies; examples are the Hanuman langur, and some of the red and black-and-white colobus of Africa. And because most cercopithecines can be maintained in captivity more readily than most colobines, there is a rich literature on cercopithecine behaviour in captive settings, and from studies of animals introduced into non-native environments (the rhesus macaques of Cayo Santiago, for instance). In contrast, almost all observations on colobine behaviour have relied on field studies.

In an effort to produce a broad review of colobine behavioural ecology with an evolutionary perspective, we have recruited specialists from several fields to help write this book. At the same time, we have tried to keep diverse material integrated by either co-authoring chapters with these specialists, or by working closely with them as editors. Even so, it has not been possible to cover all the topics we would have liked to cover. For example, we have dropped a planned chapter on foraging behaviour, although the topic is touched on in several places. While we have a considerable understanding of some aspects of colobine food selection, knowledge of the monkeys' ranging behaviour in relation to food distribution and availability is still poor, and this impedes a clear understanding of foraging strategy. We hope that this book will encourage further research to fill this and many other gaps in our knowledge.

Colobine monkeys has been a long time in preparation, as our colleagues frequently remind us. The compilation of information from many scattered sources and the preparation of new analyses have been time-consuming processes, not expedited by the two of us being on different continents (and often at quite remote field sites) for much of the time since the project began. Given the many delays there have been, we owe a great debt of gratitude to our co-authors, who have been so patient in waiting for the final product, helpful with their comments, and tolerant of final editorial changes to their work.

Final production of this book owes much to the dedicated work of those who helped to type and edit manuscripts, draw diagrams and compile bibliographies: Lydia Abiero, Jeanette Belen, Cary Anne Cadman, Milcah Karanga, Dennis Milewa, Laura Robinson and Elizabeth Siuvejas. At Cambridge University Press, Robin Pellew encouraged us to begin this project, and Alan Crowden and Tracey Sanderson helped guide it to completion. We are grateful to all of them, as well as to our colleagues who have provided photographs for the book.

In the field, our studies of colobines would have been impossible if we had not had colleagues prepared to share with us the exhilarations and hardships of field work, as well as their thoughts on the ecology of the tropical systems in which we had a joint fascination. Thanks are particularly due to Tom Struhsaker in Uganda, to Steve Green, Karen Minkowski and Rauf Ali in India, to Junaidi (John) Payne in Sabah, and to George Whitesides in Sierra Leone. For their wise counsel over many years, both in and out of the field, JFO is also very grateful to both Peter Jewell and Peter White.

Our close colleague in research at Tiwai Island, Georgina Dasilva (1955–1992), died tragically just as her contributions to our understanding of colobine ecology were being published and were gaining recognition. We remember Georgina's great commitment to her work, her patience, her selfless friendship, and her courage; this book is dedicated to that memory.

Glyn Davies and John Oates, Nairobi and New York, 1993

1

What are the colobines?

JOHN F. OATES and A. GLYN DAVIES

Introduction

The monkeys of Africa and Asia (the 'Old World monkeys') are usually regarded as belonging to one family, the Cercopithecidae, made up of two subfamilies, the Colobinae and the Cercopithecinae (see Chapter 3). Fossil evidence (reviewed by Delson in Chapter 2) suggests that the two subfamilies diverged from a common ancestor in the Miocene; since that time each group has had a complex evolutionary history, involving several distinct radiations. Today, the Colobinae include at least 30 species, which can be grouped into 4–9 genera. These colobine monkeys occupy a wide range of forest and woodland habitats in tropical Africa and in southern and eastern Asia. They vary in size from the West African olive colobus (*Procolobus verus*), which has an adult body weight of around 4 kg, to the proboscis monkey of Borneo (*Nasalis larvatus*) which has adult males weighing over 20 kg (see Table 3.2).

The relative lack of interest in and knowledge of colobines, together with the concentration of field studies on a few species, has tended to obscure the fact that the colobines are an anatomically, ecologically and socially varied group of primates. One aim of this book is to portray this diversity and analyse its pattern.

Anatomy

A diagnostic feature of the colobines, and one that probably has the most profound influence on their ecology, is their stomach morphology. The stomach is large and multi-chambered, and the forestomach (presaccus plus saccus) supports a bacterial microflora with cellulose-digesting abilities (discussed in detail in Chapter 8); in contrast, the cercopithecines have simple stomachs. Colobines are enabled by their stomach to digest plant fibres, and many include large quantities of foliage in their diet. The group as a whole is therefore sometimes labelled as 'the leaf eaters'. Not all colobines have leaf-dominated diets, however; as we shall see in this volume, several species feed very heavily on seeds. Conversely, there are folivorous monkeys that lack enlarged forestomachs, like the howlers of the Neotropics, in which leaf cellulose is partly digested by bacteria in the hindgut; hindgut fermentation is also used to a lesser extent by the colobines (Chapter 8).

Other key anatomical differences between colobines and cercopithecines are also related to food processing. Colobines possess greatly enlarged salivary glands that produce copious saliva. This saliva probably helps to buffer the acidity of the forestomach fluid (keeping it within acceptable limits for the bacterial microflora), and it may contain proline-rich proteins that can form complexes with tannins in the food and therefore nullify the potential interference of tannins with the digestive process (Chapter 9). On the other hand, cercopithecines possess cheek pouches, which colobines lack; cheek pouches have profound influences on feeding strategies, and these in turn influence cercopithecine social behaviour (Murray, 1975).

All cercopithecid molars and premolars are high-crowned, and the molars bilophodont; but, whereas cercopithecine molar teeth have low, rounded cusps, colobine molars have high, pointed cusps linked by ridges (or lophids) and separated by deeper lateral notches (lingual on lowers, buccal on uppers). Kay (1977*a*) has shown that the lophids serve as guides to bring the opposing crowns into tighter occlusion, but do not shear food as was once thought; in colobines, the sharper crests and higher cusps generally fold and slice leafy food better than do the more 'bunodont' cheek teeth of cercopithecines. Colobine incisors are narrower than those of cercopithecines of similar body size, and the lower incisors are entirely sheathed in enamel, rather than being enamel-free on the lingual surface as in cercopithecines, where this arrangement is better adapted to cutting open fruit, which forms a relatively larger percentage of the diet of most cercopithecines.

The face of colobines is generally conservative and rather gibbon-like, with wide interorbital spacing and a shorter snout than in cercopithecines. The nasal bones are short and broad in colobines and the lacrymal bone usually does not extend beyond the inferior orbit margin, although its fossa encroaches onto the maxilla; the long-faced colobines such as *Nasalis larvatus* and some extinct species are cercopithecine-like in these features. The braincase is often high and globular, but brain size is often smaller than in cercopithecines of equal body mass; sagittal crests occur in males of a few species. The mandible of colobines is often deep below M_3 and the gonion (where the ramus meets the corpus) is expanded; the median mental foramen which passes through the symphysis in cercopithecines is absent except in *Procolobus verus* and some fossil taxa. Figure 1.1 illustrates many of these features schematically.

Colobines get their name from the very reduced or absent thumbs of the African species (Greek *kolobos*, mutilated); Asian colobines have small thumbs. Although colobine thumbs are reduced or lacking, their other phalanges are relatively much longer than those of cercopithecines (Fleagle, 1988) (Figure 1.2). In the foot, the hallux *appears* to be shorter than the lateral toes, but relative to body mass the colobine hallux is actually as well developed as the cercopithecine hallux. The non-hallucal digits are considerably longer in colobines than in cercopithecines, however, when body mass is taken into account (Strasser, 1994). In addition, colobine digit-length formulas in the foot are distinctive (Washburn, 1942), with the third and fourth digits projecting considerably beyond the second and fifth (Strasser, 1988, 1994), producing an almost artiodactyl-like foot. Typically, the hindlimbs of colobines are much longer than the forelimbs, while in cercopithecines the forelimbs and hindlimbs tend to be similar in size (Strasser, 1992). Finally, almost all colobines have long tails, whereas cercopithecines display a great variety of tail lengths, including very short tails for a number of species; among the colobines, only the simakobu (*Simias concolor*) of the Mentawai Islands has a very short tail. Most of these postcranial differences between colobines and cercopithecines are probably related to the greater commitment to arboreality and leaping shown by the colobines (Strasser, 1992).

Geographical ecology

The geographical ranges of the colobines and cercopithecines show a great deal of overlap in Africa and Asia, but the present-day distribution of cercopithecines is the more extensive. Colobines are not found today in areas at the periphery of the range of *Macaca*, such as North Africa and Japan, and colobines are also absent in southern Africa and the southwest of the Arabian peninsula where *Papio* and/or *Cercopithecus* occur. As these distribution patterns imply, present-day cercopithecines generally fare better than colobines in open habitats where considerable terrestrial locomotion is required; the majority of living colobines are inhabitants of moist lowland tropical forests and are predominantly arboreal. Fossil evidence (see Chapter 2) suggests, however, that Miocene and Pliocene colobines inhabited a larger geographical area, and that many of these earlier colobines lived in relatively open woodland and were at least partly terrestrial.

Although a typical modern colobine lives in the canopy of a moist tropical

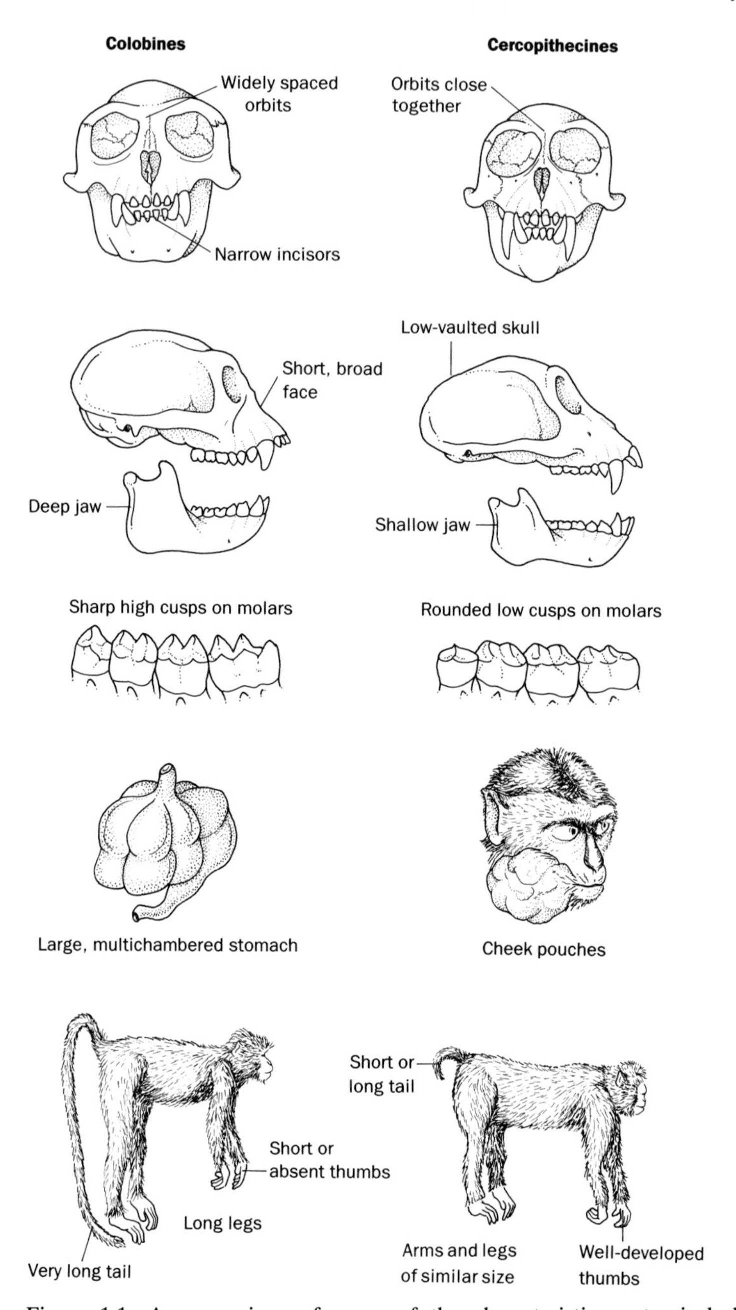

Figure 1.1. A comparison of some of the characteristic anatomical differences between colobines and cercopithecines (from Delson, 1992, with permission of Cambridge University Press).

Figure 1.2. The thumbless condition of the olive colobus (*Procolobus verus*) is shown in the hands of these animals shot by a hunter in Sierra Leone (photograph by J. Oates).

forest, there are many exceptions to this generalization, and some of the most striking of these are reviewed here because they illustrate the diversity of the group (a much fuller treatment of ecology is given in Chapters 4 and 5). Perhaps the most atypical living colobines are the 'odd-nosed' forms, which are restricted to the fringes of the Asian colobine geographical range. They have had a long evolutionary separation from other Asian colobines and they demonstrate some unusual ecological features. Three species in the subgenus *Rhinopithecus* occur in subtropical and temperate montane forests (up to an altitude of 4300 m) north of the Tropic of Cancer, in southern China, south-eastern Tibet and possibly north-eastern India (Groves, 1970; Wolfheim, 1983; Wu, 1993). Two of these species (*Pygathrix* (*R.*) *roxellana* and *P.* (*R.*) *bieti*) are at least partly terrestrial (Tan & Poirier, 1988; Wu, 1993); the third species (*P.* (*R.*) *brelichi*) spends most of its time in trees, but frequently crosses open areas on the ground (Bleisch *et al.*, 1993). The simakobu, which has the build of a macaque, is also partly terrestrial. The simakobu's closest relative, the proboscis monkey, very rarely walks on the ground, but will often swim from one part of its range to another in its mangrove- and peat-swamp habitat in Borneo; proboscis monkeys can swim underwater for up to 20 m (Bennett & Sebastian, 1988).

The Hanuman or grey langur (*Semnopithecus entellus*) of the Indian subcontinent (including Sri Lanka) occurs in open scrub forest, woodland and dry forest in the tropical zone, and in woodlands and forests north of the Tropic of Cancer, including temperate forests in the Himalayan foothills up to an altitude of 4000 m (Hrdy, 1977; Bishop, 1979). In some habitats, Hanuman langurs spend up to 80% of their day on the ground, where much of their food may be obtained (Hrdy, 1977). Members of the genus *Trachypithecus* (a sister group to *Semnopithecus*) spend most of their time in the trees, although animals will travel on the ground between food trees, despite the increased risk of predation they face in these circumstances (e.g., *T. pileatus* in Bangladesh (Stanford, 1991)). The smaller *Presbytis* species, which occupy moist forests in inland south-east Asia, hardly ever travel on the ground, although they will forage in small shrubs close to the forest floor.

In Africa, a major part of the range of one black-and-white colobus species, *Colobus guereza*, is in the highlands of East Africa (where the monkeys inhabit montane forest up to at least 3300 m) and in the dry forest and woodland zone on the periphery of the rain-forest (where the monkeys occur in riverine gallery forests). In gallery forest habitats, *C. guereza* frequently travels on the ground from one wooded patch to another (Oates, 1977*a*), behaviour also reported for *C. vellerosus* in Ghana (Booth, 1956) and for red colobus inhabiting galleries in the dry forest and woodland region of Casamance in Senegal (Gatinot, 1976). However, in closed-canopy forests in other parts of West Africa, such as Sierra Leone, black-and-white colobus and red colobus hardly ever descend to the ground, although the sympatric olive colobus frequently forages in low thickets.

Feeding ecology

Colobines differ from cercopithecines not only in how they digest food, but also in how they procure it. In foraging for fruits and insects, often along the ground, cercopithecines use their more dextrous thumbs to pick up small morsels. Colobines, in contrast, feed mostly above the ground, typically pulling leaves and unripe fruits from branches or climber stems.

Because microbial digestion in an enlarged forestomach allows colobines to exploit fibrous foods that are not readily available to many other arboreal mammals, colobines have proved useful objects for the study of food selection by large 'generalist' herbivores foraging for both foliage and fruit parts in tropical forests. The relationships found between body size, the selection of nutrients and the evasion of digestion inhibitors and toxins is discussed in Chapter 9. The ability of colobine monkeys to process relatively indigestible foliage when other foods are scarce allows them, in some situations, to attain high biomasses (approaching or exceeding 1000 kg/km^2 in certain lowland moist forest habitats). Indeed, colobine biomass is commonly higher than that of all other primates in the community combined (Oates *et al.*, 1990). Such biomass figures suggest that colobines can have a significant impact on forest vegetation, in the short term by suppressing growth (through foliage browsing) or reproduction (through flower consumption or seed predation), and in the long-term by exerting selection on plants through these actions. For example, Struhsaker (1978) has noted that in the Kibale Forest in Uganda, individuals of the tree *Markhamia platycalyx* only set fruit once in a 5-year period. At this time, the *Markhamia* flowered in synchrony, swamping red colobus predation which at other times removed all flowers from the trees; Struhsaker suggests that synchronous flowering in *Markhamia* might have evolved as a response to red colobus feeding. Where colobine biomass is low, as in the inland forests of Borneo, the monkeys probably have a relatively insignificant influence on the vegetation as a whole.

Social behaviour

Many colobines have been observed to live in relatively small social groups containing several adult females (who may often be close relatives) and a single adult male. Such groups are only one part of a variety of social structures (discussed more fully in Chapter 11), ranging from small monogamous groups in some *Presbytis potenziani* and *Simias concolor*, to groups of over 200 with many adult males in *Pygathrix* (*R.*) *roxellana, P.* (*R.*) *brelichi* and *P.* (*R.*) *bieti* (Schaller, 1985; Zhao, 1988; Bleisch *et al.*, 1993), and in some *Colobus angolensis* populations (A. Vedder, personal communication). Compared with cercopithecines, relatively low frequencies of agonistic interactions occur within groups, and dominance hierarchies are less obvious (Struhsaker & Leland, 1987). For instance, Stanford (1991) recorded 36 aggressive interactions among adult *Trachypithecus pileatus* during 75 000 scans of his main study group, and was unable to recognize any meaningful dominance hierarchy among the group members. Similarly, Davies (1984) noted only 16 instances of grooming and 43 aggressive interactions during 94 full days of observing one group of six *Presbytis rubicunda.*

Ripley (1970) and McKenna (1979) have suggested that these differences are founded upon a phylogenetic divergence in feeding strategy. They postulate that leaves are a key food resource for colobines, and that the relative abundance and even dispersion of leaves within tree crowns allow a number of animals to feed together without competition; such feeding habits, it is suggested, lessen the advantage of developing the more highly competitive social strategies common in cercopithecines, which typically exploit less abundant and more clumped food resources, such as fruits. However, studies of colobine feeding (Chapters 4, 5 and 9) have increasingly demonstrated the highly selective nature of colobine leaf-eating and the distinctly patchy distribution of high-quality foliage in forest environments, as well as the prevalence of seed consumption in colobines. In other words, colobine foods may also have a clumped distribution, though whether these are typically more or less clumped than cercopithecince foods has not been measured. Perhaps the presence of cheek pouches in cercopithecines favours a more competitive food-harvesting strategy, even in similar food clumps.

McKenna (1979) has extended his argument about the relationship between feeding strategy and within-group relationships to explain another colobine peculiarity, the prevalence of 'allomothering' (the handling of young infants by animals other than the mother). McKenna regards allomothering as facilitated by the lack of large status differences between females. In those colobine species in which allomothering is common, infants typically have coat coloration which contrasts strongly with that of adults, a phenomenon that is discussed further in Chapter 11.

The coat pattern of young infants has been used as evidence of phylogenetic relationships among Asian colobines (see Chapter 3), and so have the acoustic properties of adult male loud calls. These long, resonating calls are typical of the black-and-white colobus monkeys, of *Semnopithecus, Trachypithecus*, and *Presbytis*. They vary in pitch and tempo across genera and species, but commonly contain at least some low-pitched portion. No long, resonating male call has been described in *Procolobus, Nasalis* or *Pygathrix*, although male proboscis monkeys make loud honks which carry more than 50 m (E. L. Bennett, personal communication).

A further variable feature related to sociality and reproduction in colobines is swelling of the female circumvulvar skin. Such sexual swellings are typical of many papionin cercopithecines, and occur in a few other catarrhines (e.g. *Miopithecus, Pan*), but among the colobines prominent swellings are restricted to *Procolobus* (Pocock, 1936; Hill, 1952) in Africa and to *Simias concolor* in Asia (Tenaza, 1989).

Infant-killing by adult males in wild populations of primates was first carefully documented in Hanuman langurs (Sugiyama, 1965; Hrdy, 1974) and has subsequently been described in several populations of that species and in several other colobines (see Chapter 11). Although infanticide by adult males is therefore sometimes considered a special colobine trait, it is actually widespread among primates and other mammals (Hausfater & Hrdy, 1984; Struhsaker & Leland, 1987).

Prospective

In sum, the colobine monkeys are not only a distinct adaptive radiation, but also one that is considerably more diverse and widespread than is generally recognized. Although they have been relatively neglected by primatologists in comparison with their cercopithecine cousins, the colobines have been an important focus of research on some topics of broad interest to mammalian biologists, such as the significance of infanticide and the consequences of the interaction between herbivores and plant secondary compounds. Subsequent chapters will explore this colobine radiation from an ecological perspective, beginning with a review of colobine evolutionary history and then progressing through an analysis of variation in the structure, distribution, ecology and behaviour of the living species to a synthesis of current knowledge about what produces observed patterns of variation, especially in the density and social organization of populations.

Acknowledgements

We are very grateful to Eric Delson and Elizabeth Strasser for contributing material on skeletal anatomy to this chapter, and we thank them and Elizabeth Bennett for their comments on the chapter as a whole.

2

Evolutionary history of the colobine monkeys in paleoenvironmental perspective

ERIC DELSON

Introduction

Colobines are today rarer and less speciose than the cercopithecine subfamily of the Cercopithecidae, but this has not always been the case. At times in the past, for example in the later Miocene and Pliocene of southern Europe or in the Pliocene of eastern Africa, colobines were more numerous (at least in terms of known species) than cercopithecines. The goal of this chapter is to review the past history of Colobinae and explore the evolutionary 'reasons' for their changing diversity and distribution, especially as they may relate to the changing environments in which these animals lived. A number of new fossil samples are in the course of description and interpretation, so that only preliminary results can be reported here. After a brief review of relevant paleoenvironments and current concepts on the origin of the Cercopithecidae and the Colobinae, the fossil record of colobines is surveyed in turn from Eurasia and from Africa.

Review of Neogene paleoclimate

In order to place the fossil colobines in a more explicitly paleoenvironmental perspective, it is useful first to summarize the history of Cenozoic climate evolution. The Paleocene and Eocene epochs of the early Cenozoic, *c.* 66–34 million years ago (hereafter, Ma = Megennia) were characterized by tropical and subtropical forests across most of the continents (see Berggren & Prothero, 1992). By late in the Eocene, climatic 'deterioration' (drying and cooling) had set in, so that Northern hemisphere habitats favourable to primates were restricted to such areas as southern Europe and the southern United States and Mexico (see Gingerich, 1986). Global temperature fell sharply in the early Oligocene (*c.* 34–32 Ma), and a few million years (Myr) later, world sea level also dropped precipitously (Haq *et al.*, 1987), permitting greater interchange of faunas between Africa, Asia and Europe. Both temperature and sea-level rose slowly in the later Oligocene and early Miocene (26–17 Ma), leading to a period of stability in climate, but then both decreased again in the Late Miocene (10.5–5.3 Ma), a time of generally increasing aridity and spread of open-country environments in the Old World (Bernor, 1983; see Figure 2.1).

Around 6 Ma, plate tectonic movements of Africa with respect to Eurasia led to the closure of the Mediterranean Basin in the west (eastern closure had occurred early in the Miocene; see Thomas, 1985) and the desiccation of the seaway. After nearly 1 Myr of alternating local flooding and salt deposition, the Atlantic Ocean broke through in the Gibraltar region, refilling the Mediterranean and initiating an early Pliocene (5.3–3.6 Ma) interval of greater humidity, at least in Europe. By the later Pliocene (2.5 Ma), the combination of earth-orbit-influenced climatic cycles and surface factors such as Antarctic ice-sheet development, Himalayan orogeny and Panamanian isthmus elevation initiated a long interval of increasing climatic fluctuation (Prentice & Denton, 1989).

Through the Pleistocene, temperature cycles increased in amplitude, decreased in period (from 0.5 to 0.1 Myr) and involved worldwide glaciation and related sea-level fluctuation (Van Couvering & Kukla, 1988*a*,*b*,*c*). Several dozen major cycles have occurred since the mid-Pliocene and the pattern is expected to continue for some time unless disrupted by the effects of human pollution of the atmosphere.

Cercopithecid origins

The ancestry of the Cercopithecidae is now widely agreed to be traceable to early catarrhines similar to the Fayum propliopithecids (Delson, 1975*a*; Szalay & Delson, 1979; Fleagle, 1988). The oldest known members of the family are two genera of the Early to early Middle Miocene (20–15 Ma), *Prohylobates* and *Victoriapithecus*, usually placed in the subfamily Victoriapithecinae (see Figure 2.2). Based on a limited sample of fossils from Maboko Island, Kenya (see Figure 2.8, below), dated to the end of this interval, Delson (1975*a*) discerned two species of *Victoriapithecus*. Each was linked to a different modern subfamily by minor derived features, and this in turn was said to document the subfamilial divergence.

More recently, as a result of new finds, Benefit (1987, 1993), Benefit & McCrossin (1991, 1993) and Fleagle (1988) have discussed further the morphology of these taxa and their role in cercopithecid evolution, coming to somewhat differing conclusions, but agreeing that the victoriapithecines are

Epoch	Subepoch	Age (Ma)	Africa	Europe	Asia
Pleistocene	L	0	Co,Pc, PP		Na,Pg,Pr,PR,Se,Si,Tr
	Middle		Taung (late; Co?)		
		0.5	Andalee (Ethiopia; Co?)		Lang Trang (Vietnam; P?)
			Turkana Basin — Other areas		Xinan (China; PR)
	Early	1	(C?) — Olduvai III (Ce)		Yanjinggou (PR)
			(C?)		Gongwangling (PR)
		1.5	(Rh?,C?) — Olduvai II (Ce)		
			(Rh,Pa,Ce,C?) — Kromdraai (Ce?)		
Pliocene	Late	2	(Rh,Pa,A) — Bolt's Farm (Ce)		Upper Siwaliks (P?)
		∫∫∫ ∫∫∫	(Rh,Pa) — Sterkfontein (Ce)	Kotlovina (Do), Crag (Me)	Shamar, Atsugi (D?)
		3	(Rh,Pa,A) — Makapan (Ce)	Villafranca (Fornace; Me)	Udunga (D?)
	Early		(Rh,Pa, Ce) Hadar(Rh,A)JM90(Pa)	Baraolt, Wölfershm. (Do,Me)	
		4	Laetoli (Pa?,A?)	Perpignan (Do, Me)	
				Malusteni, Voinichevo (Do)	
Miocene	Late	5	Sahabi (Li)	Montpellier (Me)	
			Wadi Natrun (Li)	Baltavar (Me)	Yushe Mahui (P?)
		6		Ditiko (Me)	Hasnot, Domeli (P?)
			Lukeino (C?)		↑
		7	Marceau (C?)		Kotal Kund (P?)
				Pikermi,Veles,Bulgaria (Me)	↓
		8		R. Zouaves, Grebeniki (Me)	Molayan (Me)
			Nakali (Mi?)		Maragha (Me)
		9	Ngeringerowa (Mi)		
		10	Ngorora B (Mi?)	Wissberg (Me?)	
	Middle	∫∫∫ ∫∫∫			
		12			
		14			
			Maboko (Vi)		
		16	Moghara (Py),Loperot (Vi)		
	Early		Zelten (Py), Buluk (Vi?)		
		18	Napak (Vi?)		
		20			

Figure 2.1. Chronological table of localities yielding colobine and victoriapithecine fossils. The taxa present at each site are listed in parentheses, according to the following key. Dates on the left refer to the bottom of the respective line in the table, and in general, sites are dated or estimated to date between the age on their line and the next line up, but they may be in error by one line either way, except that for Kotal Kund, ↓ ↑ indicates an uncertainty of 2–3 Ma about time placement of the locality. The top line ('0 Ma') refers to the present. Note the two changes in spacing of time intervals indicated by the symbol ∫∫∫ ' ∫∫∫. Key to taxonomic abbreviations: A, colobine 'species A'; Ce, *Cercopithecoides*; Co, *Colobus*; C?, African colobine species?; Do, *Dolichopithecus*; D?, ?*Dolichopithecus*; Li, *Libypithecus*; Me, *Mesopithecus*; Mi, *Microcolobus*; Na, *Nasalis*; Pa, *Paracolobus*; Pc, *Procolobus*; PP, *P.* (*Piliocolobus*); Pr, *Presbytis*; P?, Asian colobine species?; Py, *Prohylobates*; Pg, *Pygathrix*; PR, *Pygathrix* (*Rhinopithecus*); Rh, *Rhinocolobus*; Se, *Semnopithecus*; Si, *Simias*; Tr, *Trachypithecus*; Vi, *Victoriapithecus*; ? after any taxon indicates questionable identification. Atsugi is an alternative geographic name for the locality called Nakatsu in the text.

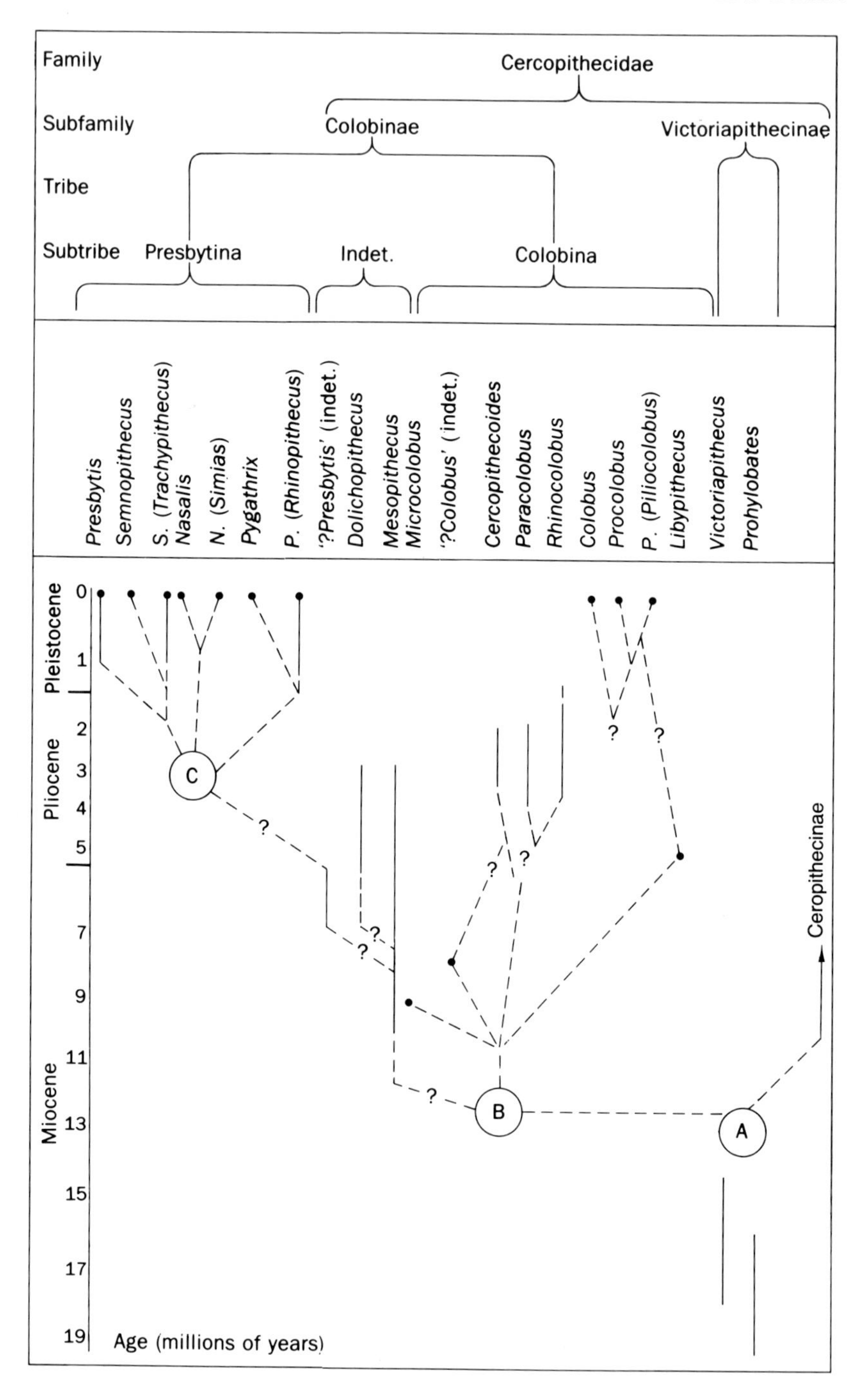
Family
Cercopithecidae
Subfamily
Colobinae
Victoriapithecinae
Tribe
Subtribe
Presbytina
Indet.
Colobina
Presbytis
Semnopithecus
S. (Trachypithecus)
Nasalis
N. (Simias)
Pygathrix
P. (Rhinopithecus)
'?Presbytis' (indet.)
Dolichopithecus
Mesopithecus
Microcolobus
'?Colobus' (indet.)
Cercopithecoides
Paracolobus
Rhinocolobus
Colobus
Procolobus
P. (Piliocolobus)
Libypithecus
Victoriapithecus
Prohylobates
Pleistocene
Pliocene
Miocene
0
1
2
3
4
5
7
9
11
13
15
17
19
C
B
A
?
Ceropithecinae
Age (millions of years)

in general conservative by comparison to all later cercopithecids. Only one species was recognized in the large dental sample of *Victoriapithecus* from Maboko. Harrison's (1989) review of the Maboko cercopithecid postcranial material further substantiated a single species. Thus, Delson's hypothesis appears to be falsified, and it seems likely that the separation between Colobinae and Cercopithecinae occurred after 15 Ma.

The Maboko sample, along with other smaller samples of Early and Middle Miocene cercopithecids, documents the nature of the earliest known Old World monkey adaptations. The typical bilophodont dental pattern was not yet fully developed: a crista obliqua was common on upper molars and the hypocone was relatively isolated; the lower first and second molars sometimes preserved a small midline distal hypoconulid, and the hypolophid mesial to that cusp was incomplete; and the buccal faces of lower cheek teeth flared out even more than in modern cercopithecids (compare Benefit, 1993; and Lucas and Teaford, Chapter 6). Postcranially, *Victoriapithecus* was moderately adapted to terrestrial locomotion, supporting the suggestions of Kay (1977*b*) and others that the differentiation of early Old World monkeys from other catarrhines was at least partially linked to terrestriality.

The environmental background to this differentiation is less clear. Andrews (1981) has suggested that there was a paleoecological 'relay' between hominoids (dominant in the earlier Miocene) and cercopithecoids (more common in the later Miocene and dominant thereafter). In his opinion, cercopithecids became distinct from early catarrhines as a result of adaptation to folivory, although he thought this occurred in an arboreal setting. Colobines further emphasized folivory, as noted below, and Andrews hypothesized that cercopithecines might have 'responded' through development of a greater tolerance for secondary compounds in fruit, an idea for which I find little evidence. Temerin & Cant (1983) expanded on this reasoning to suggest that cercopithecoids concentrated on extracting greater energy from low-quality food sources. They also inferred that early cercopithecids might have shifted from

Figure 2.2. Phylogeny and classification of Colobinae. The genera (and subgenera) of colobines recognized here are listed across the top, with higher-level taxa above; victoriapithecines are also included. The known time range for each (sub) genus is shown as a solid vertical line, with probable ranges dashed vertically and phyletic relationships indicated by oblique dashed lines. Filled circles represent living taxa and fossils known only from single sites. Lettered circles represent the uncertainty of branching sequence or date for major groups: A, the split between Colobinae and Cercopithecinae; B, the split between African and Eurasian Colobinae; and C, the divergence among Asian Colobinae. See footnote 1, page 28, as regards generic names of Asian colobines; subgenera are employed in this figure as it was prepared originally for a different book.

arboreal to more terrestrial travel, although they worried that the lack of fossil cercopithecids in forested paleoenvironments might reflect taphonomic bias. Pickford (1987) further suggested that global climatic shifts also affected the relative distribution of primate groups, finding that a major warming seemed to coincide with the more common occurrence of victoriapithecines in the early Middle Miocene (17–14 Ma) and hypothesizing that this warming was the ultimate cause of the spread of more open habitats at this time.

The paleontological evidence for such habitats is controversial. Various authors have argued that the Early Miocene of eastern Africa was typified by moist forest to woodland, and these sites have yielded almost no cercopithecid fossils (e.g. Andrews *et al.*, 1981 and Pickford, 1983; but perhaps questioned by Bestland & Retallack, 1993). Middle Miocene sites have usually been considered to sample more open habitats, but there has been little agreement about detailed paleoenvironmental parameters at the two most important localities: Maboko Island (with hundreds of *Victoriapithecus* specimens and a large fraction of primates among all mammals) and the slightly younger Fort Ternan site (lacking any cercopithecid and with five primate species accounting for about 1% of all mammals; see Harrison, 1992). Many approaches have been applied to the analysis of Fort Ternan's paleoenvironment (mammal diversity and functional morphology, gastropod index fossils, paleosol and stable isotope studies), but some workers still contend that the site sampled a Serengeti-like grassland (Retallack, 1992), while others argue for a grassy woodland (Cerling *et al.*, 1992) or as Harrison (1992, p. 517) put it: 'predominantly open to closed woodland/bushland, with scattered grassy glades and more densely vegetated areas consisting of woodland–forest mosaics' which seems to include almost everything! There has been less intense study of Maboko (but see, for example, Nesbit-Evans *et al.*, 1981), and Harrison (1992) infers that it was perhaps more densely wooded, or 'closed', than Fort Ternan, with gallery forests but fewer open glades. Yet Pickford's (1983) interpretation of the presence of gastropods was that Maboko received far *less* rainfall than Fort Ternan.

In conclusion, the earliest hominoids (26–16 Ma) probably inhabited montane or lowland evergreen forest, with few if any sympatric cercopithecids. Victoriapithecines became more common between 17–14 Ma, although it is unclear if any sites of this age sampled significantly more open woodland than earlier sites. The presumed trend to more open habitats may not be as strongly marked, at least in preserved aspects of the fossil record, than was once thought (see also below). None the less, it does seem likely that Old World monkeys probably took advantage of such microhabitats more frequently than did contemporaneous or sympatric hominoids, and it is possible that their diversification and biomass increase was delayed until such habitats became more common.

At present, there is no clear evidence of cercopithecids between about 14 and 11 Ma. Szalay & Delson (1979) suggested that the subfamilies had diverged by this time and that the cercopithecine tribes differentiated within or soon after that interval, although the lack of fossils makes this impossible to test. By 11 Ma, colobines appeared not only to have diverged from ancestral cercopithecids but also to have differentiated geographically. In contrast, the oldest known cercopithecine fossils (macaque-like teeth from Algeria) are roughly dated at 8–7 Ma.

Eurasian fossil colobines

Late Miocene

The earliest well-known colobine, *Mesopithecus pentelicus*, was not African but Eurasian. Although mainly recovered over 100 years ago, it is still the best represented, both in number of fossils and range of skeletal parts (Figures 2.3 and 2.4). The largest sample of the species is from Pikermi, near Athens, but small numbers of fossils are known from sites further north in Greek, Bulgarian and Yugoslavian Macedonia, and to the east in Ukraine, Iran and Afghanistan (see Figure 2.5). All of these sites probably date to about 8.5–6 Ma. One upper premolar of a colobine the size of *M. pentelicus* (and referred to that species) was reported by Delson (1973, 1975*b*) from Wissberg (Germany), a site which appears to be 2–3 Myr older than the others; it is possible, however, that the fauna from this locality was mixed, including elements from two distinct time intervals (see Andrews *et al.*, 1994).

Several authors (e.g. Heintz *et al.*, 1981) have suggested that more than one species may be represented in these sites, but that view was not formalized until de Bonis *et al.* (1990) described *M. delsoni* from the Ravin des Zouaves locality in Greek Macedonia. The diagnostic features of this species were said to involve slight differences in size and proportions of M_3 and the mandibular corpus (only three specimens, all lower jaws, are yet known). Zapfe (1991) considered it to be merely a large *M. pentelicus*, and my recent analysis of *t*-tests on a range of dental variables indicates that, while all the *Mesopithecus* specimens from the Macedonian region are somewhat large, the sample is only significantly different from the Pikermi sample in two or three measures. The species *M. delsoni* is thus considered to be merely a local variant of *M. pentelicus* (see Andrews, *et al.*, 1995). In turn, Zapfe (1991) named a new subspecies *M. p. microdon* for a single mandible from

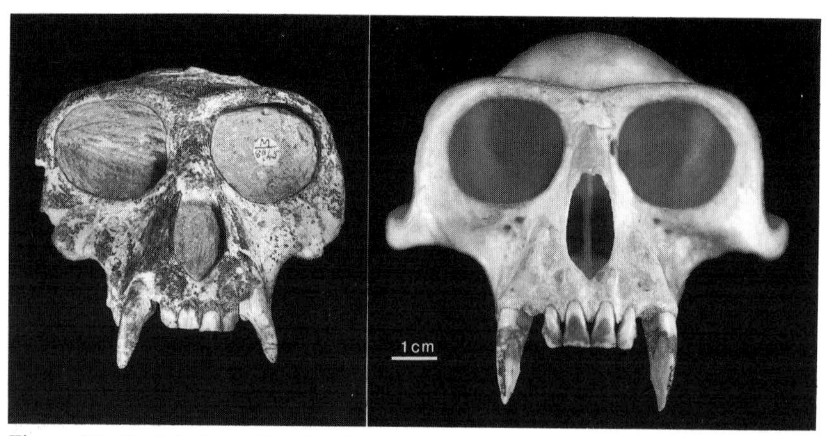

Figure 2.3. Facial view of male *Mesopithecus pentelicus* (left) and *Semnopithecus entellus* (right) at same scale.

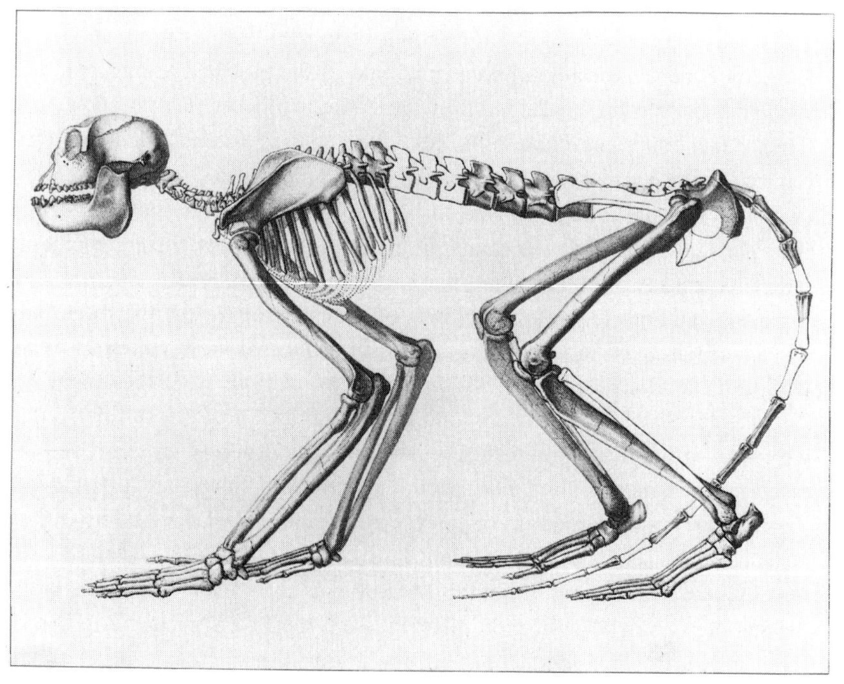

Figure 2.4. Reconstruction of skeleton of *Mesopithecus pentelicus* based on unassociated remains of several individuals; from Gaudry (1862).

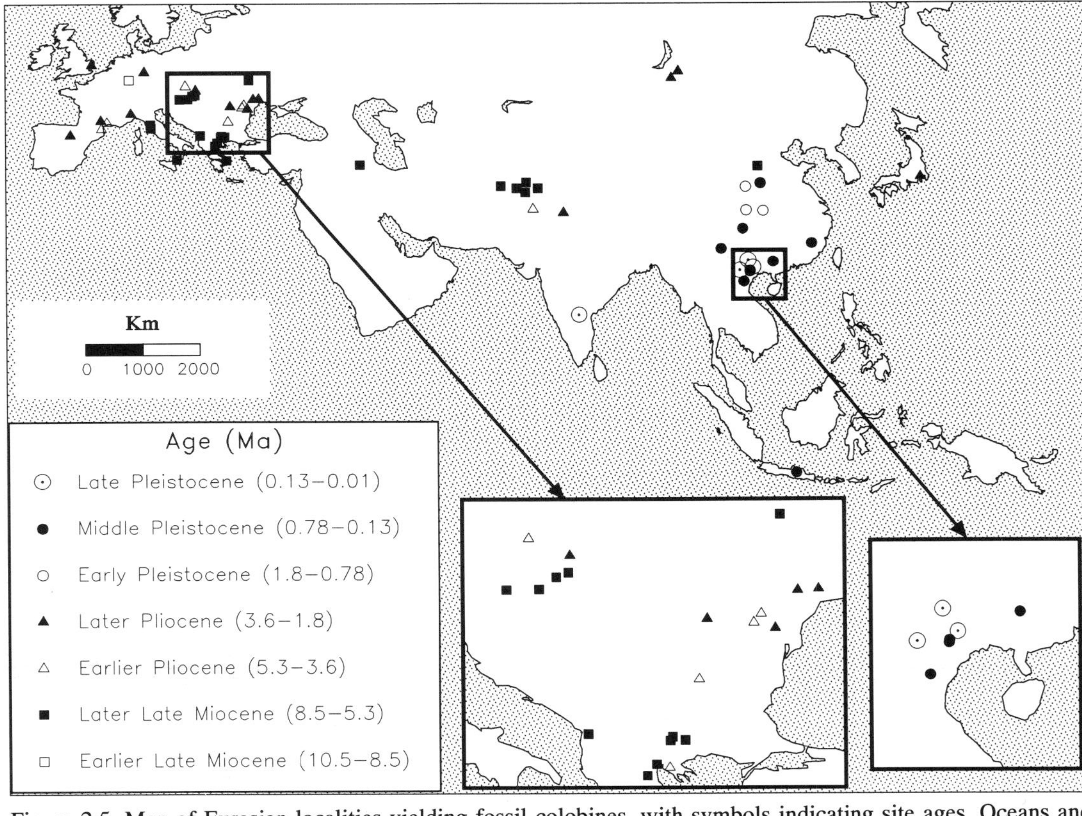

Figure 2.5. Map of Eurasian localities yielding fossil colobines, with symbols indicating site ages. Oceans and major lakes stippled; insets provide enlarged views of central European and south Chinese–Vietnamese regions.

the Chomateri locality near Pikermi, which may be slightly younger than the main horizon, but its distinctions also appear insufficient to warrant a formal name, at least until a larger sample is known. Several jaws and isolated teeth are known from younger Miocene localities, and where their taxonomic allocation is reasonably secure, they are best allocated to *M. pentelicus*.

The warm and relatively moist environments of the Middle Miocene apparently underwent a trend toward drying in the Late Miocene. Broad surveys (e.g. Delson, 1975*b*; Bernor, 1983) and some more detailed regional analyses (e.g. von Orgetta, 1979; Demarcq *et al.*, 1983; Kovar-Eder, 1987; Bernor *et al.*, 1988) allow the delineation of at least the outlines of climatic evolution. Over much of Europe, mixed woodland characterized the Vallesian (10.5–8.5 Ma). This was generally followed by decreased humidity, more open steppes and restriction of forest to gallery bands in the Turolian (8.5–5.3 Ma). In Greece, for example, paleobotanical work has suggested mixed grassland and coniferous forest in the earlier Turolian, while farther north, a Vallesian mesophytic forest may have changed little into the Turolian. Sites in the eastern Mediterranean often have even more arid faunas and floras late in the Turolian. De Bonis *et al.* (1992) have compared faunal lists and body-size distributions of mammals in these localities and a variety of modern and fossil assemblages using a range of techniques. They conclude that Pikermi, Samos and the older Macedonian sites with *Mesopithecus* were broadly similar and relatively open, while the younger Macedonian localities near Ditiko may have been more forested. It would therefore seem that *Mesopithecus* species were limited to habitats with mixed forest (at least gallery-type) and grassland, but not adapted to life in more arid steppes.

A different view of the Greek Miocene paleoenvironment, with broader provincial implications, has been put forward by Solounias & Dawson-Saunders (1988). They studied the morphology of the chewing apparatus in ruminant artiodactyls from Pikermi and the supposedly more arid Samos localities, which are together viewed as a single chronofauna. Somewhat surprisingly, almost all of the extinct bovids were morphologically comparable to modern browsers or intermediate browser–grazers; only one was probably a grazer. Rather than inhabiting an open savanna like those of East Africa, the Samos–Pikermi species would have lived in a forest–woodland similar to those found today in Sichuan (China), south-east Asia or Kanha (India). By reanalysing the pollen data reported from Pikermi by von Orgetta (1979), Solounias & Dawson-Saunders (1988, p. 169) suggested that the local environment might have been 'a warm temperate riparian woodland or forest with mixed evergreen and broad-leaf deciduous elements. This woodland was likely to include substantial undergrowth', presumably herbaceous shrubs.

The woodland on plains and low hills might have been broken up by rivers, lakes and bogs, as well as low-lying meadows. Seasonality in temperature or humidity would probably have been low. In this reconstruction, *Mesopithecus* might have lived in more closed habitats than previously thought.

The morphology of *M. pentelicus* has been discussed by several authors since Gaudry's (1862) revision of the Pikermi fauna, most recently by Szalay & Delson (1979) and Zapfe (1991). The species was comparable in size to *Semnopithecus entellus* (see Figures 2.3 and 2.4) and was at least as terrestrial, as shown especially by studies of the elbow and foot. Strasser & Delson (1987) reported that the proximal cuboid-ectocuneiform facet was absent, as otherwise seen only in African colobines. Delson (1973), however, suggested that *Mesopithecus* was more similar cranially to smaller Asian colobines. The combination of overall postcranial similarity to *Semnopithecus entellus* and the new suggestion of paleoenvironmental similarity of Pikermi to Kanha forest, which might represent a persisting area of the original habitat of *S. entellus*, reinforces the probable ecological equivalence between *Mesopithecus* and that modern langur.

The record of eastern Asian colobines is much less extensive and is only now becoming reasonably well documented. The oldest Asian colobine east of Afghanistan is represented by only a half-dozen partial jaws from the 'Dhok Pathan' zone of the Pakistan Siwaliks, dated by Barry (1987) between 7–5 Ma. These fossils were originally described as *Cercopithecus* or *Macaca* species in the last century, but Simons (1970) and Delson (1975*a*) showed that only a single colobine species was involved. This has been termed ?'*Presbytis*' *sivalensis* to indicate that it is an Asian colobine of uncertain affinities. Only dental morphology is sufficiently preserved to allow comparative study, but colobine teeth are generally quite homogeneous (see Chapter 1), and results to date are inconclusive. There is no clear distinction from (nor linkage with) *Mesopithecus, Semnopithecus* or *Presbytis*. As Barry (1987) has summarized, there are a number of faunal differences between the biogeographic 'provinces' on either side of the Baluchi Range (see also Brunet *et al.*, 1984). The Afghan mandible of *M. pentelicus*, found only 300 km to the west, appears to be typical of its species, but not enough gnathic detail is available from the Siwaliks to determine the identity of *sivalensis*.

Pending the recovery of additional fossils or the identification of new morphological characters, a possible source of information might be provided by a still-occluded mandible and lower face, apparently of a small colobine, from the Siwalik locality of Kotal Kund, near the Hasnot site of most ?*P. sivalensis*. If the mandible can be removed from the maxilla and the teeth correspond to those already known, more precise systematic indications should be forthcoming. Barry (1987) has estimated the age of this specimen as between 10–4 Ma. Barry & Flynn (1990) indicated that climate in the Siwalik region was 'variable' between 12–8 Ma, based on isotopic studies, but with cool episodes between 11–9.5 and 9–8 Ma. It was warmer from 8–6.5 Ma, after which there was marked cooling. It was not clear to Barry & Flynn whether this cooling was related to the local disappearance of primates, but Cerling *et al.* (1993) have documented a strong link between faunal turnover (e.g. hominoids – colobines – no primates) and the development of savannas in the Siwaliks and elsewhere between 7–5 Ma.

Further eastward yet, Miocene colobines (and cercopithecines) were even rarer, as might be expected if they were spreading from a south-west Asian source. Three isolated cercopithecid teeth have recently been recovered from a horizon of apparently latest Miocene age (Mahui Formation) in the Yushe Basin of Shanxi province, China, south-west of Beijing. Two teeth are cercopithecine upper molars, and a third appears to be a colobine M_3. Comparison is in progress with modern and extinct taxa, but this would seem to afford solid documentation of the arrival in China of both subfamilies at least by 6–5 Ma, if not necessarily contemporaneously. These would be the oldest cercopithecines and among the oldest colobines in Asia, roughly equivalent in age to the Siwalik material, and probably younger than the better-dated Afghan or Iranian *Mesopithecus*.

Pliocene and Pleistocene

Following the refilling of the desiccated Mediterranean Basin in the earliest Pliocene, southern Europe became densely forested, with some floral evidence suggesting a nearly monsoonal climate – warm and wet with strong seasonality. Around 4 Ma, this situation again began to deteriorate, with dry seasons (especially summer) dominating the yearly cycle, for example in south-eastern France (Bessedik *et al.*, 1984). After about 2.5 Ma, further cooling (on a global scale) led to local floristic extinctions and reductions in forest extent in southern Europe. Again, detailed paleoenvironmental reconstructions are not yet available for most of eastern Asia, but ongoing work in China should provide clarification. Glacial conditions of fluctuating temperatures and widespread loess deposition were common across most of northern (and central) Eurasia after about 1 Ma, rendering the region inhospitable for colobines.

The European Pliocene was characterized by two further colobine species, sometimes found together and/or in association with a macaque similar or identical to *Macaca sylvanus. Mesopithecus monspessulanus* occurs in Early to Middle Pliocene localities from France into Romania and Ukraine and north to Hungary, Germany and England (and perhaps in the latest Miocene of Northern Greece). Mainly known from mandibles and isolated teeth as well as a few fragmentary limb bones, this species is marginally distinguishable from *M. pentelicus* morphologically, but appears to have been more of a forest dweller. The English specimen, an isolated tooth from the Red Crag of East Anglia, is apparently the youngest record of a European colobine, if it is indeed associated with the other Red Crag taxa. Most of the latter are estimated to date about 2.3 Ma, but Hooker (1989) indicated that some specimens may be reworked from older deposits.

Far more interesting is the large-bodied *Dolichopithecus ruscinensis*, from the latest Miocene to mid-Pliocene. This species is also mainly known from a single sample, that from the 4 Myr-old type Ruscinian at Perpignan, southern France, but other Pliocene specimens come from Spain, Germany, Romania, Ukraine (see Maschenko, 1991, who recognized a second species on what seem to be insufficient grounds) and most recently from Greece (Koufos *et al.*, 1991). An ulna from Pestszentlörinc, Hungary, which is indistinguishable from Perpignan specimens was thought to date to the Late Miocene, but has now been shown to be earlier Pliocene in age (L. Kordos, personal communication). This redating removes a key bit of 'evidence' supporting the hypothesis of the origin of *Dolichopithecus* in apparently forested environments in the Late Miocene of Hungary and Austria. If *Dolichopithecus* is derived from *Mesopithecus* (see below), it is still possible that a peripheral population of the latter underwent reproductive isolation and adapted to life on the forest floor, spreading widely across southern Europe when its prime habitat expanded early in the Pliocene, but no paleontological support for this view remains.

The morphology of *D. ruscinensis* was first discussed by Depéret (1890) and reviewed most recently by Szalay & Delson (1979). Several partial female crania permit a restoration of the skull of this *Rhinopithecus*-sized species (Figure 2.6). The face was relatively long for a colobine, and the interorbital pillar correspondingly narrow, perhaps analogous to *Nasalis larvatus*. Strasser & Delson (1987) noted that, as in the mid-foot of most Asian colobines, the cuboid preserved a small proximal contact facet for the ectocuneiform. Most striking, however, was the degree of terrestrial adaptation as evidenced by aspects of the elbow joint and the short phalanges, features commonly seen in cursorial living cercopithecines. In light of its forested habitat, it is likely that *Dolichopithecus* foraged both terrestrially

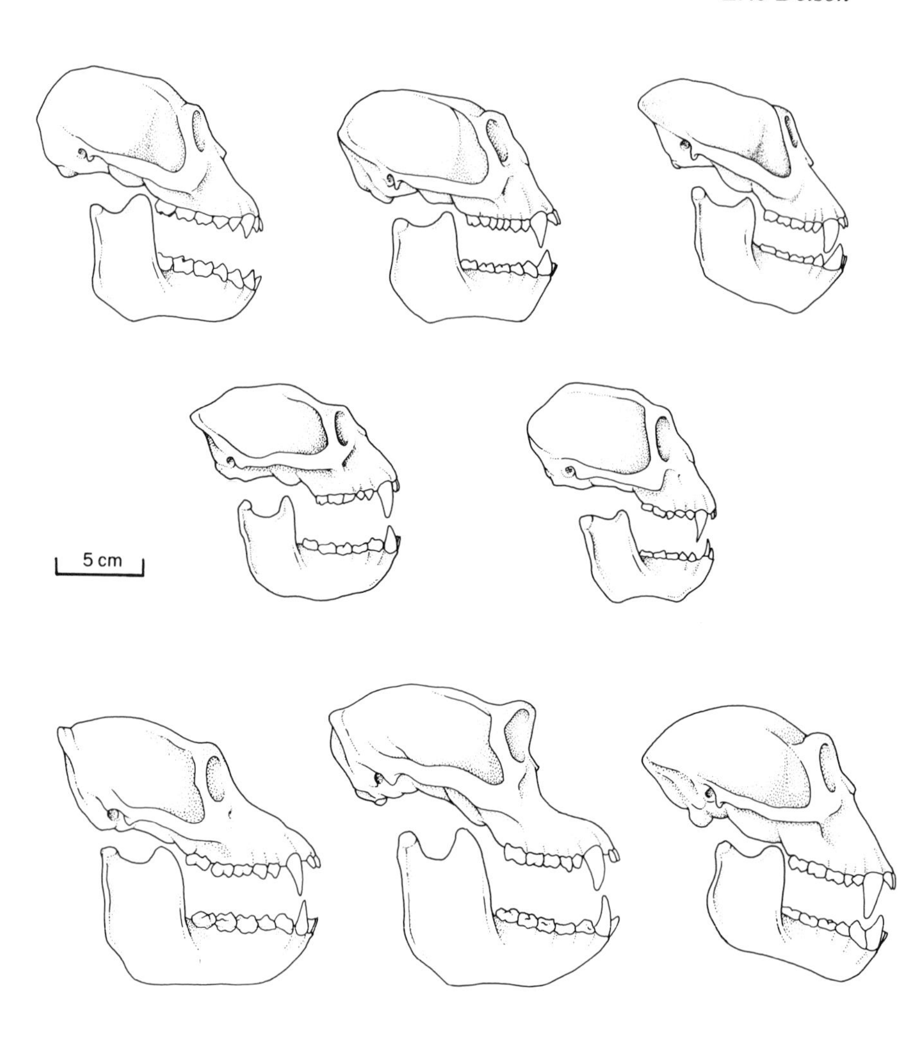

Figure 2.6. Reconstructed crania (male except as noted) in right lateral view of extinct (and selected extant – middle row) colobines: top row *Dolichopithecus ruscinensis* (female), *Mesopithecus pentelicus, Libypithecus markgrafi*; middle row *Colobus polykomos* (F), *Nasalis larvatus* (F); bottom row *Paracolobus chemeroni* (F), *Rhinocolobus turkanaensis* (F), *Cercopithecoides williamsi*. All to same scale; skulls marked (F) redrawn after Fleagle (1988), with permission; others from Szalay & Delson (1979).

and perhaps in the lower arboreal strata; analogies may be suggested with the larger colobines from south-east Asian and, although they have a different diet, the African mandrills, *Papio* (*Mandrillus*) species.

Previously, Delson (1973, 1975*a,b*, 1977; Szalay & Delson, 1979) had suggested that because *Mesopithecus pentelicus* was in all known features relatively 'primitive' by comparison to *D. ruscinensis*, and since it appeared that the two genera were each other's closest relatives, there might have been an actual ancestor-descendant relationship between them (as even earlier indicated by Gaudry, in Depéret, 1890). Strasser's discovery that *Mesopithecus* was more derived than *Dolichopithecus* in having a shorter tarsus suggested closer links between *Mesopithecus* and African colobines (Strasser & Delson, 1987). Another possibility is that the two European species, each represented by a single pedal element, fall within the variation range of some modern Asian colobines (see Strasser, 1988) and are indeed closely related.

With the spread of more open conditions in the Middle Pliocene, *c.* 2.5 Ma, *Dolichopithecus* apparently died out, about the same time as *Mesopithecus*. Macaques, which had been rarer components of the southern European mammal faunas of the earlier Pliocene, became more common, and the larger terrestrial cercopithecine *Paradolichopithecus* also spread between Spain and Central Asia, perhaps ecologically replacing *Dolichopithecus* in some ways before becoming extinct by the Early Pleistocene.

No definite extinct colobines are known from the Indo-Pakistan region after the Miocene, although a problematic population is represented by a mandibular corpus with P_4–M_3 and a fragment with M_3 (see Delson, 1980) from the 'Upper Siwaliks' of India. Barry (1987) has dated these in the range of 3.2–1.7 Ma (Late Pliocene). Originally termed *Semnopithecus palaeindicus* by Lydekker (1884), they were transferred to *Macaca* by Delson (1975*a*), but have recently been suggested to be colobine (Jablonski & Pan, 1988). Pending further study, they are accepted here as cercopithecine (as Jablonski, personal communication, has agreed for at least one of the two). Other than some latest Pleistocene specimens of *Semnopithecus entellus* (see e.g. Badam, 1979), no other colobine fossils are known in south Asia.

One of the most surprising finds of recent years was the occurrence of a rather large colobine in the later Pliocene (*c.* 3–2.5 Ma) of north-eastern Asia. Borissoglebskaya (1981) described as *Presbytis eohanuman* two mandibles, a distal humerus and most of an ulna lacking the distal quarter and part of the olecranon process, from the locality of Shamar, Mongolia. These fossils are quite close in detail to corresponding elements of *Dolichopithecus ruscinensis*, and it is perhaps best to refer the species to that genus (as suggested by Delson, 1988), accepting the important zoogeographical implications of such a move. In this case, there appears to be sufficient morphology preserved to document the close similarity between the two species, which is (as noted above) not yet true for the Miocene Siwalik monkey. More recently, Kalmykov & Maschenko (1992) reported some maxillary dentition of the same species from slightly farther north, at Udunga (near Lake Baikal) in Siberian Russia. They named the new genus *Parapresbytis* for this species, based on features which appear to relate to large body size and incisor development (and wear?) but are not significantly distinct from *Dolichopithecus*.

Even more intriguing is the report by Hasegawa (1993, in a semi-technical journal) of the recovery of a well-preserved partial face of a large colobine in Japan (see Figure 2.7). This fossil, from the Nakatsu locality near Yokohama, could belong to the same species based on tooth size and morphology, and the face is not dissimilar to that expected for a male *Dolichopithecus*. The site is dated to about 2.5 Ma, based on marine invertebrates, and also yielded proboscidean, deer and hippopotamus, as well as marine taxa such as whale, walrus, squid and turtle. Presumably it represents a coastal deposit with some forest indicated by the deer and proboscidean. It would thus appear that some eastern population of *D. ruscinensis* extended north-eastward into

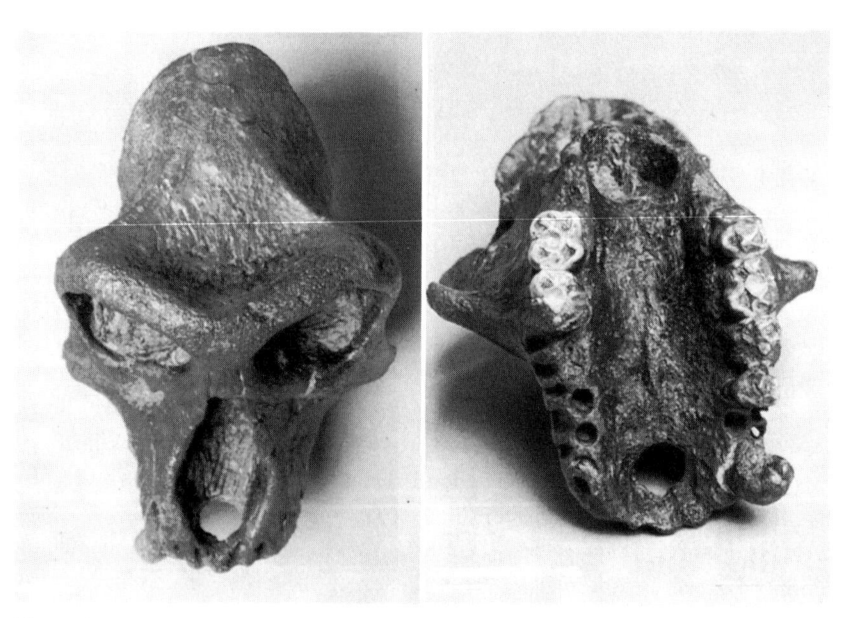

Figure 2.7. Superior (left) and palatal views of new large fossil colobine from Nakatsu, Japan, tentatively referred to *?Dolichopithecus eohanuman*. Courtesy of Dr Y. Hasegawa.

central and then north-eastern Asia during the Pliocene. It is conceivable that movement was in the opposite direction, but that appears unlikely, given the early occurrences of *Dolichopithecus* in the west and the lack of other colobines in the eastern Asian Pliocene, where sites are not uncommon.

A variety of colobines are known in the Pleistocene of China as well. *Pygathrix* (*Rhinopithecus*) was first reported from the early Middle Pleistocene fissure fillings of Yanjinggou (Yenchingkou, Wan County), in Sichuan province. Matthew & Granger (1923) named the sub-adult skull and partial jaws *Rhinopithecus tingianus*, Colbert & Hooijer (1953) identified the population as a subspecies of *R. roxellana*, while Groves (1970) allocated them to *Pygathrix (R.) brelichi*. An adult cranium, crushed laterally but quite complete (Figure 2.8), has recently been recovered from apparently Middle Pleistocene sediments in Henan province, well north of the current range of any 'golden monkey' (Gu & Hu, 1991). This cranium is comparable in size and facial length to both the Sichuan fossils and the living *P. (R.) roxellana* and has been referred to *R. r. tingianus*.

Jablonski and colleagues (Jablonski & Gu, 1988, 1991; Jablonski & Pan, 1988; Gu & Jablonski, 1989) have suggested that a crushed mandible (Figure 2.8) and other jaws from the Gongwangling locality at Lantian may also be colobine. Lantian is the source of a partial cranium of *Homo erectus*, and the fauna was described in some detail by Hu & Qi (1978). Recent paleomagnetic studies demonstrate a Matuyama age (> 0.78 Ma) for these fossils, perhaps close to 1.2 Ma (An & Ho, 1989). Hu & Qi named the taxon *Megamacaca lantianensis* for the Gongwangling cercopithecid fossils, with the mandible as holotype. The name apparently referred to the great depth of the corpus, as the teeth are not distinctly larger than those of *Macaca anderssoni* (= *M. robusta*) from Zhoukoudian and other sites; part of this depth may be due to crushing and plastic deformation. Jablonski & Gu (1991) referred this species to *P.* (*Rhinopithecus*), reducing *Megamacaca* to a synonym. Its northern location, similar to that of the Henan cranium noted above, was at the time probably covered by warm and moist forest, as documented by both faunal and floral remains. However, An & Ho (1989) noted that the fossiliferous horizon at Gongwangling was a thick silty loess with carbonate concretions, which indicated a dry and cold glacial climate; they suggested that local tectonic uplift might have isolated the region from the northern faunal province, and this does seem to be a satisfactory resolution of the conflicting climatic indications.

Fragmentary remains probably referable to *P.* (*Rhinopithecus*) are known from a variety of southern Chinese cave sites, such as Lingyian cave, in Liujiang county, Guanxi province; most are probably of later Pleistocene age.

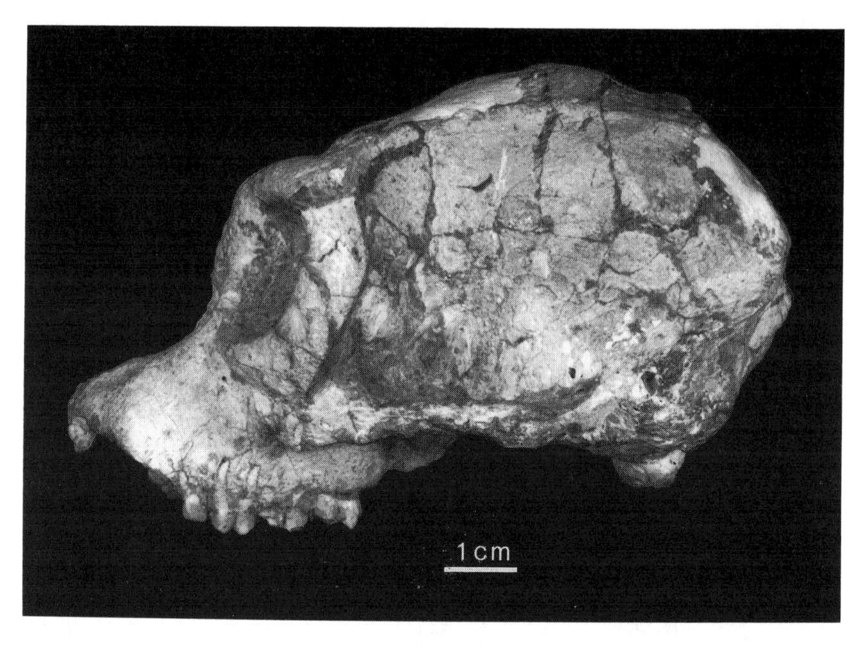

As indicated by Pan & Jablonski (1987), no fossil remains of smaller colobines (*Presbytis* or *Trachypithecus*)[1] have yet been reported from China.

The reverse situation holds for south-east Asia, where fossils have been reported for the two common living genera, but not the 'odd-nosed' colobines, *Nasalis* and *Simias*[1]. Hooijer (1962) described specimens allocated to several species of both *Presbytis* and *Trachypithecus* from Middle Pleistocene caves in Java, and Holocene (late Late Pleistocene) sites in both Java (only the latter genus) and Sumatra. Niah Cave on Borneo has also yielded both genera, in deposits ranging between 40 000 and 5000 years ago. Kahlke (1973) has mentioned *Trachypithecus* fossils from several Middle Pleistocene caves farther north, in Vietnam, and recent work there may lead to publication of further details (Ciochon *et al.*, 1990). In terms of paleoenvironments, Indonesia is the best known, but the details are still unclear for all of south-east Asia.

De Vos and colleagues have recently reviewed the collections of mammalian fossils made by Dubois in the early 1890s, concluding that new correlations are required among the various faunal units at Trinil, Sangiran and

[1] Although I prefer to rank *Trachypithecus* and *Simias* as subgenera of *Semnopithecus* and *Nasalis*, respectively, in the interests of consistency throughout this volume, I have agreed to rank them here as full genera; see discussion in Chapter 3, page 57.

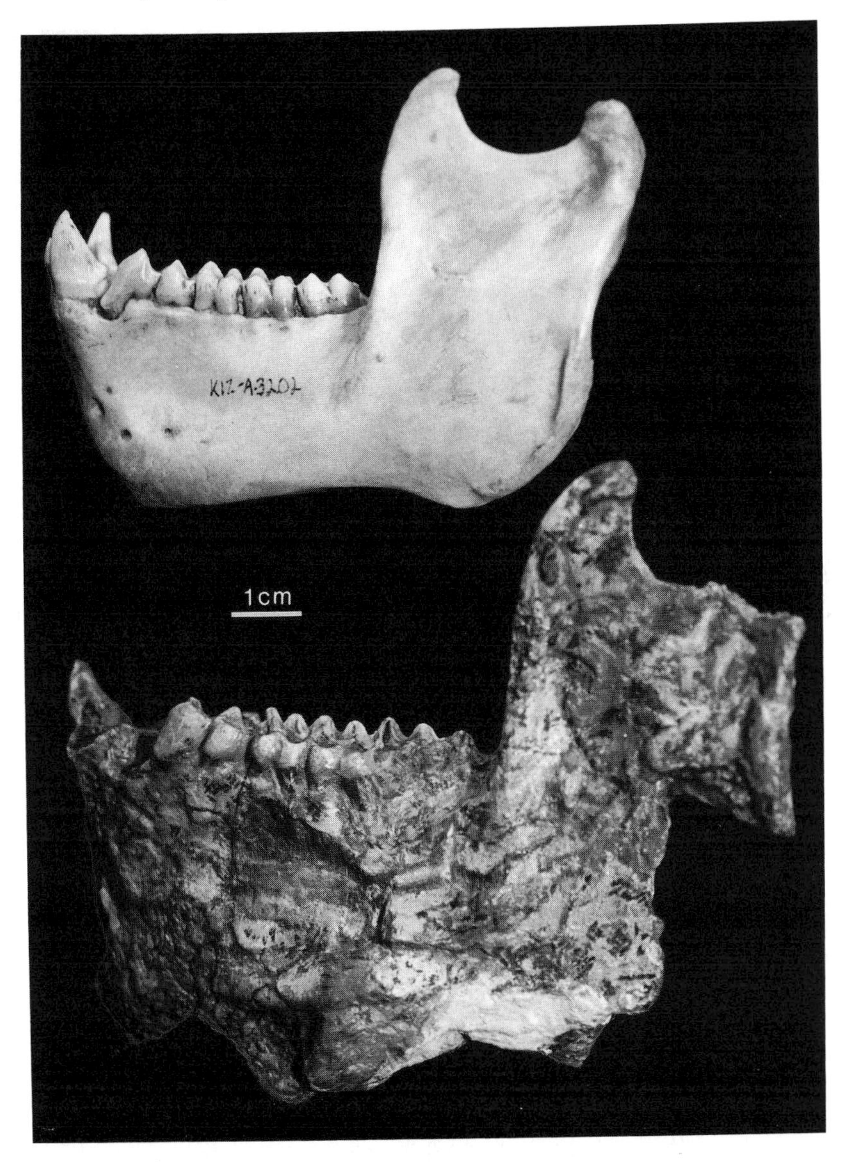

Figure 2.8. Left lateral views of Chinese *Pygathrix (Rhinopithecus)*: left, *P.* (*R.*) *roxellana ?tingianus* cranium from Henan Province; right, mandibles of modern *P.* (*R.*) *brelichi* above *P.* (*R.*) *lantianensis* from Gongwangling (last two courtesy of N. G. Jablonski).

elsewhere. De Vos (1989) summarized some of this work and argued that the main Trinil collection was indeed from a single broad Middle Pleistocene horizon. From these fossils, de Vos suggested that: (1) the low number of species indicates an isolation of the island from the mainland (more species might be expected if later forms were mixed in, as had been suggested by others); and (2) the presence of several bovids might imply an open woodland landscape, and in turn a colder climate. Pope (1988), however, noted the absence in Indochina and Indonesia of any indisputable open-country mammals such as camelids, giraffids or equids. He suggested that the number of specimens upon which de Vos and colleagues based their interpretation of open woodland habitats in Indonesia was too small for definite determination of paleoenvironment, and argued that the region was basically forested throughout the Pleistocene.

African fossil colobines

Miocene

As opposed to the record of apparent slow diversification of colobines in Eurasia to a modern peak, the pattern in Africa was one of increase into the Pliocene and then reduction (at least at the genus level) to the present. This is presumably tied to regional climatic change and perhaps the rise of competition from cercopithecines, among other factors. Unfortunately, the Late Miocene is not well known in Africa, even in the eastern zone. Hill (1988; Hill *et al.*, 1985, 1991) has discussed the Baringo Basin sequence in central Kenya, which extends discontinuously from Middle Miocene through Middle Pleistocene. A flora from early in this sequence (about 12 Ma) indicates forested conditions, based on distribution of identified taxa and the common occurrence of entire margins and acuminate tips. By about 10 Ma, equids and other grassland mammals (such as an increasing variety of bovids) appear in the record. Hill *et al.* (1985) suggested that the modern African open-country fauna was well developed by the end of the Miocene (5–6 Ma). Northern Africa was perhaps less densely forested than the east, with development of the Sahara as a zoogeographic barrier during the Late Miocene (see Delson, 1975*b*; Bernor, 1985; but compare Geraads, 1982, 1987).

Cerling (1992) has examined this problem from an entirely different approach. Most plants utilize one of two photosynthetic pathways, known as C_3 or C_4, which can be distinguished by differences in the relative amounts they contain of the carbon isotope ^{13}C. Most trees, shrubs and cool-season grasses are C_3 plants, with a value of $\delta^{13}C$ between −2.3 and −2.9%; grasses and dwarf shrubs of open country or savanna (warm season) are C_4 plants, with $\delta^{13}C$ of about −0.2%. Soil carbonates of modern or fossil soils have values which reflect the dominant plant types: C_3 leading to −1.1 ± 0.1% and C_4 averaging +0.2%. Intermediate values reflect a mixture of the two photosynthetic pathway types. Values at two Middle Miocene East African sites were in the pure C_3 range, confirming the essentially forested (closed-canopy) nature of these environments. At the Ngeringerowa locality in the Baringo sequence (as yet not dated directly, but *c.* 10–9 Ma), a $\delta^{13}C$ value of around −0.8% indicated the presence of C_4 plants in moderate frequency. Other results are discussed in sequence below.

The oldest definitely colobine fossil from Africa (Figure 2.9) is a mandible with nearly complete dentition (but damaged corpus and no rami) from Ngeringerowa. Benefit & Pickford (1986) named this specimen *Microcolobus tugenensis* and showed that it combines clearly colobine teeth with several distinctive features of the mandible: the lack of an inferior transverse torus of the symphysis, the lack of a median mental foramen, a fairly even inferior border of the corpus below the molars and a long and steep planum alveolare. Moreover, the mandible and teeth are smaller than those known in any other colobine, only slightly larger than those of *Miopithecus talapoin*. One isolated lower molar of similar size is slightly younger, and a premolar somewhat older, but no other colobines from East Africa are definitely pre-Pliocene.

In the Maghreb, the locality of Menacer (previously named Marceau) has yielded several dozen cercopithecid teeth, of which about seven are colobine. Originally, Arambourg (1959) allocated all the monkeys to a species of macaque, but Delson (1973 *et seq.*) showed that the type of Arambourg's species and a variety of isolated molars belonged to a medium-sized colobine, which he termed '*?Colobus*' *flandrini* (Arambourg), using the modern generic name as a 'form-genus' to indicate African colobines of uncertain affinity. The teeth seem most similar in proportion to those of *Cercopithecoides* (see below).

Pliocene and Pleistocene

About eight taxa of colobines are known in the African Pliocene and Pleistocene, and the co-occurrence of up to five of these in the same region indicates that a diverse radiation of colobines occurred at this time (Figure 2.6). The global cooling discussed above for Eurasia also affected African climate over the past 5 Myr, although there was greater buffering in the more equatorial regions. Major floral and faunal changes have been described in the late Pliocene (*c.* 2.5–2.0 Ma), especially in eastern Africa, which appear to reflect

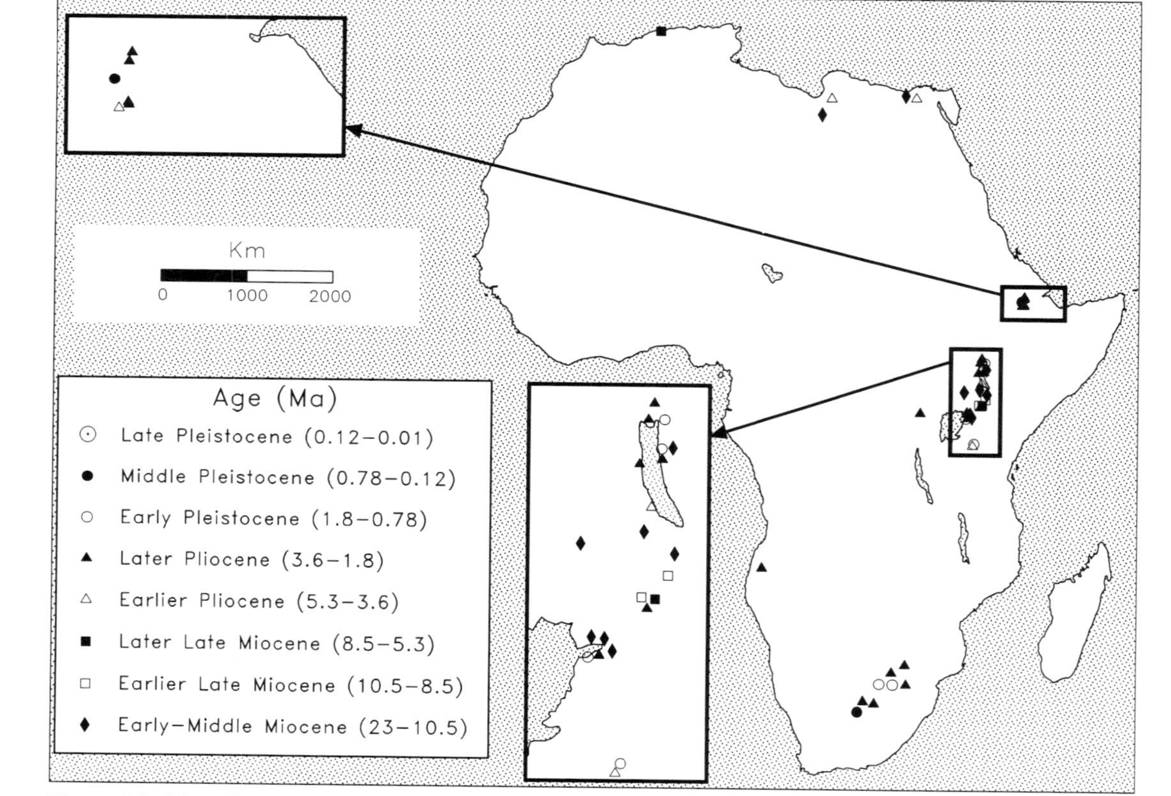

Figure 2.9. Map of African localities yielding fossil colobines and victoriapithecines, with symbols indicating site ages. Oceans and major lakes stippled; insets provide enlarged views of Awash Valley (Ethiopia) and central Rift Valley (Ethiopia–Kenya–Uganda–Zaire) regions.

this cooling and a related increase in relative aridity (see Bonnefille, 1984; Coppens, 1989; and many papers in Coppens, 1985). The work of Bonnefille and collaborators (among others) on palynology has allowed careful reconstruction of paleoenvironments at a number of East African sites yielding Plio-Pleistocene hominids (and colobines).

Paleoenvironmental reconstructions

At Laetoli, Tanzania, deposits aged 3.75–3.45 Ma seem to document a mainly open savanna environment, dominated by a variety of grasses indicating a slightly warmer and drier climate than today's (Bonnefille & Riollet, 1987). About 10–15% of the pollen represents tree species, mainly of the Afromontane flora of the East African mountains, of an altitude similar to that of the region today (1500–1800 m). Most (75%) of the plant species are found locally today, indicating that Pliocene floras were essentially modern in composition and diversity. Most individual reports (in Leakey & Harris, 1987) on elements of the Laetoli mammalian fauna, as well as the summary by Harris (1985), agreed with the essentially savanna nature of the assemblage, but Andrews (1989) has reanalysed the data and concluded that the high species diversity suggests a more wooded paleoenvironment than any present in the Laetoli/Serengeti region today. He has further suggested that some faunal mixing might be involved, given the 0.3-Myr span of the deposits, but the high number of primate, giraffe and rodent species still indicates the presence of woodland, perhaps with shifting ecotonal margins. This is not entirely at odds with the work of Bonnefille & Riollet (1987), as they recognized significant tree pollen but thought it represented a restricted ecozone. Cerling (1992) reported $\delta^{13}C$ values between −0.4 and −0.8% for paleosol carbonates from the Laetoli Beds, which again suggests a flora of mixed C_3 and C_4 type, i.e. grassy woodland with regional variation in canopy closure.

The Hadar deposits (Ethiopia; see Aronson & Taieb, 1986; Tiercelin, 1986) sample lakeside situations dating between 3.35 and 3.0 Ma. Faunas are generally dominated by forest-dwelling mammals in the lower (SH) and upper (KH) horizons but suggest more open environments in the middle (DD) member dated 3.2–3.1 Ma. Bonnefille *et al.* (1987) described a discontinuous series of pollen samples which mainly indicate montane forest and evergreen bushland in the SH and KH members. These floras suggest seasonal climates and environments similar to those found today above 1600 m (about 1000 m higher than the current elevation), with annual rainfall above 800 mm. In the

DD member, more arid pollen spectra were recovered, indicating extensive savanna with scattered acacia trees and a much lower rainfall.

The Lake Turkana Basin includes a number of fossiliferous regions, of which the most important are the Lower Omo Valley deposits (southern Ethiopia), Koobi Fora (= East Turkana, ex-East Rudolf) and West Turkana (Nachukui Formation), both in northern Kenya. Geology, paleogeography and geochronology were reviewed recently by Harris *et al.* (1988), Brown & Feibel (1989, 1991) and Feibel *et al.* (1989), while paleoenvironments were explicitly evaluated by Feibel *et al.* (1991). Pollen records are poor for these long sequences (mainly dating between 4 and 1 Ma), but Bonnefille & Vincens (1985) summarized their most recent findings. The Omo deposits, mainly riverine and deltaic sediments, record mostly woodland and wooded grasslands of broadly Sudano-Zambezian character. The region may have been relatively moister around 3 Ma and after 1.7 Ma, relatively drier between 2.6–2.1 Ma. Contrary to Bonnefille & Vincens (1985), the more recent dates on this sequence (Feibel *et al.*, 1989) show that the relatively cold and dry interval recorded at Gadeb, in the Ethiopian highlands, between 2.5 and 2.35 Ma coincided with, rather than predated, the Omo drying. From Koobi Fora, lakeshore, riverine and floodplain deposits between 2 and 1.4 Ma yielded far more pollen spectra, representing both local Sudano-Zambezian wooded grassland floras and regional Afromontane forest floras. Before 1.9 Ma, the pollen rain was almost entirely of Graminae (grasses), indicating a relatively more open and perhaps drier environment than found after 1.9 Ma, when the area was characterized as a sub-desertic landscape even somewhat drier than that found locally today. Cerling *et al.* (1988) analysed the isotopic composition of Koobi Fora carbonates, finding that a mixed C_3/C_4 plant assemblage ($\delta^{13}C$ −1.0 to −0.5%) probably dominated before 1.8 Ma, with a change to mostly C_4 plants ($\delta^{13}C$ of −0.6–0.0%, wooded or dwarf-shrub grassland) after. They further found that major climatic instabilities may have occurred around 3.4–3.1 and 1.8 Ma on the basis of oxygen isotope values. Williamson (1985) had suggested the spread of rainforest into eastern Africa about 3.4–3.3 Ma on the basis of snail distributions at East Turkana, but Bonnefille *et al.* (1987) accepted only the increased evidence of humidity, as their data for Hadar suggested seasonality at that time. No palynological work has yet been reported for West Turkana.

Many authors have employed interpretations of mammalian fossil assemblages to offer paleoenvironmental reconstructions, but Vrba (1980, 1985, 1989) has concentrated on relative percentages of various bovid tribes by reference to their modern adaptations. Shipman & Harris (1989) modified and extended this work to survey all three Turkana Basin sequences. For the

Lower Omo Valley sequence, they found no fossil assemblages equivalent to modern open/arid zones, but instead found mainly closed environments with differing degrees of moisture. The oldest layers studied, between 3 and 2.8 Ma, were moderately wet, followed by drier deposits between 2.8 and 2.3 Ma, then slightly more moist between 2.3 and 2.0 Ma, and younger levels significantly wetter (2.0–1.4 Ma); the resulting climatic pattern agrees well with that reported by Bonnefille & Vincens (see above). Among the most interesting results (not indicated by the authors) is that, except for the driest interval of 2.5–2.4 Ma, most of the levels between 2.8 and 2.0 Ma have no modern counterparts. Almost all of the localities studied from Koobi Fora (and the few of West Turkana) were indicative of a closed and moist habitat, which is rather at odds with the pollen work cited above.

Feibel *et al.* (1991) reviewed these and other lines of evidence in a detailed investigation of local paleoenvironments around Lake Turkana, especially in the Koobi Fora (East Turkana) region. In the older horizons at Koobi Fora, differing results are suggested by the several types of data analysed (flora, mammals, geochemistry), perhaps implying local variation and the presence of both woodland and grassland patches around the lake. Later horizons may indicate several decreases in Omo River outflow (caused by deltaic silting) and temperature increase, leading to more savanna-like conditions.

From the lacustrine deposits of Olduvai Gorge (northern Tanzania), Bonnefille & Vincens (1985) reported a wooded grassland between 1.8 and 1.75 Ma, followed by a very dry interval around 1.75–1.7 Ma and the spread of moist woodlands ('closed/wet' habitats) around 1.6 Ma. Shipman & Harris (1989) discerned the dry interval and considered some of the early levels fairly wet, but they found little evidence for the later forested period, despite having fossils from the same site as the pollen. Cerling (1992) discussed the Olduvai paleoenvironment in terms of $\delta^{13}C$ values, which showed great variation (−0.8 to −0.1%, probably grassy woodland with mixed canopy closure) in the 1.75–1.7 Ma interval, then a change to more open conditions (−0.5–0.0%). Evidence for high C_4 proportions was clearly marked at 1.2 Ma and 0.6 Ma, as well as during the last 200 thousand years.

The East African results reveal that comparisons between even relatively nearby regions are fraught with difficulty. Analyses of the paleoenvironments for the South African cave sites that have yielded both hominids and cercopithecids are even more problematic. Cadman & Rayner (1989) provided one of the few palynological studies, in which they examined a sequence of samples from the Makapansgat site, probably spanning part of the 3.1–2.9-Ma interval. Low in the section there was an increase of arboreal elements, mainly bushveld trees with varying if generally moisture-loving habitats. A period of open grasslands followed, in turn replaced by dry bushveld, and then a dramatic increase in wet forest elements in the first layers that yielded mammalian fossils. As Cadman & Rayner reported, studies of the mammals have generally implied a relatively moist but open environment in this interval.

In the Sterkfontein (or Blaaubank) Valley to the south, several collapsed caves have also yielded important fossil collections. Vrba (1980, 1985, 1989) has analysed these in terms of relative frequencies of bovid groups, finding a general trend toward more open and dry habitats in a broadly bushveld regime. Thus the Sterkfontein Member 4 site unit, dated roughly at 2.7–2.4 Ma, was thought relatively more closed than the younger Swartkrans Member 1 and Kromdraai A units (estimated to date at 1.9–1.5 Ma). The Kromdraai B unit was thought to be intermediate in age on the basis of the less derived australopith and relatively wet on the basis of the bovids and presence of a monkey (see below). Delson (1984, 1989) questioned the age distinction from the younger group and the monkey-based climatic inference, but the presence of the bovids might thus imply local variation in habitat within that time range. Shipman & Harris (1989) found little difference among the Sterkfontein and Swartkrans subunits they analysed, placing all as open-arid habitats, but Vrba (1989) noted that their analysis was effectively less fine-grained than hers, which included more taxa and compared site units within the admittedly narrow range of habitats in the valley, seeking local differences rather than regional or subcontinental comparisons. Overall, their results appear compatible.

Colobine radiation

With that detailed framework available, it is now possible to examine fossil colobine species and their distributions and look for patterns of habitat utilization. One of the largest of these species was *Rhinocolobus turkanaensis*, known from long-faced male and female crania (see Figure 2.6), numerous jaws, and rare postcranial elements which suggest a rather arboreal habitus (Leakey, 1982) and body size perhaps comparable to that of modern larger species of *Pygathrix* (*Rhinopithecus*). The best representation is in the Omo sequence (M. G. Leakey, 1987), with several good specimens (including partial postcranial elements associated with a mandible fragment) known from Koobi Fora (Leakey, 1982). Unpublished specimens from Hadar (SH and DD members), including four jaws and a partial humerus, appear referable to this species, in the 3.35–3.0 Ma range.

Paracolobus chemeroni (and the quite similar *P. mutiwa*) were shorter-faced than *Rhinocolobus*, but seem to have had larger teeth and perhaps skulls (see Figure 2.6). Until better comparisons of limb bones and of tooth-to-skull size are available, it is unclear which genus had a larger body size, but they were probably in the same range. A nearly complete skeleton (Figure 2.10) of *P. chemeroni* from the type locality in the Chemeron Formation (probably close to 3.2 Ma – see Delson & Dean, 1993) indicates a mainly arboreal adaptation in the hindlimb and superficially more cursorial-appearing forelimb (Birchette, 1982). *P. mutiwa* is represented all around the Turkana Basin by fragmentary jaws and isolated teeth between 3.5 and 1.9 Ma (M. G. Leakey, 1982, 1987) and by a new partial face and skeleton from West Turkana (*c.* 2.5 Ma; Harris *et al.*, 1988). A smaller and unnamed species is represented only by dentognathic material at Laetoli (Leakey & Delson, 1987). Laetoli is famous for its mammalian footprints preserved in volcanic ash, and several cercopithecid trails are known (M. D. Leakey, 1987). Skelton (1990) has suggested that one print with a short thumb impression (according to Leakey & Hay, 1979) might represent a colobine, presumably *Paracolobus* sp., which would thus have been at least occasionally terrestrial. However, I

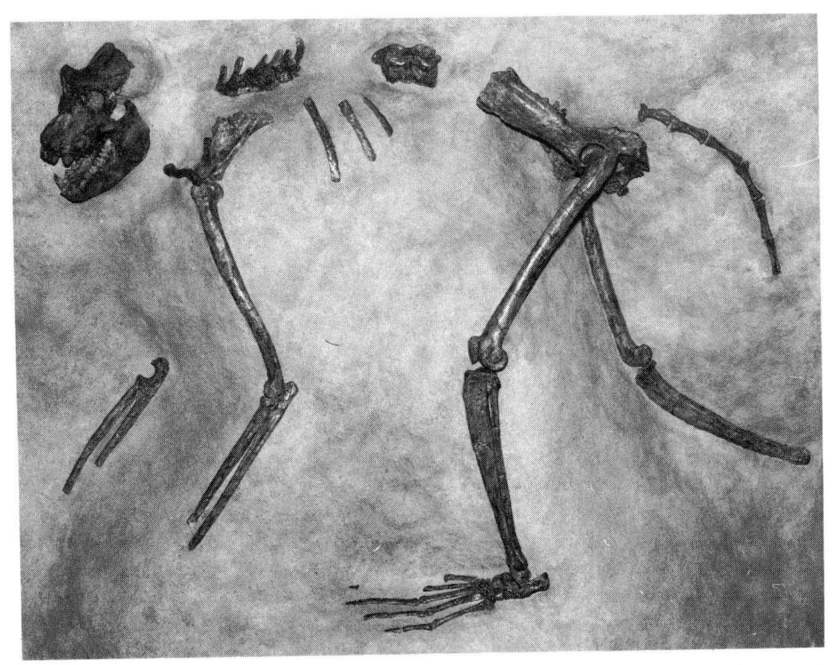

Figure 2.10. Mounted reconstruction of partial skeleton of *Paracolobus chemeroni* from Loc. JM 90, Chemeron Formation, Kenya. Designed (with casts) for Hall of Mammals and Their Extinct Relatives in the American Museum of Natural History.

question the distinctiveness of this trail by comparison with the others, with supposedly longer thumbs.

The third large extinct African colobine, *Cercopithecoides*, is the most common and widespread. In South Africa, *C. williamsi* is represented by numerous craniodental elements (see Figure 2.6) at Makapansgat (*c.* 3–2.9 Ma), Sterkfontein Member 4 (*c.* 2.7–2.4 Ma) and Bolts Farm (estimated 2.2–2.0 Ma); a single mandible is also known from the Leba fissure filling in Angola (Delson, 1984). Jaws and a partial associated skeleton apparently of the same species are known from Koobi Fora in the 2.0–1.9 Ma range (Leakey, 1982). They indicate an extreme terrestrial adaptation (Birchette, 1981), supported by the high tooth wear indicative of a diet including gritty food items found at or below the ground surface. No postcranial elements from South Africa have been identified for this species, but given the lack of associations with teeth, they may have been misidentified as baboons. The species was probably comparable in body size to a large *Nasalis larvatus* or smaller *P. (Rhinopithecus)*.

A slightly larger 'variant' of *C. williamsi* has been recovered from the later Sterkfontein Valley sites (?2.0–1.5 Ma) of Kromdraai (A and B and nearby Cooper's A) and Swartkrans (provenance and/or identification ambiguous) (Delson, 1984, 1989). The still larger *C. kimeui* (perhaps approaching *Rhinocolobus* in body size) is less frequent but also has heavy dental wear (Leakey, 1982). It is represented by jaws and teeth from Koobi Fora over a long but discontinuous range: *c.* 3.4–3.3 Ma and 2.0–1.7 Ma, and at Olduvai Gorge by a partial skull from *c.* 1.65–1.3 Ma and an isolated tooth potentially younger than 1 Ma.

The habitat preferences of these three genera are less clear than might be expected. The Laetoli and earlier Hadar and Omo intervals suggest a wooded and well-watered landscape verging on denser forest, where *Rhinocolobus* and *Paracolobus* occur predictably, given the arboreal adaptations of these taxa, but rarely sympatrically. The presence of *Rhinocolobus* in the more open Hadar DD Member is unexpected, but given the low number of specimens it is not possible to tell if the changing climate led to local disappearance; the species is not recognized from the later KH member which saw a return to woodland habitat. The lack of *Paracolobus* at any Hadar level is surprising, given its locomotor adaptations, perhaps in some ways comparable to (if not as cursorial as) *Semnopithecus entellus*. Both taxa do occur sympatrically in the relatively open habitats of the middle Turkana horizons, with no apparent reduction in frequency at Omo through the cool and dry phase of 2.5–2.1 Ma. Based upon an adaptation to terrestrial locomotion and gritty diet, *Cercopithecoides* would be expected to occur in more open habitats, such as at Koobi Fora after 2 Ma, in what is characterized by some as open to sub-desertic savanna. Although there may be a trend toward drier conditions through time, all indications point to the southern African Pliocene sites as also representing broadly open and poorly watered habitats. *C. kimeui* may have had a wider habitat tolerance, as it is known from relatively wetter conditions in the early Omo horizons and the later Olduvai levels; I might predict a less cursorial locomotor adaptation on this basis, but no postcrania are yet reported. The general scarcity of primates at Omo in post-2-Ma time renders distribution here less meaningful, although the presence of *Rhinocolobus* in the later forested span is reasonable.

Delson (1973, 1975*a*; Szalay & Delson, 1979) suggested that *Cercopithecoides* and *Paracolobus* were close relatives, based on preliminary analyses of craniofacial shape, but M. G. Leakey (1982, 1987) has rejected this view. In 1987, she suggested that *Paracolobus* and *Rhinocolobus* shared more recent ancestry and were closely linked to the living colobines, while the more terrestrially adapted *Cercopithecoides* might be specially related to the (semi-) terrestrial European colobines. I would agree with the former hypothesis, and in fact my earliest views combined these two taxa before *Rhinocolobus* was formally named. On the other hand, I would reject the second suggestion without better evidence, as I would Leakey's (1982) conjecture that *Rhinocolobus* might be phyletically linked to other 'long-faced' colobines like *Nasalis* or *Libypithecus*.

Both facial elongation and postcranial adaptations to terrestriality have occurred several times within the Cercopithecidae and are not themselves sufficiently unique derived features on which to link taxa. Instead, I would suggest that *Cercopithecoides* is definitely an African colobine, perhaps part of a Pliocene radiation which preceded the differentiation of the modern genera, and thus a possible sister-taxon to *Paracolobus* and *Rhinocolobus* (see Figure 2.2). Interpretation of 'species A' (below) and the Marceau colobine is required before these alternatives can be tested carefully.

Two other extinct forms are smaller than the preceding but larger than most extant African colobines. *Libypithecus markgrafi* is known from a single partial cranium at the North African site of Wadi Natrun (Egypt; see Figure 2.6). It is most comparable to male *Procolobus badius* (Szalay & Delson, 1979), in terms of size and especially sagittal crest development, which is rare in smaller cercopithecids. A colobine of similar dental size was reported at nearby Sahabi, Libya (Meikle, 1987), where apparently colobine postcranial elements are neither strongly terrestrial nor as arboreally adapted as those of living African colobines. Both sites are probably earliest Pliocene in age, *c.* 5 Ma, and sampled the interface between woodland and savanna.

Another taxon, as yet unnamed and known informally as 'species A', is represented by a partial cranium and robust associated skeleton from the Hadar region, which seems to suggest a (semi-) terrestrial habitus. This site (Leadu) is not well-dated, but closely similar jaws and a humerus fragment were identified from the main Hadar SH Member (3.35–3.1 Ma). None of these fossils have yet been fully published (see Delson, 1984). Some workers have suggested that this probably new species might best be placed in the genus *Libypithecus*, but although a natural endocast is preserved at Leadu, the dorsal surface of the neurocranium is lacking and it is impossible to tell whether a sagittal crest was present. Specimens of comparable size have been tentatively referred to this taxon, although identification is obviously uncertain on size alone, from Laetoli, Omo (*c.* 3.0–2.5 Ma), and Koobi Fora (*c.* 2.0–1.8 Ma) (see M. G. Leakey, 1987; Leakey & Delson, 1987). Most of the site units mentioned (except for Koobi Fora and perhaps Laetoli) are relatively tree- or bush-covered and well-watered.

Finally, smaller teeth, comparable to those of living colobines, have been reported from Turkana (*c.* 1.8–1.0 Ma) and the Kanam East region (Kenya, perhaps 3.5–3.0 Ma; Szalay & Delson, 1979; M. G. Leakey, 1987; Harris & Harrison, 1991). These intervals would all appear to represent at least partially forested habitats.

From the Taung region (northern Cape Province, South Africa), an apparently later Pleistocene collection of about 50 isolated cercopithecid teeth includes a half-dozen which are identifiably colobine, comparable in size to those of the living taxa. It is not yet possible to allocate them to species or even genus, but from a distributional point of view, their presence in a relatively arid area in association with *Cercopithecus* cf. *aethiops* and *Papio hamadryas* cf. *ursinus* implies the past presence of moister habitats, at least gallery forest, and a significant range extension.

Summary and conclusions

The colobines presumably differentiated from cercopithecines in Africa but exited to Eurasia early in the Late Miocene, perhaps via a wooded savanna 'corridor'. Colobines did not reach eastern Asia until the end of the Miocene. The European genera represent the most terrestrial radiation, if indeed they are closely related to each other. *Dolichopithecus* demonstrates that a terrestrial or cursorial locomotor adaptation does not imply open country habitats, as most fossils can be linked to woodland or subtropical forest; Andrews (personal communication) has noted that because subtropical and temperate forests have generally simple canopies, a large primate would probably be forced to travel on the ground due to the lack of suitable arboreal pathways between feeding areas. A member of this genus may have reached north-eastern Asia in the later Pliocene, when possible relatives are known in Mongolia, Siberia and Japan. The oldest representative of a modern genus is *Pygathrix* (*Rhinopithecus*) *lantianensis* from central China, perhaps just over 1 Myr old.

African Miocene colobines are rather distinctive but poorly known; *Microcolobus* was probably the smallest colobine ever. The Pliocene witnessed a major differentiation of large colobines, with the arboreal *Rhinocolobus*, terrestrial *Cercopithecoides* and perhaps intermediate *Paracolobus, Libypithecus* and 'species A'. They inhabited a range of environments, including grasslands and woodlands, although no true moist forest habitats are yet known in the African Plio-Pleistocene record. Fossils possibly referable to modern genera may occur in the Pleistocene.

As opposed to the cercopithecines (see Szalay & Delson, 1979), most extinct colobine taxa cannot be linked to living forms. They are thus archetypally catarrhine: whereas platyrrhine generic lineages extend back into the Middle or even Early Miocene, no catarrhine genus can be traced back beyond 12 Ma (*Pongo* to *Sivapithecus*). Extant cercopithecines and humans, on the other hand, can be linked to later Miocene or Early Pliocene taxa (*Macaca*; *Papio* and *Theropithecus* to *Parapapio*; *Homo* to *Australopithecus*). Colobines apparently underwent 'adaptive radiations' in Europe and Africa, producing a variety of distinctive species that died out, to be replaced by collateral relatives that survive today.

Colobines were never dominant members of the primate fauna except in Europe; in Africa, *Theropithecus* and *Parapapio* were always more common, as was *Macaca* in eastern Asia. The last genus supplanted colobines during the Pleistocene in Europe, probably during early episodes of glacial climatic regime, and in Africa the large colobines of the Pliocene also disappeared by the Early Pleistocene. It is unlikely that early human hunting or habitat interference was related to this disappearance, such effects becoming important only later in the Pleistocene. Only in eastern Asia did colobine variety increase, and there is as yet no clear evidence of the antiquity of that radiation, nor its phyletic fine-structure.

On the other hand, it might be possible to determine whether the African colobines truly represent a monophyletic radiation. The living taxa (see Chapters 1 and 3) are linked by their reduced thumb and mid-tarsal shortening. If animals as different in locomotor adaptation as *Cercopithecoides* and *Rhinocolobus* were found to share these features, it would be a strong confirmation of the monophyly hypothesis. Moreover, it was noted above that terrestriality and facial elongation had apparently evolved independently several times among cercopithecids, including colobines. Although this pair of features occurs together commonly in cercopithecines (*Papio*, some *Theropithecus* and *Macaca*, and *Paradolichopithecus*), they are typically disjunct in colobines: only *Dolichopithecus* presents both, with the highly terrestrial *Cercopithecoides* being relatively short-faced and the more arboreal *Rhinocolobus* and *Nasalis* more 'snouty'. The relatively terrestrial extinct colobines are distinctive in comparison to their modern relatives, and thus they appear to be more common in the fossil record than was probably the case. This apparent frequency is further exaggerated by the prevalence of relatively open habitats sampled, although the Eurasian *Dolichopithecus* implies that forested regions might have been inhabited by more terrestrial colobines (or cercopithecines) as well.

In light of recent discussions (Happel, 1988; Davies, 1991) of the likely diet of ancestral cercopithecids and especially colobines, it is reasonable to suggest that some dietary specializations away from an eclectic cercopithecid pattern characterized the earliest members of the subfamily. Dental and presumably digestive modifications probably followed soon after, so that by 11–10 Ma, the earliest colobines in Africa and Europe were recognizably colobine in those systems. Thumb reduction appears most likely to have been homologous among colobines; if so, it had also begun by the same time, as *Mesopithecus* is characterized by a thumb relatively shorter than in any cercopithecine, but longer than in living colobines. At present it is not possible to determine the time of origin of such other diagnostically colobine traits as aunting behavior and contrasting natal coats.

Recent paleoenvironmental reappraisals summarized here (Andrews, 1989; Solounias & Dawson-Saunders, 1988; Cerling, 1992) have demonstrated that two of the most widely-cited examples of Late Miocene and Pliocene open-country faunas may be more accurately characterized as woodland assemblages. The Pikermi (and Samos) 'savannas' of Late Miocene southeastern Europe were shown to be probably mixed riparian woodlands on the basis of both palynology and functional morphology of fossil bovids (but compare de Bonis *et al.*, 1992). The mid-Pliocene Laetoli assemblage from Tanzania has indicators of species diversity typical of discontinuous seasonal forested environments, and its paleosol carbonates yield $\delta^{13}C$ values indicative of mainly C_3 plants, i.e. woodland. It appears from these studies that open, nearly treeless plains are quite a recent feature of the landscape in Europe as well as Africa (although not farther east, where C_4 grasslands are known in Pakistan back to 6 Ma – Quade *et al.*, 1989; Cerling *et al.*, 1993). Some of the increase in open environments is presumably due to human activity, especially burning. As a typical catarrhine, *Homo* replaces many contemporaneous primates one way or another.

Acknowledgements

I thank Drs Peter Andrews and John Oates for considerable discussion on the development of this chapter and John for his patience during the course of its production. Conversations with or written comments from Drs Glyn Davies, Meave Leakey and Nikos Solounias were also of much help. Dr Mitsuo Iwamoto drew my attention to the Nakatsu colobine; Elena Cunningham, Reiko Matsuda and Ivy Rutsky kindly helped with translations; and Margaret Heffernan, Lorraine Meeker and Chester Tarka assisted with illustrations. I thank them all. Figures 2.5 and 2.9 were prepared with the Atlas-GIS computer mapping software; Figures 2.2 and 2.6 were originally prepared by Cambridge University Press to illustrate Delson (1992), and their permission to reprint them here (slightly modified) is gratefully acknowledged. The research reported here was financially supported, in part, by Grant Nos. 667370, 668540 and 669381 from the PSC-CUNY Faculty Research Award Program and by the Committee on Scholarly Communication with the People's Republic of China.

3

The diversity of living colobines

JOHN F. OATES, A. GLYN DAVIES
and ERIC DELSON

Introduction

Taxonomically, the diversity of living colobines is portrayed by the recognition in most classifications of some 30 species (see e.g. Thorington & Groves, 1970; Delson *et al.*, 1982; Napier, 1985). As is discussed below, such arrangements probably underestimate the number of 'good' species in the subfamily.

The living colobines have been grouped into between four and nine genera, and these genera are often arranged in two clusters, one African and one Asian. Napier (1970) argued for an ancient phylogenetic separation of these two clusters, a view supported by Szalay & Delson (1979) in their recognition of two subtribes, Colobina and Semnopithecina (= Presbytina of Delson, 1975*a*) (note also Figure 2.2, this volume). Szalay & Delson acknowledged, however, that the Asian colobines are less clearly united than the African species, and Groves (1989) has argued that *Nasalis* groups with other Asian species largely through sharing a set of retained primitive features. Groves therefore proposed that *Nasalis* (with *Simias*) should be regarded as a sister group to all other colobines, and treated Nasalinae and Colobinae as subfamilies within the family Colobidae.

This chapter examines the generic- and species-level diversity of the living colobines, in the main following the order used by Napier (1985), and retaining the traditional African and Asian clusters. Table 3.1 summarizes the classification that we follow, and Table 3.2 presents information on body weights.

African colobines

Africa's colobus monkeys appear to be a monophyletic group, sharing a vestigial thumb, mid-tarsal shortening, and some other postcranial and dental fea-

Table 3.1. *The species of living colobines*

	Common name
Colobus polykomos (Zimmerman, 1780)	Ursine colobus
Colobus vellerosus I. Geoffroy, 1830	White-thighed colobus
Colobus guereza Rüppell, 1835	Guereza
Colobus satanas Waterhouse, 1838	Black colobus
Colobus angolensis Sclater, 1860	Angolan colobus
Procolobus (*Piliocolobus*) *badius* (Kerr, 1792)	Red colobus
Procolobus (*Procolobus*) *verus* Van Beneden, 1838	Olive colobus
Pygathrix (*Pygathrix*) *nemaeus* (Linnaeus, 1771)	Douc
Pygathrix (*Rhinopithecus*) *roxellana* (Milne Edwards, 1870)	Sichuan snub-nosed monkey
Pygathrix (*Rhinopithecus*) *bieti* (Milne Edwards, 1897)	Yunnan snub-nosed monkey
Pygathrix (*Rhinopithecus*) *brelichi* (Thomas, 1903)	Guizhou snub-nosed monkey
Pygathrix (*Rhinopithecus*) *avunculus* (Dollmann, 1912)	Tonkin snub-nosed monkey
Nasalis larvatus (Wurmb, 1784)	Proboscis monkey
Simias concolor (Miller, 1903)	Simakobu, pig-tailed monkey
Presbytis melalophos (Raffles, 1821)	Banded leaf-monkey
Presbytis comata (Desmarest, 1822)[a]	Javan leaf-monkey
Presbytis frontata (Müller, 1838)	White-fronted leaf-monkey
Presbytis rubicunda (Müller, 1838)	Red (or maroon) leaf-monkey
Presbytis potenziani (Bonaparte, 1856)	Mentawai leaf-monkey
Presbytis hosei (Thomas, 1889)	Hose's leaf-monkey
Presbytis thomasi (Collett, 1893)	Thomas's leaf-monkey
Trachypithecus vetulus (Erxleben, 1777)[b]	Purple-faced langur
Trachypithecus auratus (E. Geoffroy, 1812)	Lutung
Trachypithecus cristatus (Raffles, 1821)	Silvered langur, or lutung
Trachypithecus johnii (Fischer, 1829)	Nilgiri langur
Trachypithecus obscurus (Reid, 1837)	Dusky langur
Trachypithecus pileatus (Blyth, 1843)	Capped langur
Trachypithecus phayrei (Blyth, 1847)	Phayre's langur
Trachypithecus francoisi (Pousargues, 1898)	Francois's langur
Trachypithecus geei (Gee, 1956)	Golden langur
Semnopithecus entellus (Dufresne, 1797)	Hanuman or grey langur

[a]Weitzel & Groves (1985) have explained why the more familiar name for the Javan leaf-monkey, *P. aygula*, is not valid under the rules of nomenclature.
[b]In the behavioural literature, *T. vetulus* is usually referred to as *Presbytis senex*. Napier (1985) has explained why the name *senex* is probably invalid; it was apparently used by Erxleben to describe an albino monkey from an unknown locality.

Table 3.2. *Body weights of adult colobine monkeys. Mean weight given in kilograms, with sample size in parentheses, where available*

Species	Female	Male	Source of data
Colobus polykomos	8.3 (10)	9.9 (5)	Oates *et al.* (1990)
Colobus vellerosus	6.9 (5)	8.5 (3)	JFO
Colobus guereza guereza	9.2 (4)	13.5 (3)	WLJ and JFO
Colobus guereza matschiei	7.9 (7)	10.1 (4)	WLJ
Colobus satanas		10.9 (2)	Eisentraut (1973); M. J. S. Harrison, personal communication
Colobus angolensis	7.4 (6)	9.8 (4)	WLJ; Napier (1985)
Procolobus badius badius	8.2 (16)	8.3 (9)	Oates *et al.* (1990)
Procolobus badius tephrosceles	7	10.5	Struhsaker & Leland (1979)
Procolobus badius kirkii	5.5 (1)	5.8 (1)	Napier (1985)
Procolobus verus	4.2 (14)	4.7 (20)	Oates *et al.* (1990)
Nasalis larvatus	10.0 (14)	21.2 (13)	WLJ
Simias concolor	7.1 (1)		Napier (1985)
Pygathrix (Rhinopithecus) bieti	9.0 (1)	13.0 (1)	Jablonski & Pan (1991)[a]
Presbytis melalophos	5.8	5.9	Waterman *et al.* (1988)
Presbytis rubicunda	5.7 (21)	6.2 (18)	WLJ
Presbytis potenziani	6.4 (4)	6.5 (5)	Tilson & Tenaza (1976)
Presbytis hosei sabanus		6.2 (6)	Davis (1962)
Trachypithecus vetulus	5.9 (3)	8.2 (3)	Napier (1985)
Trachypithecus cristatus	5.7 (25)	6.6 (13)	WLJ
Trachypithecus johnii	10.9 (1)	12.7 (4)	Leigh (1926)
Trachypithecus obscurus	6.6 (22)	7.3 (12)	Napier (1985)
Trachypithecus pileatus	10.0 (3)	12.8 (2)	Oboussier & von Maydell (1959)
Trachypithecus phayrei	6.9 (5)	7.9 (8)	Napier (1985)
Trachypithecus geei	9.5 (1)	10.9 (4)	Oboussier & von Maydell (1959)
Semnopithecus entellus entellus	11.2 (22)	18.3 (3)	Hrdy (1977) (females listed as 'parous')
S. entellus schistacea	15.6 (5)	19.8 (3)	Hrdy (1977) Bishop (unpublished data); Napier (1985)
S. entellus thersites	6.7 (4)	10.6 (8)	Napier (1985)

WLJ, information from specimens in the US National Museum (Washington, DC), the Natural History Museum (London) and the Museum of Comparative Zoology (Harvard), collated by W. L. Jungers (personal communication); JFO, information from museum specimens in the Natural History Museums of London and Paris, collated by J. F. Oates.

[a]Male weight stated to be of 'young adult'.

tures (Szalay & Delson, 1979; Strasser & Delson, 1987; Delson, Chapter 2, this volume). As Pocock (1936) and Kuhn (1967) recognized, the colobus monkeys fall into two distinct subgroups, each of which merits generic status. The red and olive colobus, *Procolobus*, are united by a set of anatomical characteristics that distinguish them from members of the black-and-white colobus group, *Colobus*. Female *Procolobus*, unlike *Colobus*, have sexual swellings; male *Procolobus* have separate rather than united ischial callosities and a sagittal crest, and young males have a perineal organ; *Procolobus* have a four-chambered stomach, compared with a three-chambered arrangement in *Colobus*; and *Colobus* have a sub-hyoid sac (absent in *Procolobus*) and a large rather than small larynx (Napier, 1985; Strasser & Delson, 1987). The phylogenetic polarity of these differences is unclear, but Strasser & Delson (1987) have suggested that the discontinuous callosities, perineal organ and four-chambered stomach are derived features of *Procolobus*, while the large larynx, sub-hyoid sac and lack of female swellings are derived features of *Colobus*. Adult male *Colobus* make resonant, low-pitched loud calls that are not heard in *Procolobus* (presumably a correlate of their laryngeal

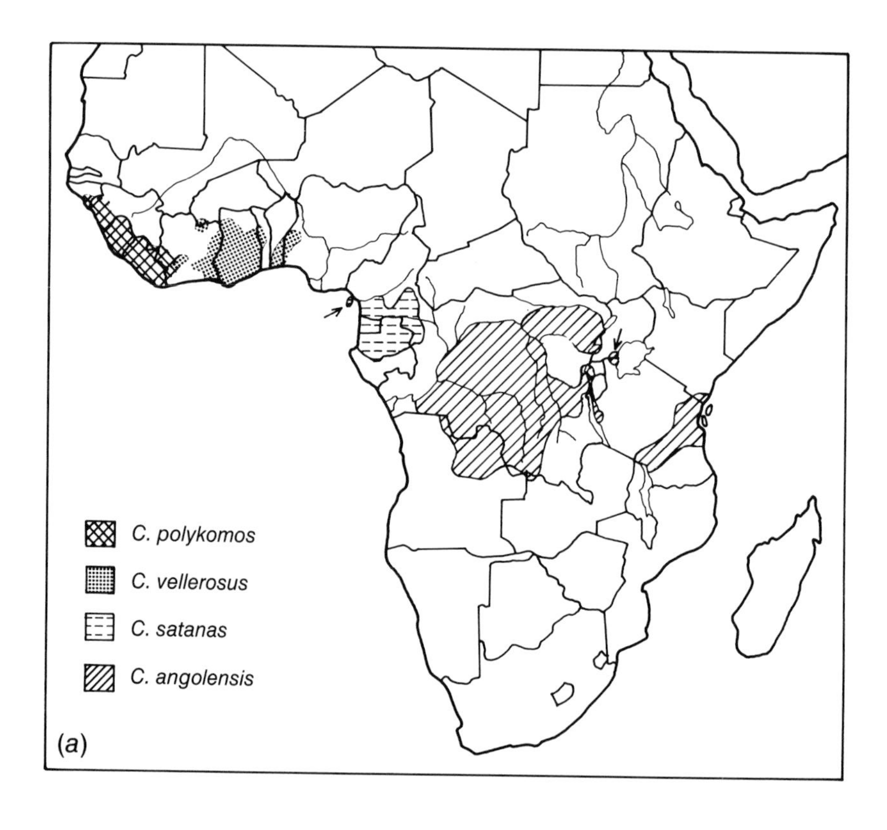

differences), and *Colobus* mothers allow their infants to be handled by other group members, a behaviour that is very rare in *Procolobus*.

Both *Colobus* and *Procolobus* have a wide distribution in tropical African forests, each occurring from the Atlantic coast of the far west to the Indian Ocean coasts of Kenya and Tanzania.

Genus Colobus, *black-and-white colobus monkeys*

These monkeys have long black or black-and-white pelage. Although all populations of black-and-white colobus monkeys have sometimes been grouped into one species (see e.g. Schwarz, 1929), they are now widely regarded as a diverse group of four or five species. Here we follow Oates & Trocco (1983) in recognizing five forms: *C. satanas*, *C. polykomos*, *C. vellerosus*, *C. guereza* and *C. angolensis*. Figure 3.1 shows the distribution of these species.

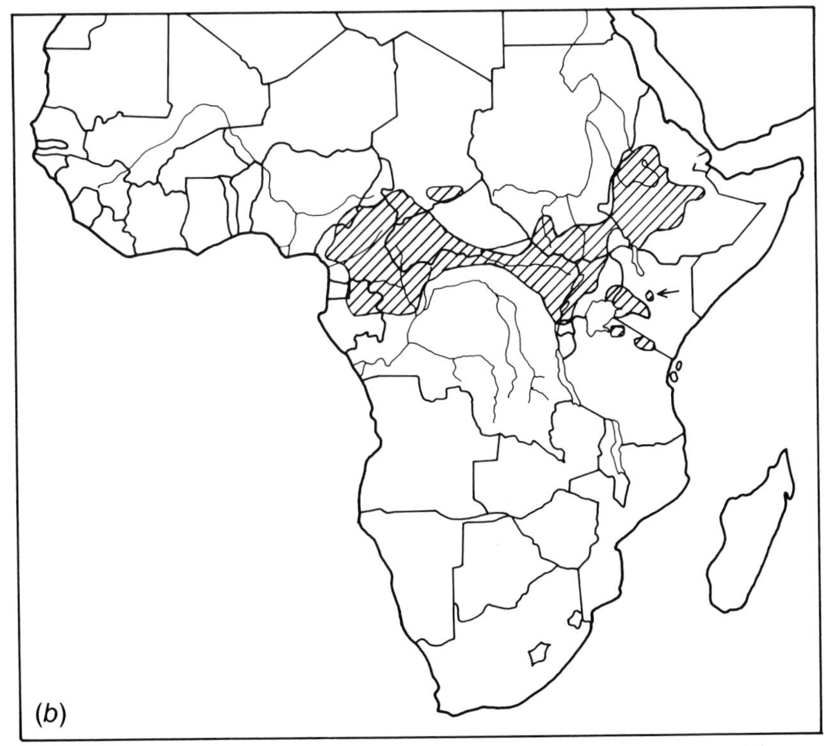

Figure 3.1. Distribution of black-and-white colobus monkeys: (*a*) *Colobus polykomos*, *C. vellerosus*, *C. satanas* and *C. angolensis*; (*b*) *C. guereza*. Based on Oates & Trocco (1983), Colyn (1991), Machado (1969), Mitani (1990), Rahm & Christiaensen (1960) and Rodgers (1981). Arrows draw attention to isolated populations.

Each has a different coat pattern: *C. satanas* is all black, *C. guereza* (Figure 3.2) has a characteristic white peridorsal mantle, and the other three have differing combinations of white or grey markings on the tail, thighs, shoulders and/or head (illustrated by Rahm, 1970). A craniometric study by Hull (1979) showed four highly distinct clusters: *polykomos* + *vellerosus*; *angolensis*; *guereza*; and *satanas*. Of these, *guereza* was most distinct, especially in the teeth (e.g. small incisors, longer molars, large female canines), and face (e.g. broader nasal apertures). In the other three clusters, *satanas* had the largest incisors, and *polykomos* (+ *vellerosus*) had larger jaws than *angolensis*. A study of loud call pitch and tempo by Oates & Trocco (1983) found three clusters: *vellerosus* + *guereza*; *polykomos* + *angolensis*; and *satanas*. The low-pitched male loud call of *vellerosus* could not be discriminated from that of *guereza*, and *satanas* had a very distinct high-pitched roar. This evidence led Oates & Trocco (1983) to agree with Dandelot (1971) that the status of *vellerosus* 'is perhaps more nearly specific than subspecific'. The specific status of *C. vellerosus* has been supported in a recent analysis by Groves *et al.* (1993). Relative to other black-and-white colobus, *C. guereza* appears to have the largest number of derived features, while *C. satanas* may be the

Figure 3.2. Group of *Colobus guereza occidentalis* in gallery forest by the River Nile, Murchison Falls National Park, Uganda (photograph by J. Oates).

most primitive member of the group (Grubb, 1978; Oates & Trocco, 1983). *Colobus satanas* is the only form in which the infant is not born with a pure white coat; instead, the neonate is brown (though still noticeably different from the mother's black coat) (M. J. S. Harrison, personal communication).

Unlike the other forms, which have the majority of their populations in moist lowland forest, much of the range of *C. guereza* is in the deciduous forest and savanna woodland zone north of the moist forest, and in the montane forest zone of East Africa, including the Ethiopian highlands. This ecological divergence from other members of the group is discussed in Chapter 4. As Hull (1979) has suggested, the divergence of *C. guereza* from the others in its teeth and jaws is likely to be related to dietary differences. Such adaptive divergence apparently allows *C. guereza* to coexist with other black-and-white colobus species in areas where it enters the moist forest zone; sympatry with *C. angolensis* certainly occurs in eastern Zaire (Thomas, 1991), and distribution records and hunters' reports suggest that until recently it coexisted with *C. satanas* in West Africa (Schwarz, 1929; Rahm, 1970; Oates & Trocco, 1983; Mitani, 1990). *Colobus guereza* may have expanded its range within these moist forest areas in relatively recent times, along with humans. Thomas (1991) reported that, in the Ituri Forest, *guereza* exhibits a strong preference for roadside secondary forest, while *angolensis* occurs mainly in primary forest; in Uganda's Kibale Forest, *guereza* is the only monkey found at higher density in heavily logged than in unlogged or lightly logged forest (Skorupa, 1988).

Genus **Procolobus,** *red and olive colobus monkeys*

Brandon-Jones (1984), Strasser & Delson (1987) and Groves (1989) have followed Hill & Booth (1957) and Kuhn (1967) in arguing that the red and olive colobus monkeys are more closely related than is either to *Colobus*, and should be united in the genus *Procolobus*, but separated into two subgenera: *P.* (*Piliocolobus*) for the red colobus and *P.* (*Procolobus*) for the olive. That arrangement is followed here, but with the recognition that the olive colobus is a very distinctive animal. Several authors (e.g. Verheyen, 1962) have given equal generic or subgeneric status to the red, olive and black-and-white colobus, but such an arrangement ignores the set of shared, derived features uniting the red and olive species.

Groves (1989) has noted that red and olive colobus are also united by the absence of two features present in the black-and-white group: infant-handling by group members other than the mother, and colour contrast between young infants and adults. Although the first point is consistent with field observations, the latter is not. Struhsaker (1975) reported that newborn red colobus of the *tephrosceles* subspecies are black dorsally, grey ventrally, and totally lacking in red or brown, making them 'very distinct from older monkeys'. Struhsaker also noted differences between neonatal and adult coloration in four other forms of red colobus that he observed. On the other hand, olive colobus infants are not very different in colour from their parents (Oates, personal observation).

Subgenus Piliocolobus, *red colobus monkeys*

Red (or bay) colobus monkeys occur from Senegambia in West Africa across the continent to Zanzibar (Figure 3.3(*a*)). However, their distribution within this area is patchy and they are completely absent from a large area of the western equatorial forest (e.g. Gabon and mainland Equatorial Guinea). The nominate form of red colobus, *Procolobus badius badius* from West Africa, is bright reddish-brown on the venter and lower limbs, while the dorsal sur-

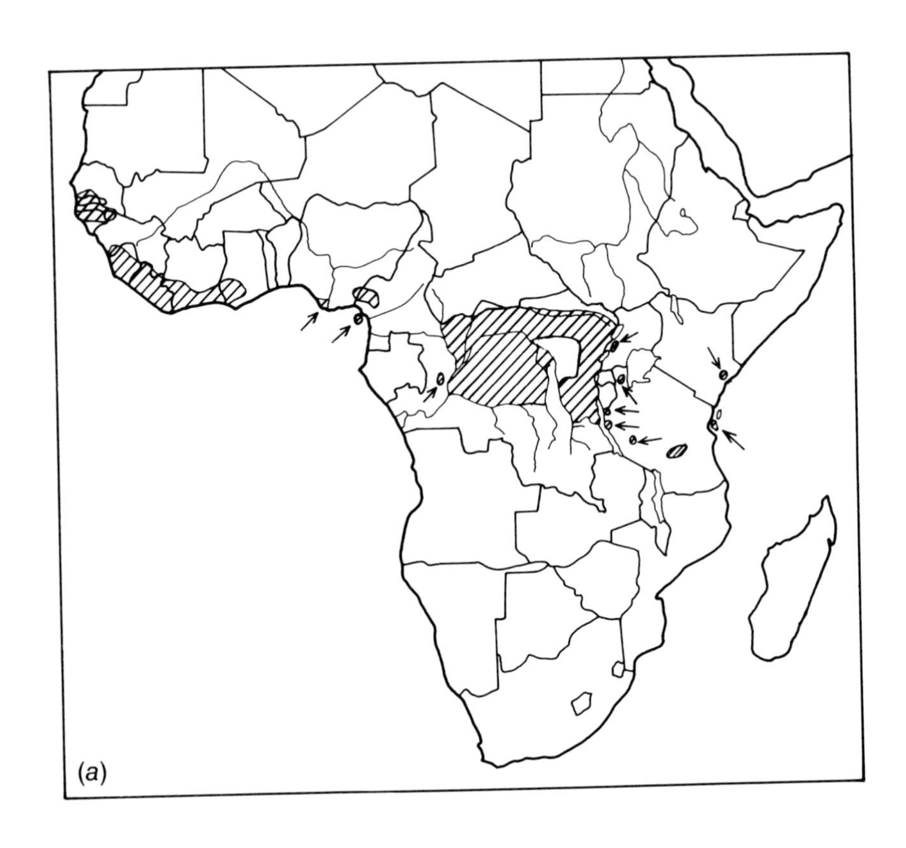

face is black. Other populations display varying permutations of black, brown, red and (in some cases) white hair. Red colobus have a complex, graded repertoire of vocalizations (Hill & Booth, 1957; Marler, 1970; Struhsaker, 1975), and wild animals call frequently (as Hill & Booth put it (p. 312), 'the species is notable for its inability to remain silent for any length of time'). The vocal system is very different from that of black-and-white colobus; for instance, there is no distinctive low-frequency adult male loud call. Probably related to this is the relatively small larynx of adult males, which is less than half the length and width of the larynx of *Colobus polykomos* (Hill & Booth, 1957).

A number of widely-followed classifications recognize 14 distinct and allopatric forms of red colobus (see e.g. Rahm, 1970; Dandelot, 1971; Napier, 1985). These are (in the order in which they were named): *badius, temminckii*

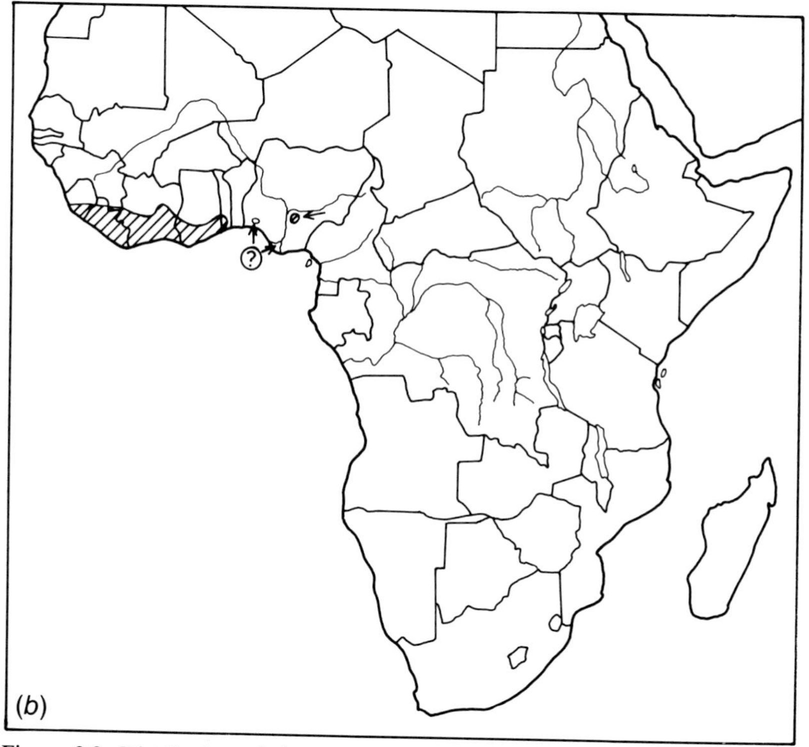

Figure 3.3. Distribution of the genus *Procolobus*. (*a*) Red colobus monkeys, *P. badius*; (*b*) olive colobus monkeys, *P. verus*. Based on information from Oates (1981, and unpublished data collated from museum collections), Colyn (1987, 1991), Rahm (1970), Rodgers (1981) and Wolfheim (1983). Isolated populations arrowed.

(Figure 3.4), *pennantii, kirkii, rufomitratus, tholloni, bouvieri, foai, gordonorum, preussi, oustaleti, tephrosceles, ellioti,* and *waldroni*. This arrangement has recently been questioned, however, by Colyn (1991) who has reviewed the zoogeography and taxonomy of monkeys of the Congo Basin. Colyn has recognized a new subspecies from E. Zaire, *parmentieri* (described by Colyn & Verheyen, 1987), and argued that the populations usually labelled *ellioti* actually comprise four distinct subspecies (*langi, lulindicus, foai* and *semlikiensis*) together with hybrids between them.

Several taxonomists have recognized that the different red colobus forms display more variation (in their pelage, vocalizations, and cranial morphology) than can easily be contained within a single-species concept. However, there is considerable disagreement about how many different red colobus species might therefore be recognized. Rahm (1970) recognized one species, Dandelot (1971) recognized five 'good' species and three 'potential' species, Delson *et al.* (1982) recognized four species and Napier (1985) recognized two. Table 3.3 summarizes these arrangements. An analysis by

Figure 3.4. Red colobus (*Procolobus badius temminckii*) at Abuko, The Gambia (photograph by D. Starin).

Table 3.3. *Classifications of red colobus monkeys*

Rahm (1970)	Dandelot (1971)	Delson *et al.* (1982)	Napier (1985)	Oates (1986)	Colyn (1991)
Colobus badius badius	*C. badius badius*	*C. badius badius*	*C. badius badius*	*Procolobus badius badius*	
C. b. temminckii	*C. b. temminckii*	*C. b. temminckii*	*C. b. temminckii*	*P. b. temminckii*	
C. b. waldroni	*C. waldroni*	*C. b. waldroni*	*C. b. waldroni*	*P. b. waldroni*	
C. b. preussi	*C. preussi*	*C. preussi*	*C. b. preussi*	*P. pennantii preussi*	
C. b. pennantii	*C. pennantii pennantii*	*C. pennantii pennantii*	*C. b. pennantii*	*P. p. pennantii*	
C. b. bouvieri	*C. p. bouvieri*	*C. p. bouvieri*	*C. b. bouvieri*	*P. p. bouvieri*	*C. badius bouvieri*
C. b. oustaleti	*C. rufomitratus oustaleti*	*C. p. oustaleti*	*C. b. oustaleti*	*P. rufomitratus oustaleti*	*C. b. oustaleti*
C. b. ellioti	*C. ellioti*	*C. p. ellioti*	*C. b. ellioti*	*P. rufomitratus ellioti*	*C. b. langi, C. b. semlikiensis,* and hybrids
C. b. foai	*C. rufomitratus foai*	*C. p. foai*	*C. b. foai*	*P. rufomitratus foai*	*C. b. foai and C. b. lulindicus*
C. b. tholloni	*C. tholloni*	*C. p. tholloni*	*C. b. tholloni*	*P. rufomitratus tholloni*	*C. b. tholloni*
C. b. tephrosceles	*C. rufomitratus tephrosceles*	*C. p. tephrosceles*	*C. b. tephrosceles*	*P. rufomitratus tephrosecles*	*C. b. tephrosceles*
C. b. gordonorum	*C. r. gordonorum*	*C. p. gordonorum*	*C. b. gordonorum*	*P. gordonorum*	
C. b. rufomitratus	*C. r. rufomitratus*	*C. rufomitratus*	*C. b. rufomitratus*	*P. rufomitratus rufomitratus*	
C. b. kirkii	*C. kirkii*	*C. pennantii kirkii*	*C. kirkii*	*P. kirkii*	
					C. b. parmentieri

Struhsaker (1981*a*) of vocalizations showed major differences between West African populations (i.e. *temminckii*, *badius* and *preussi*) and some from central and eastern Africa; Struhsaker did not, however, propose a new taxonomy.

Because there is no broad consensus on the most appropriate classification of the red colobus monkeys, we treat them in this book as a single species, *Procolobus* (*Piliocolobus*) *badius*, consisting of different local forms named according to Rahm's (1970) subspecific arrangement. A resolution of phylogenetic relationships within the group must await further research; Colyn's study (1991) is a useful start, but covers only one geographical region and relies largely on traditional craniometric and skin-colour evidence. A broader study should look at vocalizations in populations additional to those examined by Struhsaker (1981*a*); DNA studies would also be very helpful, perhaps using museum skins and/or faecal extracts.

Subgenus Procolobus, *olive colobus monkeys*

With a body weight of about 4.5 kg, the olive colobus is the smallest living colobine (see Table 3.2), and at a glance it looks like a small, slender version of the red colobus. More careful examination reveals a number of differences in addition to the size disparity. For instance, the olive colobus coat lacks any distinctive colour pattern – it is light reddish brown above and light grey below – although there is a short but noticeable sagittal crest on the crown of the head; the external ear has a hairy lateral surface; the glans penis bears minute horny papillae (unique among primates); and both the fundus of the stomach and the rectum are sacculated (Hill, 1952). Uniquely among monkeys, young olive colobus infants are carried in their mother's mouth, rather than clinging to her trunk (see Chapter 4). Only one species of olive colobus is recognized, *Procolobus* (*Procolobus*) *verus*. This species occurs only in Guinean coastal forests of West Africa (Figure 3.3(*b*)) and shows no evident subspecific variation, even between populations in Ghana and eastern Nigeria separated by two well-known zoogeographic boundaries, the Dahomey Gap and the Niger River (Menzies, 1970; Oates, 1981). This may be partly a consequence of the monkey's common association with riverine forest, because a tenuous gallery-forest network bridges the Dahomey Gap (Oates, 1988*a*).

Asian colobines

We have noted above that there is debate as to how recently the Asian colobines have shared a common ancestry. Delson (Figure 2.2, this volume) tentatively places their ancestor in the middle Pliocene, but Groves (1989) has suggested that *Nasalis* (with *Simias*) separated from other Asian forms more anciently and that it forms a sister-group to all other living colobines. *Nasalis* apart, Groves notes that the Asian colobines can be distinguished from the African species by a number of shared, derived features, including a shorter face and the presence of a suborbital fossa in the skull.

One well-known classification (Napier & Napier, 1967) has recognized five genera of Asian colobines: *Presbytis*, *Rhinopithecus*, *Pygathrix*, *Nasalis* and *Simias*. Although this classification is widely used, at least in the behavioural literature, it is not accepted by most authorities on Asian colobine systematics. In particular, it is clear that the group of monkeys commonly lumped together as *Presbytis* includes a number of distinct lineages. On the other hand, a generic-level division of the other ('odd-nosed') Asian colobines may not be the best way of expressing their relationships; Groves (1970), for instance, has presented evidence for regarding *Rhinopithecus* as a subgenus of *Pygathrix*, and for uniting *Simias* with *Nasalis*.

A division of *Presbytis* (*sensu lato*) into several genera and/or subgenera has a long history; for instance, Reichenbach (1862) recognized four subgenera of what he called *Semnopithecus*. Hill (1934) and Pocock (1935, 1939) used the coloration of infants, skull morphology, and features of the female external genitalia to separate *Semnopithecus*, *Trachypithecus* and *Kasi* as genera from *Presbytis* (*sensu stricto*). Brandon-Jones (1984) (tentatively followed by Strasser & Delson (1987)) proposed a two-genus arrangement into *Presbytis* and *Semnopithecus* (with the latter including *Trachypithecus* as a subgenus).

Groves' most recent (1989) review of Asian colobine taxonomy used a five-genus arrangement: *Nasalis, Pygathrix, Presbytis, Trachypithecus* and *Semnopithecus*. We feel that in general this arrangement well reflects both existing phylogenetic evidence based on morphology, and patterns of variation in behaviour and ecology. We therefore follow it with one exception; we agree with Medway (1970) and Napier (1985) that the great differences in the external appearance and ecology of *Nasalis* and *Simias* warrant their generic separation. The distribution of the six Asian genera follows a complicated pattern. The odd-nosed genera occupy areas on the northern and eastern boundaries of Asian colobine distribution, with *Pygathrix* occurring in the mountains of southern China and in eastern Indo-China, *Nasalis* in Borneo and *Simias* in the Mentawai Islands. *Presbytis* dominates moist inland forests in South-east Asia, south of the Isthmus of Kra, and *Semnopithecus* occupies the Indian subcontinent (including Sri Lanka). *Trachypithecus* is the most widely distributed genus, occurring from Sri Lanka to Java, with several species occupying a large area of mainland South-east Asia, including dry forests and mangrove swamps.

Trachypithecus occurs sympatrically with *Presbytis* in many parts of Malaysia and Indonesia, with *Nasalis* in coastal Borneo, and with *Pygathrix* in Indochina. *Simias* is sympatric with *Presbytis* on the Mentawai Islands.

Genus **Pygathrix,** *snub-nosed monkeys*

We follow Groves (1970) and Napier (1985) in regarding the Chinese snub-nosed monkeys (*Rhinopithecus*) as a subgenus of *Pygathrix*. This genus is notable for its nasal peculiarities; small flaps of skin are present on the upper borders of the nostrils and the nasal bones themselves are reduced or absent. These monkeys are unusual among colobines in having forelimbs almost as long as their hindlimbs, and ischial callosities are separate in both males and females (Napier, 1985).

Jablonski & Peng (1993) have analysed 178 characters in *Pygathrix* and *Rhinopithecus*, and found that all the *Rhinopithecus* species are closer to each other than any of them is to *Pygathrix nemaeus*, which has the largest number of primitive features but also some unique specializations (for instance, in its locomotor apparatus and gut). On this basis, Jablonski & Peng propose the generic separation of *Rhinopithecus* from *Pygathrix*. We regard a subgeneric separation as more sensible, given the clear affinities of these forms.

Subgenus Pygathrix

The douc (or douc monkey), *P. nemaeus*, occurs in the forests of eastern Indo-China (Figure 3.5). Two subspecies of this strikingly coloured colobine are usually recognized, each with a different pattern of grey, white, red and black hair (Figure 3.6). Brandon-Jones (1984) treated the southern form, *nigripes*, as a separate species from the northern *nemaeus*. Doucs have little or no sexual dimorphism in body size (Napier, 1985).

Subgenus Rhinopithecus

These snub-nosed monkeys have extremely reduced nasal bones, and the flaps of skin on the upper borders of the nostrils stand erect, as twin peaks (Napier 1985; Caton, 1991). They occur in northern Vietnam and south-eastern China (including Tibet) (Figure 3.5). There are four allopatric forms, but there has been disagreement on how many species these comprise. Napier & Napier (1967) followed Ellerman & Morrison-Scott (1951) in recogniz-

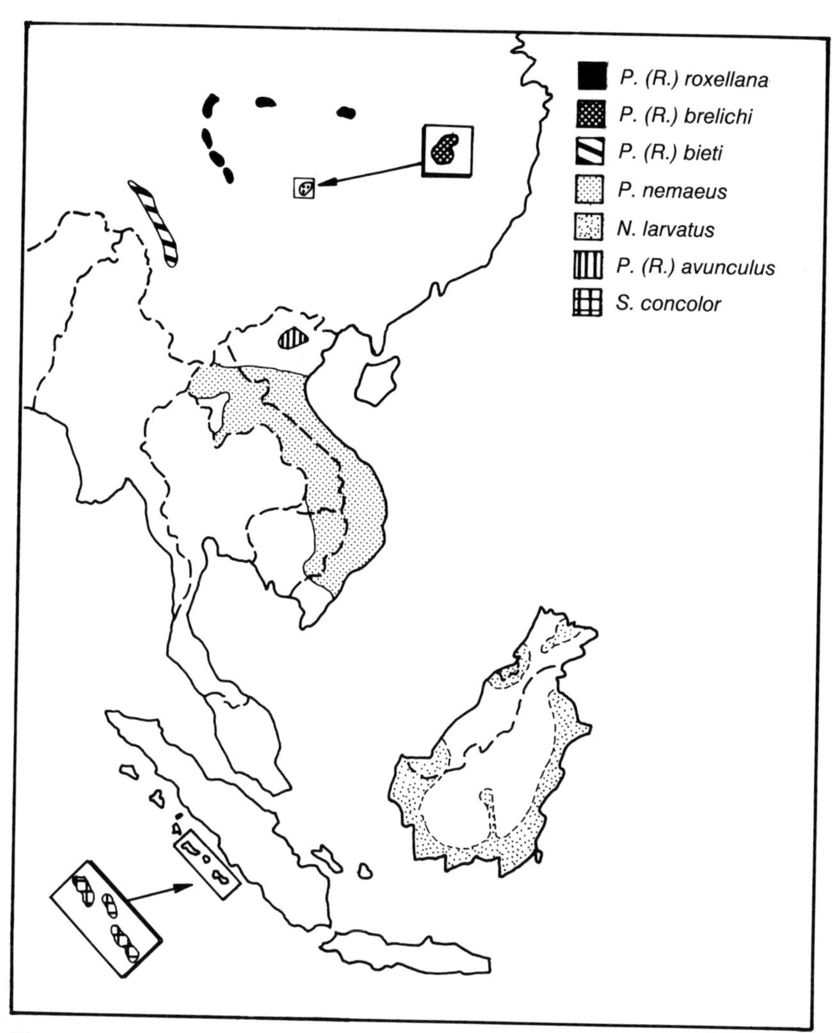

Figure 3.5. Distribution of the genera *Pygathrix* (with subgenus *Rhinopithecus*), *Nasalis* and *Simias*. Based on information from E. Bennett (personal communication), Bodmer *et al.* (1991), Long *et al.* (1994), MacKinnon & MacKinnon (1987) and Wolfheim (1983).

ing two species: *Rhinopithecus roxellanae* containing the three Chinese forms (*bieti, brelichi* and *roxellanae*), and *R. avunculus* containing the Vietnamese monkey. Subsequently, Groves (1970) elevated *brelichi* to species status, and has now (1989) accepted *bieti* also as a full species, based on the studies of Chinese zoologists (see e.g. Li *et al.*, 1982). Such a four-species arrangement

Figure 3.6. The southern form of the douc (*Pygathrix nemaeus nigripes*) (photograph by N. Rowe).

has also been supported by the analysis of Jablonski & Peng (1993), and we therefore accept these four species here: *P.* (*R.*) *roxellana*, *P.* (*R.*) *bieti*, *P.* (*R.*) *brelichi*, and *P.* (*R.*) *avunculus*. According to Napier (1985), *roxellana* and not *roxellanae* is the correct name for the Sichuan form, which is often known as the golden monkey. Jablonski & Peng separate *P.* (*R.*) *avunculus* as the subgenus *Presbytiscus*, implying a phyletic separation from the other three species, but we do not think such a subdivision is necessary.

P. (*R.*) *roxellana* has a greyish-brown back ornamented with long golden strands, and its yellowish-white underparts become golden with age; *P.* (*R.*) *bieti* and *P.* (*R.*) *brelichi* (Figure 3.7) are darker in colour; *P.* (*R.*) *avunculus* is dark brown on the back, and has yellowish-white underparts and an orange throat patch (Napier, 1985). While *P.* (*R.*) *avunculus* inhabits lowland forests in northern Vietnam, the other three forms are restricted (at least today) to montane habitats in southern China. Of the three, *P.* (*R.*) *bieti* occurs at the highest altitudes (between 3000 and 4300 m), in temperate fir-larch forest subject to harsh winter conditions; this species has 70% of its activity on the ground (Li *et al.*, 1982; Zhao, 1988; Long *et al.*, 1994; Wu, 1993).

Genus **Nasalis**, *proboscis monkey*

This genus contains the single species, *N. larvatus*, the proboscis monkey. The species takes its common name from the large, pendulous, fleshy nose of the adult male, a feature unique among primates (Figure 3.8). Proboscis monkeys occur only in Borneo, where they are typically associated with coastal swamp forests, including mangroves, although they also occur in riverine forest far inland (Chivers & Burton, 1988). They are unusual among primates in that they regularly swim. This monkey is also notable for its extreme sexual dimorphism; adult males have an average body weight of 21 kg, twice the size of females (Table 3.2).

Genus **Simias**, *pig-tailed monkey, or simakobu*

This is another very unusual colobine. The single species *S. concolor* is found only in the Mentawai Islands, and is the only colobine with a markedly short tail. Both male and female simakobu have short, turned-up noses. In body proportions, simakobu resemble macaques; they have relatively short arms and legs of similar length, and short, bare tails (Napier, 1985). They show little sexual dimorphism in size, and two colour phases occur, not related to sex (Tilson, 1977) (Figure 3.9). Most observed social groups are small (3–8 monkeys), and sometimes monogamous (Tilson, 1977; Watanabe,

Figure 3.7. The Guizhou snub-nosed monkey *Pygathrix* (*Rhinopithecus*) *brelichi* in the Wuling Mountains of south-central China (photograph by N. Rowe).

Figure 3.8. Adolescent male proboscis monkeys (*Nasalis larvatus*), part of an all-male group in Tanjung Puting National Park, Kalimantan Tengah, Borneo (photograph by T. Blondal and C. Yeager).

Figure 3.9. Light-phase juvenile simakobu, *Simias concolor* (photograph by R. Tilson).

1981), although Watanabe found two primary forest groups to contain about 20 individuals each. Tenaza (1989) has observed prominent pink sexual swellings in females on S. Pagai Island, a unique feature among Asian colobines. Simakobu may descend to the ground and flee when disturbed by hunters (Tilson, 1977).

Groves (1970) first suggested that simakobu were closely related to proboscis monkeys and classified them as *Nasalis concolor*. Delson (1975*a*) agreed that the simakobu skull was an excellent structural intermediate between those of *Nasalis* and *Presbytis*, and ranked the proboscis monkey and simakobu as monotypic subgenera of *Nasalis*, given the many differences between the two species. Delson continues to strongly support this argument, which takes a middle course between the simple congeneric association of Groves (1989) and the generic distinction of others. This generic distinction is followed in this volume and favoured by Davies and Oates, because of the ecological distinctiveness of the simakobu.

Genus Presbytis, *leaf-monkeys*

Although these monkeys have often been grouped in the same genus with the langurs (*Semnopithecus* and *Trachypithecus*), they may be distinguished from them by several craniodental features: a short face with weakly developed brow ridges, a convex nasal profile, deep and consistent underbite, relatively broad homomorphic incisors, thick dental enamel and a reduced or absent hypoconulid on the lower third molar (Brandon-Jones, 1984; Napier, 1985; Groves, 1989). Newborn infants are white or whitish and as the coat darkens during development they pass through a stage which displays a dark cruciform pattern on the back and upper head (Pocock, 1928). Compared to *Trachypithecus*, *Presbytis* have relatively longer hindlimbs, leap more (Figure 3.10) and use quadrupedalism less (Fleagle, 1976; Strasser, 1992), and they have relatively smaller stomachs (Chivers, Chapter 7, this volume). Adult *Presbytis* are typically less dimorphic than *Trachypithecus*, with adult males only slightly larger than females (Table 3.2).

The genus *Presbytis* is restricted to rain forests in southern Thailand, the Malay Peninsula, Borneo, Sumatra (including the Mentawai Islands), Java, Bali and the Lomboks (Figure 3.11). We recognize seven species in this region: *P. melalophos*, *P. comata*, *P. frontata*, *P. rubicunda*, *P. potenziani*, *P. hosei*, and *P. thomasi*. Napier (1985) made the same division of this group, modifying Chasen's earlier (1940) arrangement. The seven species are differentiated largely by their coat patterns; skeletally they are difficult to distinguish (Medway, 1970). The distinctive nature of the male loud call was used

Figure 3.10. Banded leaf-monkey *Presbytis melalophos* leaping, West Malaysia (photograph by J. Fleagle).

by Wilson & Wilson (1975) to justify a separation of *P. melalophos* into three species: *femoralis, cruciger* and *melalophos*. However, Bennett & Bennett (1988) have found that differences in the calls of *femoralis* and *melalophos* are not consistent in the wild, and in that light we revert here to regarding *P. melalophos* as one (highly variable) species. We also retain *siamensis* as a subspecies of *melalophos*, although Brandon-Jones (1984) gave it species status.

Brandon-Jones (1977) claimed that *P. potenziani* is the most primitive member of the group, but provided no convincing support for this claim. He implied that he regarded its dark coloration as a significant primitive feature. However, this is by no means an all-black monkey; Brandon-Jones (1984) and Napier (1985) made it clear that the coat carries white and reddish markings. Furthermore, Tilson (1976) observed that *P. potenziani* infants are white at birth, but soon develop the 'cruciger' pattern.

Among the species we recognize, sympatry occurs between *P. hosei* and *P. rubicunda* in north Borneo, and between *P. rubicunda* and *P. frontata* in western and central Borneo.

We refer to the members of this genus as 'leaf-monkeys' because this term has wide usage. The name is not entirely satisfactory, because these monkeys

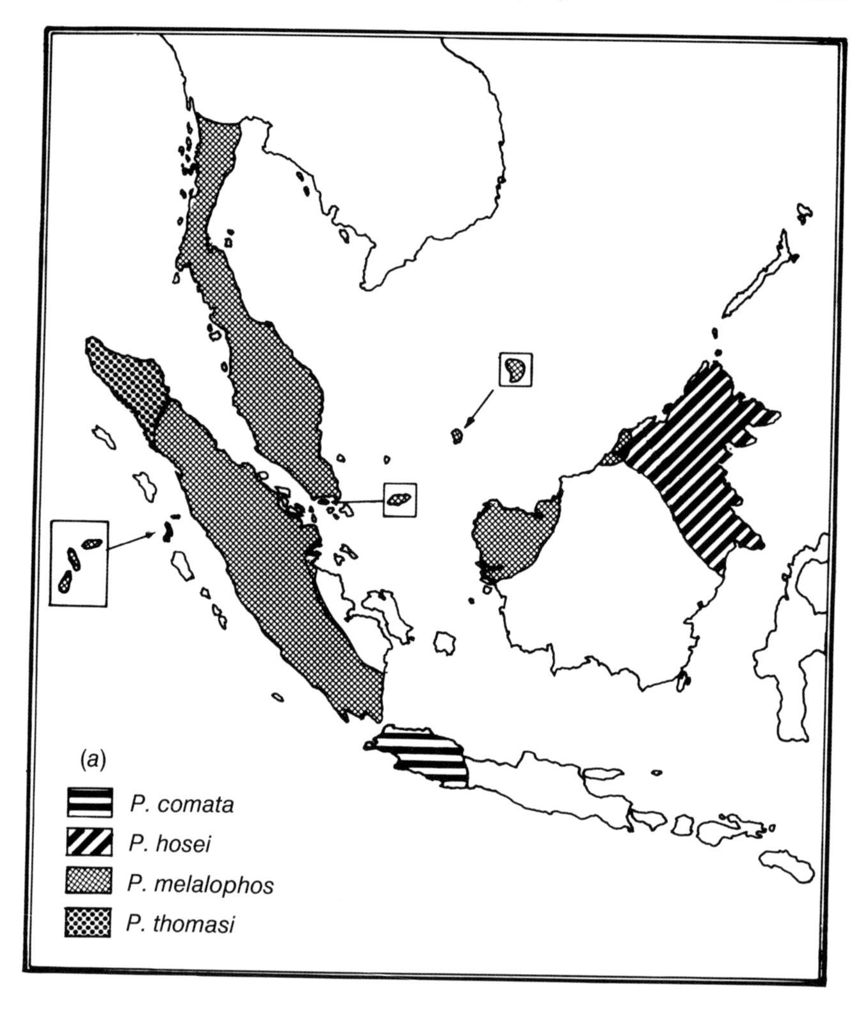

often include large quantities of seeds in their diet (see Chapter 5), but we think it is more suitable than 'sureli', the name used by Brandon-Jones (1984). *Sureli* is a Javan word and is used for just one, peripheral, member of the genus.

Genus **Trachypithecus,** *langurs*

Members of this genus have heteromorphic incisors with an edge-to-edge bite, and the hypoconulid on the lower third molar is unreduced (Weitzel, 1983; Napier, 1985); their newborn young are orange, brown or grey in col-

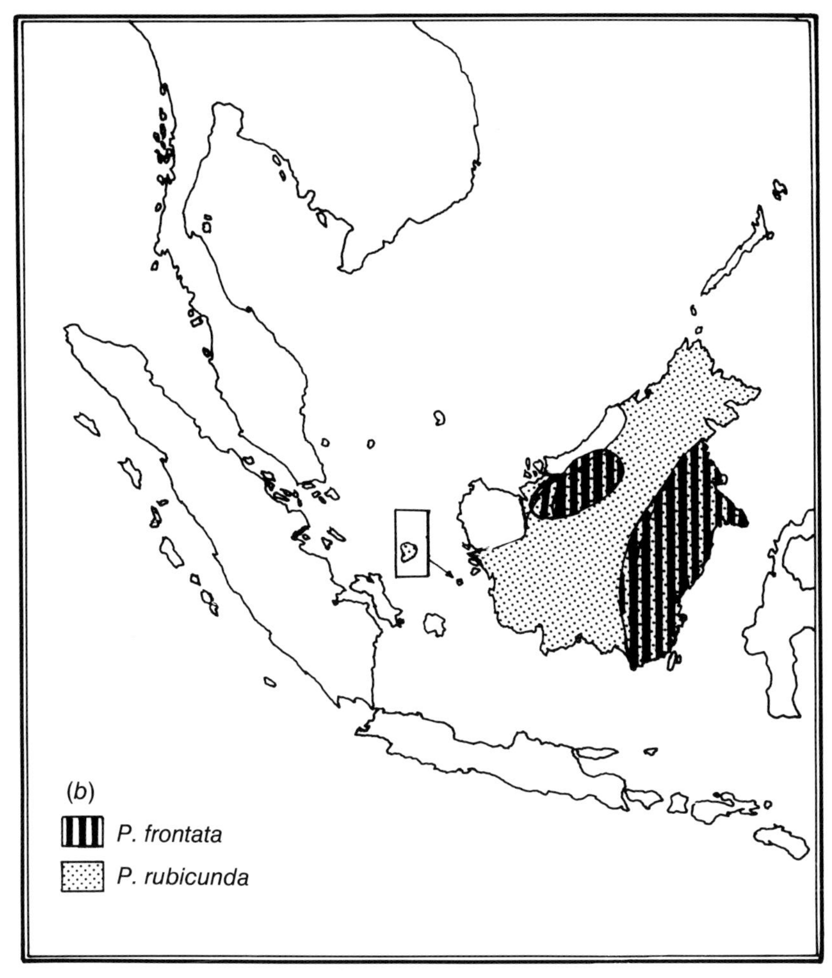

Figure 3.11. Distribution of the genus *Presbytis*. (*a*) *P. melalophos*, *P. thomasi*, *P. comata* and *P. hosei*; (*b*) *P. frontata* and *P. rubicunda*. Based on E. Bennett (personal communication), Payne *et al.* (1985) and Wolfheim (1983).

our. Compared with *Presbytis*, they have relatively shorter hindlimbs, engage in more quadrupedal walking and running (Fleagle, 1977), and have relatively larger stomachs in relation to their body size (see Chapter 7). *Trachypithecus* species occur in a wider range of conditions than *Presbytis*; in addition to moist and wet lowland forests, they are found in dry deciduous forests, coastal mangrove swamps, and montane broad-leaved forests. Their range includes Sri Lanka and South India, eastern India and Bangladesh, Burma,

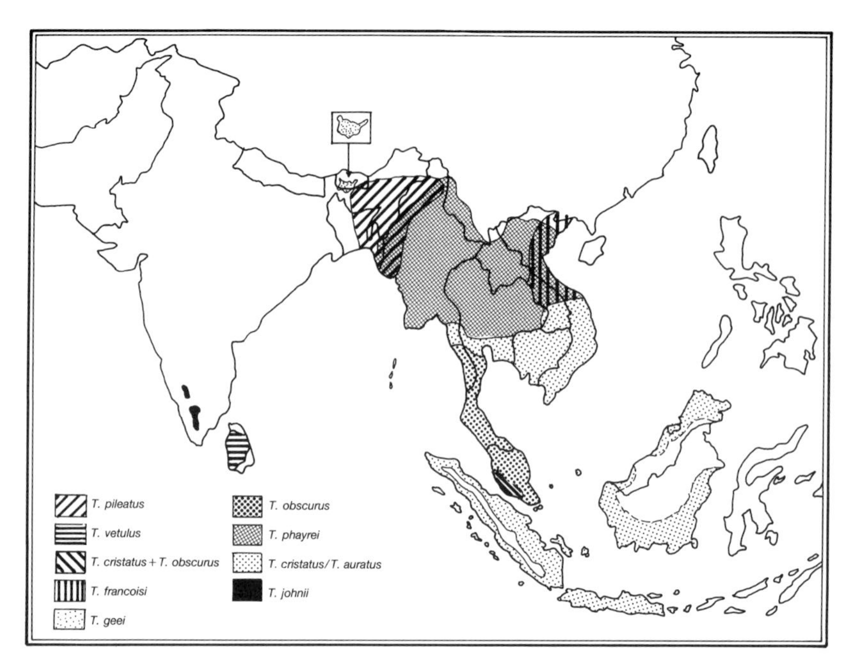

Figure 3.12. Distribution of the genus *Trachypithecus*. Based on E. Bennett (personal communication), Choudhury (1988, 1992), Fooden (1976), Green (1978), MacKinnon & MacKinnon (1987), Medway (1970), Oates (1979), Payne *et al.* (1985), Wilson & Wilson (1977) and Wolfheim (1983).

Indo-China and southern China, the Malayan Peninsula and the Sunda Islands (see Figure 3.12).

There is no entirely appropriate common name for this group of monkeys, but we feel that the widely-used Hindi name 'langur' (originally applied to *Semnopithecus entellus* and referring to the monkey's long tail (Hrdy, 1977)) is a suitable term, both because four species of *Trachypithecus* occur in India itself, and because *Semnopithecus* and *Trachypithecus* appear to be sister taxa (an issue discussed below).

We recognize nine species in this group. These species are the eight members of Napier's (1985) *Presbytis cristata* group (i.e. *T. cristatus*, *T. vetulus*, *T. johnii*, *T. obscurus*, *T. pileatus*, *T. phayrei*, *T. francoisi* and *T. geei*) (Figure 3.13) together with *T. auratus*. *T. auratus* was included by Napier and others within *T. cristatus*, but Weitzel & Groves (1985) have shown that this Javan population has distinctive cranial morphology.

Brandon-Jones (1984) questioned the unity of *T. francoisi*, a species from northern Vietnam and an adjacent area of southern China. He regarded the

Figure 3.13. Dusky langur (*Trachypithecus obscurus*) (photograph by J. Caldecott).

subspecies *francoisi*, *leucocephalus* and *delacouri* as full species, apparently on the basis of pelage differences, and recently Ratajszczak (1988) has reported that Chinese zoologists have seen *leucocephalus* and *delacouri* living sympatrically without interbreeding. Brandon-Jones listed another subspecies, *T. francoisi poliocephalus* (from Cat Ba Island in the Gulf of Tonkin), as a subspecies of the South Indian *T. johnii*. We see no cogent reason to accept such a grouping, in the absence of a published analysis; Ratajszczak listed *poliocephalus* as a form of *francoisi*.

Groves (1989) considered *geei* to be a 'well-differentiated offshoot' of *T. pileatus*, although Brandon-Jones (1984) kept it as a species. Brandon-Jones (1984) also placed *phayrei* in *T. obscurus* (but did not justify this), and recognized the form *barbei* as a separate species. Napier (1985) has discussed the position of *barbei* (a species tentatively proposed by E. Blyth in 1847 based on two specimens which are from an uncertain locality), and concluded that its status is in doubt. Gupta, who has studied *T. phayrei* in Tripura, India, considers that the *barbei* specimens belong to *phayrei* (A. K. Gupta, personal communication).

Pocock (1939) separated *T. vetulus* and *T. johnii* (Figure 3.14) from the other *Trachypithecus* and placed them in the genus *Kasi*. The adults of these

Figure 3.14. Nilgiri langur (*Trachypithecus johnii*) at Kakachi, India (photograph by J. Oates).

two species are somewhat larger in body weight than the other species and their newborn infants are not orange but grey (*vetulus* of Sri Lanka) or predominantly reddish-brown (*johnii* of South India). However, parts of the neonatal coat of *johnii* have an orange tinge (J. F. Oates, personal observation), and the skull anatomy and diet of both species are similar to other *Trachypithecus*; we therefore include them here, pending further study.

Apart from a very limited area of sympatry between *T. cristatus* and *T. obscurus* on the west coast of the Malay Peninsula (Marsh & Wilson, 1981), the only sympatry between members of this genus is between *T. phayrei* and other species in the east (*pileatus*) and west (*francoisi*) of its range (see Figure 3.12).

Genus **Semnopithecus**, *Hanuman or grey langur*

Groves (1989) has noted that, while the Hanuman langur shares many characteristics with *Trachypithecus*, it can also be clearly separated by the blackish-brown coat coloration of young infants and by skull morphology (e.g. heavy horizontal brow ridges with a marked depression posteriorly (Napier, 1985)). In this volume we follow Groves in recognizing *Semnopithecus* as a monotypic genus, containing only the species *S. entellus*. Strasser & Delson (1987), however, followed Brandon-Jones (1984) in regarding *Semnopithecus* as also including the subgenus *Trachypithecus*, a course which Delson still prefers and one which has the merit of equalizing to some degree the variation seen among colobines; the African genera *Colobus* and *Procolobus* are about as distinct as *Presbytis* and *Trachypithecus*, while the subgenera within *Procolobus* are about as similar as are *Semnopithecus* and *Trachypithecus*. Davies and Oates, on the other hand, are impressed by the considerable ecological divergence between the Hanuman langur and the *Trachypithecus* species. Hanuman langurs do not inhabit closed-canopy tropical forests, they exhibit considerable terrestriality (Figure 3.15), and they have very flexible diets that are rarely as folivorous as those of *Trachypithecus*.

S. entellus is remarkably variable and adaptable, occurring from Sri Lanka north to the Himalayas, from sea level to an altitude of 4000 m (map, Figure, 3.16), and in a great range of habitats from dry tropical scrub jungle to montane coniferous forest (Roonwal & Mohnot, 1977; Wolfheim, 1983). It is commonly associated with human settlements where it has traditionally been tolerated due to its association with the monkey-god Hanuman in Hindu mythology (Hrdy, 1977). Many populations are highly terrestrial in their behaviour. Given the wide geographical spread and the range of habitats occupied by Hanuman langurs, it is not surprising that there is a great deal

Figure 3.15. Hanuman langurs (*Semnopithecus entellus*) drinking at a boulder pool at Kanha, India (photograph by P. Newton).

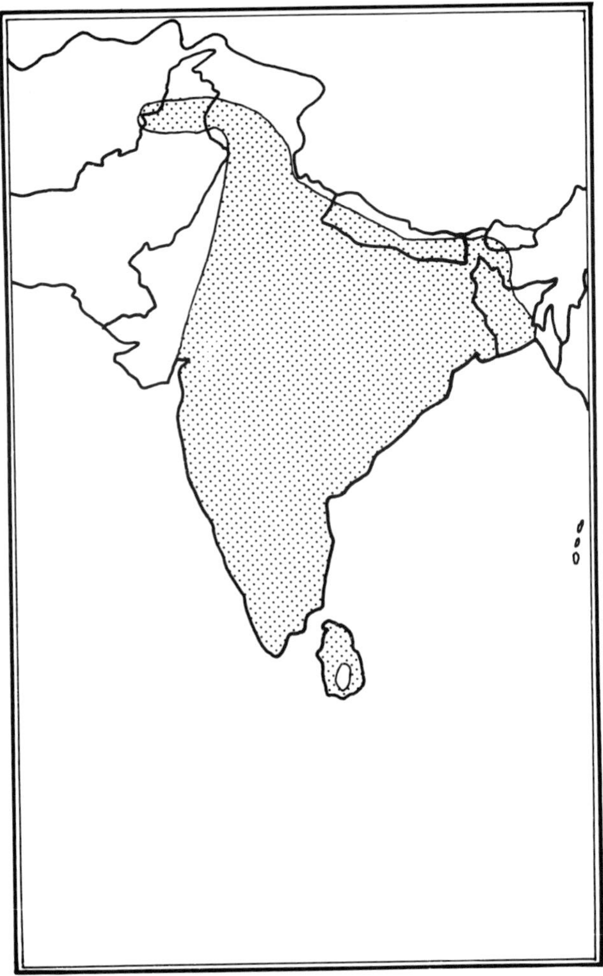

Figure 3.16. Distribution of *Semnopithecus entellus*. Based on Napier (1985), Roonwal (1981) and Wolfheim (1983).

of interpopulational variability in size (adult females from Sri Lanka average 7 kg, those from the Himalayas over 15 kg) and coloration (from predominantly pale grey to dark chocolate brown) (Napier, 1985); this has led to the description of numerous subspecies. Roonwal & Mohnot listed 16 subspecies, while Napier listed 15 in four groups (based largely on pelage colour and pattern). Brandon-Jones (1984) regarded the south-west Indian form *hypoleucos* as a distinct species. This form lives in relatively moist habitats and has particularly dark hair; Brandon-Jones appears to believe that this dark coat is a primitive feature, justifying separation at the species level. In the absence of a careful assessment of other characteristic we prefer to follow the traditional arrangement, and recognize only a single species of grey langur.

Following similar logic to that of Brandon-Jones, Groves (1989) argued that *Semnopithecus* is more primitive than *Trachypithecus*, because of the blackish colour of newborn infants. We are not convinced that dark coloration alone is a strong indicator of primitiveness.

Acknowledgements

We thank William Bleisch, Ardith Eudey and Nina Jablonski for guiding us to some of the information reviewed in this chapter, and Elizabeth Bennett for commenting on an early draft.

4

The natural history of African colobines

JOHN F. OATES

Introduction

Africa's colobines – the colobus monkeys – are often thought of as rare, specialized forest dwellers, with limited geographical distributions, monotonous diets and dull social lives. Although some kinds of colobus monkeys are among the most localized and endangered of African primates, others are widespread and relatively common. In fact, the colobus monkeys are a diverse group of primates that display considerable ecological and social flexibility, and have some intriguing special features.

As explained in Chapter 3, there are three distinct subgroups of colobus monkeys, the black-and-white, the red, and the olive. The black-and-white and red groups each contain many different forms, but how many of these merit recognition as distinct species is disputed; all authorities recognize only a single species of olive colobus, however. Although colobus monkeys are restricted to wooded habitats, they occupy a great diversity of such habitats. Until quite recently, colobus monkeys of some kind occurred throughout most of the African moist lowland forest zone, as well as in many montane forests and in gallery forests deep into the savanna zone. Even today, despite the widespread destruction of their habitats and, often, intense hunting pressure, these colobus populations occur from Senegal in the west to Zanzibar in the east, and from the Ethiopian Highlands to the southern edge of the Congo Basin. Colobus monkeys can achieve some of the highest biomasses recorded for any primates anywhere (for example, red colobus in Uganda's Kibale Forest). Their diets may be dominated by tree leaves (black-and-white colobus in the Kibale Forest), or by seeds (black colobus at Douala-Edéa, Cameroon, and at Lopé, Gabon), or may be very diverse (many red colobus populations).

The social groups of colobus monkeys range in size from some of the largest found in any forest-living primates (for instance, groups of Angolan black-and-white colobus in the Nyungwe Forest of Rwanda may contain over 300 individuals) to some of the smallest (groups of olive colobus monkeys in West African forests often number less than half-a-dozen). Some populations of East African guerezas exhibit a one-male harem group structure, while many other colobus have multi-male groups. Red colobus are one of the few kinds of monkeys in which adolescent (and adult) females transfer out of their natal group more frequently than do males, and olive colobus females are the only monkeys which carry their infants in their mouths rather than having them cling to their thorax.

This chapter reviews findings from field studies of African colobine natural history, with a focus on diet, habitat use and social structure. Five major research sites, which are mentioned frequently in subsequent chapters, will be described in more detail than others.

General features of African colobine habitats

African forests are different in several significant respects from those of the two other major tropical forest areas inhabited by primates: southern Asia, and Central and South America. Tropical Africa covers a larger land area than tropical Asia or America and has a generally drier climate. Many African forests experience prolonged and pronounced dry seasons, and only small areas have annual rainfalls exceeding 2500 mm. Owen-Smith (1989) ascribes the strong seasonality of African climates to the continent's extensive high-altitude interior plateau.

Most of the African moist forest lies within 10° of the equator, in a zone where two rainfall peaks are normally separated by two relatively dry periods, one short and one long. Local factors, however, produce a single peak of rainfall in many areas within this zone (see Figure 4.1). Even more arid conditions have affected the African tropics in the past, as shown by a wealth of evidence on pollen distribution, lake level fluctuations and wind-blown sand deposits; for instance, wind-blown Kalahari sands underlie much of the Congo River Basin (Moreau, 1966). This may be one factor that has acted to produce a lower tree-species richness in African than in Asian and American forests (Richards, 1952).

Some extensive areas of the African moist forest zone, particularly in the north-eastern part of the Congo Basin, are dominated by a single tree species. These dominants are often caesalpinioid legumes such as *Gilbertiodendron dewevrei* and *Cynometra alexandri*. In this group of leguminous trees, symbiotic ectomycorrhizas are common in the root system (Hogberg, 1986). Soil conditions may contribute to the production of these 'monodominant' forests,

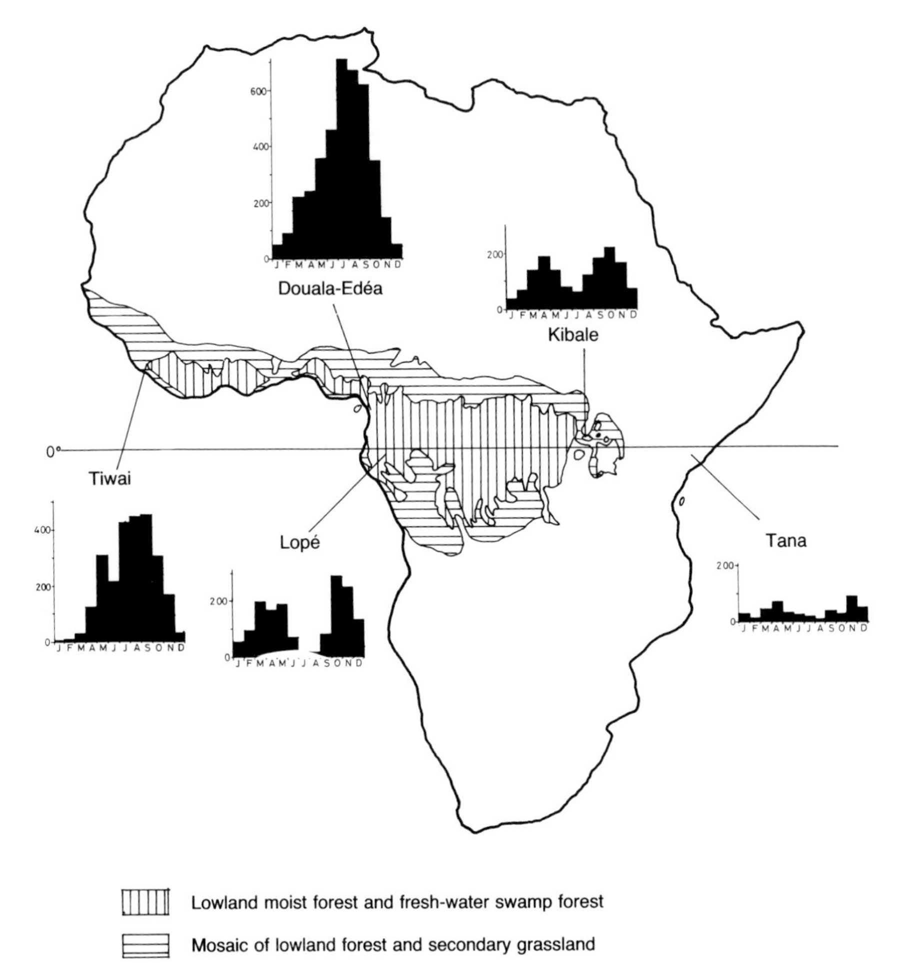

Figure 4.1. Location of five colobus study sites in relation to the African lowland moist-forest zone, and histograms of average monthly rainfall (in mm) for these sites. Rainfall data for Tiwai were collected by the author and colleagues; data for Douala-Edéa from McKey (1978*a*); for Lopé from L. J. T. White (personal communication); for Kibale from Oates (1974); and for Tana from Marsh (1976).

but it seems that long-term local environmental stability may be a more important determinant of their occurrence (White, 1983; Hart *et al.*, 1989). Recent studies in eastern Zaire and northern Congo suggest that *Gilbertiodendron* forests support only low densities of colobus and other monkeys (Thomas, 1991; Mitani, 1992); however, in riverine *Gilbertiodendron* forest in northern Congo, red colobus, although not common, were the most abundant monkey species (Mitani, 1992).

The tree-species richness of tropical African forests may be lower than that of the richest Asian and American sites, but faunal diversity (of birds and mammals, at least) is similar. Any one area of African moist forest appears to harbour at least as many mammalian species as an equivalent area in Asia or South America, while bird species-richness in African forests is intermediate between Asian and American levels (Bourlière, 1983). Primate species-richness is greatest at some African sites. For instance, 15 primate species occur together in some parts of eastern Zaire (Allen, 1925; Hart *et al.*, 1986), compared with 13 in the richest New World locality (Terborgh, 1983), and eight at the richest Asian site (Bourlière, 1985). However, no more than three colobine species occur sympatrically in Africa, and even those cases are rare; typically, only one or two colobine species are found at any one site; this is discussed further in Chapter 12. Even after allowing for the underestimation of African colobine diversity by traditional classifications, the Asian colobine fauna as a whole is evidently richer in species than that of Africa (see Chapter 3). This is most probably related both to the ecological diversity of the forests inhabited by Asian colobines, and to the marked discontinuity of the Asian forest zone (which is especially striking in the Sundaland archipelago).

The Hominidae probably evolved in wooded environments in Africa (Leakey, 1963; Andrews, 1989), and today human influences are pervasive in African forest ecosystems. Human hunters are active, or have been active in the recent past, in most of the moist forest zone. Monkeys, including colobus, are usually included among the most frequently or preferentially hunted animals, whether by traditional hunters (e.g. the pygmies of Lobaye, Central African Republic (Galat, 1977)), or by organized commercial groups using vehicles and shotguns (e.g. in Sierra Leone (Davies, 1987*a*)). Habitats themselves have been widely modified, in most areas by shifting cultivation and small-scale tree cutting, and in many areas by plantation agriculture or commercial forestry. In West Africa, forest that has never been cultivated exists only in a few special situations, such as swamps and steep rocky slopes (Richards, 1952).

Colobus study sites

Long-term, ecologically oriented studies of colobus monkeys, employing systematic observational sampling of habituated groups, have been conducted in most of the major African forest regions, as well as at many outlying sites. A notable exception is the Congo (or Zaire) Basin, the largest forested region on the continent, where several forms of colobus occur. So far only one field study extending over a complete year has been completed in the Congo Basin and this study (Maisels *et al.*, 1994) did not involve fully habituated animals.

Five colobus study sites will be given special attention in this chapter. At these sites (Tiwai, Douala-Edéa, Lopé, Kibale and Tana River), similar methods were used to study habituated groups over many months, with an emphasis on feeding behaviour. Except at Tana River, information on the chemistry of plants and soils was also collected at these sites, whose key ecological features are compared in Tables 4.1–4.3, and which are mapped in Figure 4.1.

Tiwai Island

Tiwai is a 12 km^2 island in the Moa River of southern Sierra Leone. It is the largest and most northerly in a series of Moa islands in a zone of geological faulting about 80 km inland from the Atlantic Ocean. When river levels are low at the end of the dry season, parts of Tiwai are separated from the mainland by only narrow, shallow water channels (Figure 4.2), while tree bridges provide connections to downstream islands that are adjacent to the Gola West Forest Reserve. Gola West is part of a forest ecosystem area that extends east to the Liberian border (30 km from Tiwai) and that once stretched from southern Sierra Leone through Liberia to western Côte d'Ivoire. This region probably contained a distinct forest refuge during dry phases of the Pleistocene (Booth, 1958), and evidence from the distribution patterns of bird species (Diamond & Hamilton, 1980) suggests that this refuge might have been located in Sierra Leone itself.

Tiwai has a gently undulating terrain. Shallow sandy soils are underlain by granite and laterite, which outcrop in places. Several shallow valleys are drained by small perennial streams and become flooded in the wet season. Most of the high annual rainfall (averaging 2800 mm) is concentrated in the period May–October, and there is very little rain from mid-December to March.

At one time, most of Tiwai must have supported high-stature moist lowland forest that displayed spatial variation as a result of edaphic factors, especially soil moisture. Tiwai has been uninhabited by humans for many years, but has long been subject to farming (especially rice cultivation) and tree cutting by local Mende people from mainland villages. At present, the island supports a mosaic of young secondary forest (regenerating on old farmland), *Raphia*-palm swamp, old secondary forest and riverine forest. Old secondary forest covers about 60% of the island; in this forest, leguminous trees are

Table 4.1. *Soil characteristics of four colobus study sites*

		Particle composition (%)			Mineral Concentrations				
					(ppm)	(mequiv./100 g)			
Site	pH	Sand	Silt	Clay	P	K	Ca	Mg	Reference
Tiwai Island	4.3	77	14	9	4.5	0.017	0.394	0.044	Oates *et al.* (1990)
Douala-Edéa (transect A)	3.4	85	12	3	22.4	0.255	0.497	0.283	McKey (1978*b*); Gartlan *et al.* (1980)
Lopé, de Brazza Hills	4.5	54	23	23	<4	0.113	0.258	0.203	M. Harrison (personal communication)
Kibale, Kanyawara	5.6	60	28	13	14.0	0.195	4.550	1.833	McKey (1978*b*); Gartlan *et al.* (1980)

Table 4.2. *Vegetation of selected colobus study sites*

Site	Vegetation classification	Top three tree species in canopy[a]	Percentage contribution of top three species to canopy[a]	Percentage contribution of legumes to canopy[b]	Phenology: Peak of young leaf production	Phenology: Peak of fruit production	Reference
Tiwai	Semi-deciduous lowland moist forest	*Pentaclethra macrophylla* *Funtumia africana* *Piptadeniastrum africanum*	30	41	Mar–Jul	Oct–Mar	Oates *et al.* (1990); Dasilva (1989)
Douala-Edéa	Evergreen lowland rain forest	*Protomegabaria stapfiana* *Lophira alata* *Librevillea klainei*	48	18	No obvious peak	Jun–Aug	McKey *et al.* (1981)
Lopé	Semi-deciduous lowland moist forest	*Aucoumea klaineana* *Scyphocephalium ochocoa* *Pentaclethra macrophylla*	35	24	Oct–Nov, Jan–Feb	Dec–Feb	M. Harrison (personal communication)
Kibale	Semi-evergreen medium-altitude moist forest	*Diospyros abyssinica* *Markhamia platycalyx* *Celtis durandii*	48	3	Apr–May, Oct–Dec	No obvious peak	Struhsaker (1975); Oates (1974); Waser (1975)
Tana River	Semi-evergreen gallery forest	*Diospyros mespiliformis* *Sorindeia obtusifoliolata* *Albizia gummifera*	50	27	Nov–Dec, April	Jan–Mar	Marsh (1978*b*, 1981*b*); Kinnaird (1992)

[a]Based on these measures: Tiwai, basal area of trees ≥ 5 cm dbh; Douala-Edéa, basal area of trees ≥ 16 cm dbh; Lopé, basal area of trees > 50 cm dbh; Kibale, stem frequencies of trees > 10 m height; Tana, estimated crown volume of trees > 10 m height.
[b]Members of the Caesalpinioideae, Mimosoideae and Papilionoideae.

Table 4.3. *Anthropoid primate faunas of selected colobus study sites*

Taxonomic group	Tiwai	Douala-Edéa	Lopé	Kibale	Tana River
Colobinae	*Procolobus badius* *Procolobus verus* *Colobus polykomos*	*Colobus satanas*	*Colobus satanas*	*Procolobus badius* *Colobus guereza*	*Procolobus badius*
Cercopithecinae, Papionini	*Cercocebus atys*	*Cercocebus albigena*	*Cercocebus albigena* *Mandrillus sphinx*	*Cercocebus albigena* *Papio anubis*[a]	Cercocebus galeritus Papio cynocephalus
Cercopithecini	*Cercopithecus diana* *Cercopithecus campbelli* *Cercopithecus petaurista* *Cercopithecus aethiops*[a]	*Cercopithecus nictitans* *Cercopithecus mona* *Cercopithecus pogonias* *Cercopithecus erythrotis*	*Cercopithecus nictitans* *Cercopithecus pogonias* *Cercopithecus cephus*	*Cercopithecus lhoesti* *Cercopithecus mitis* *Cercopithecus ascanius*	*Cercopithecus albogularis*
Pongidae	*Pan troglodytes*	*Pan troglodytes*	*Pan troglodytes* *Gorilla gorilla*	*Pan troglodytes*	

[a]Visitor to main colobus study area.

Figure 4.2. One of the many Moa River channels that in places separate Tiwai Island (right) from other islands and the mainland (photograph by J. Oates).

abundant. The most common emergent tree, reaching heights of > 45 m, is *Piptadeniastrum africanum* (Leguminosae, Mimosoideae) (Oates *et al.*, 1990).

With the exception of *Cercopithecus nictitans stampflii* (which has a very localized distribution in Liberia and Côte d'Ivoire), Tiwai supports populations of all the Guinea Coast forest-zone primates, including three colobines: *Procolobus verus*, *P. badius badius*, and *Colobus polykomos polykomos* (see Table 4.3).

In addition to farming, which is believed to have been more intensive 60–100 years ago, there has also been hunting on Tiwai. The intensity of this hunting has been relatively light, however, in part because of the river barrier and in part because of local antipathy towards Liberians who have, in recent years, been the chief hunters of monkeys in southern Sierra Leone. In 1987, Tiwai became a game sanctuary in which all hunting is prohibited.

Primate studies began at Tiwai in 1982 (see Figure 4.3). My own study of the olive colobus was conducted at intervals from 1983 to 1986, and studies on black-and-white colobus (by G. L. Dasilva) and red colobus (by A. G. Davies) took place between 1984 and 1986.

Figure 4.3. Part of the Tiwai Island Field Research Station, photographed in 1984. Forest in the west study area is behind the thatched building (photograph by J. Oates).

Douala-Edéa

The Douala-Edéa Forest Reserve covers approximately 1300 km^2 at the mouth of the Sanaga River on the coast of Cameroon. The reserve is underlain by deep marine sediments deposited from Cretaceous times onwards by north-flowing South Atlantic currents. Soils are therefore extremely sandy and very acidic (see Table 4.1). The surface is dissected by many low-lying streams and swamps, and parts become inundated in the wet season. Rainfall is very high and, as at Tiwai, May to October are generally the wettest months. Only January and February usually have less than 100 mm of rain (McKey, 1978*a*). The Douala-Edéa landscape is probably of comparatively recent origin, for the reserve area would have been inundated during interglacial periods of raised sea levels.

When D. B. McKey began primate studies at Douala-Edéa in 1973, human interference with the vegetation was localized to areas with large settlements, near the major rivers. McKey's studies were concentrated in Lombe, an area remote from population centers (Figure 4.4). Lombe is in the eastern part of the reserve, south of Lake Tissongo between the Sanaga and Nyong Rivers. McKey (1978*a*) estimates that about 60% of the area between the Sanaga

Figure 4.4. Douala-Edéa Forest Reserve, viewed across Lake Tissongo. Exposed along the shore are beaches of white-sand soil typical of this reserve (photograph by G. Whitesides).

and Nyong supports mature high forest, the coastal Cameroonian variant of Lower Guinean (or Congolian) rain-forest. Emergents such as *Lophira alata*, *Sacoglottis gabonensis*, and *Klainedoxa gabonensis* reach to around 40 m, with the general canopy at 19–30 m. Swampy areas support lower-stature forest (McKey, 1978*a*).

Seven species of anthropoid primates occur in Douala-Edéa, but their population densities (and therefore biomass) are relatively low (Oates *et al.*, 1990). The black colobus is the only colobine present. The low densities and absence of certain primate species have been ascribed by McKey *et al.* (1978) and McKey (1978*b*) to the prevalence of plant foods which, as a result of poor soils, have high levels of chemical defences, especially in their leaves (see Chapter 9).

Up to the time when McKey's studies began on the black colobus, there had been relatively little recent human interference with the vegetation of the Douala-Edéa forest – given its poor soils and paucity of commercially valuable trees. Black colobus and other animals (including elephants, bushpigs and antelopes) had been subject to some hunting, but this hunting was probably light in the colobus study area (McKey, 1978*a*,*b*, and personal communication). Since the time of McKey's study, there has been oil exploration (but not exploitation) in the reserve, and there has probably been an increase in hunting (McKey, personal communication).

Lopé

The Lopé Reserve covers an area of 5000 km^2 south of the Ogooué River in central Gabon. The reserve is a mixture of forest and savanna. The savanna grows in areas with poor, sandy soils, whereas the forest occurs in more hilly areas with thick clay-rich soils weathered from underlying schists and quartzite. *Aucoumea klaineana* (Burseraceae) is the most common large tree in the Lopé forest (White, 1992). *C. satanas* is the only colobine at Lopé, and M. J. S. Harrison studied this species in one of the hilly areas, the de Brazza Hills, in 1983–84 (Harrison, 1986, and personal communication; Harrison & Hladik, 1986) (Figure 4.5). The de Brazza Hills undulate steeply between altitudes of 22 and 600 m, and have a dense network of small valleys. Climate is strongly seasonal, with most of the annual rainfall concentrated in March–May and October–November. Little rain falls between June and August. Compared with Douala-Edéa, soils are clay-rich, but have lower levels of mineral nutrients, including a very low phosphorus content (Table 4.1).

Figure 4.5. *Colobus satanas* in the Lopé Reserve, Gabon, entering a cave to feed on sodium-rich soil (photograph by M. Harrison). Similar behaviour has been recorded in many colobine populations, but its significance has not been fully resolved (see Oates, 1978; Davies & Baillie, 1988; and Chapter 8, p. 247).

Small areas in the north of Lopé were logged in the 1960s, and logging was recently resumed. About one-third of the reserve has been affected, but there is a proposal to safeguard a core unlogged area of 1000 km^2. Until recently, hunting and cultivation were at low levels (C. E. G. Tutin, personal communication).

Kibale Forest

Kibale Forest Reserve is located in western Uganda near the eastern edge of the Western Rift Valley and just north of the equator. It covers approximately 560 km^2, of which about 60% is high forest. Descriptions of the reserve are given by Wing & Buss (1970), Struhsaker (1975), and Ghiglieri (1984); this summary is based on those accounts and my own observations.

Kibale lies on a southward tilted part of the uplifted western edge of the Ugandan plateau. Two major rivers (the Dura and Mpanga) flow south through the reserve in hilly terrain which drops from an average elevation of around 1600 m in the north to about 1200 m in the south. The forest grows on nutrient-rich sandy loams and sandy clay loams (Table 4.1); much of it is concentrated in the northern part of the reserve, and in the valleys of the Mpanga and Dura and their tributary streams. Many hilltops are grassy, probably resulting from past forest clearance and cultivation by humans, while swamp vegetation occupies many of the valley bottoms. These features give Kibale a distinctly mosaic character.

Annual rainfall at the nearby town of Fort Portal is a moderate 1485 mm, unevenly distributed, but typically falling in all months. Most rain falls in March–May and September–November, with less than 100 mm of rain in the months of December–February and June–July. Kibale's high elevation and consequent low temperatures produce year round conditions of high relative humidity.

The nature of the high forest in Kibale changes along the north–south altitudinal gradient. The most prominent large tree in the north is *Parinari excelsa* (Rosaceae), and in the south *Cynometra alexandri* (Leguminosae; Caesalpinioidae) is abundant; along the lower Dura River, *Cynometra* occurs in places in nearly monodominant stands. In the Kanyawara area, in the northern sector of the reserve, where many primate studies have been conducted, the forest has a discontinuous upper canopy at a height of 25–50 m. In addition to *Parinari*, common upper-canopy trees are *Strombosia scheffleri, Mimusops bagshawei, Newtonia buchanani, Olea welwitschii, Celtis africana, Aningeria altissima*, and several species of *Ficus* (Figure 4.6).

Eight anthropoid primates occur in Kibale (Table 4.3), including red colo-

Figure 4.6. *Colobus guereza* study area at Kanyawara, Kibale Forest, with abundant *Celtis* (photograph by J. Oates).

bus (*P. b. tephrosceles*) and black-and-white colobus (*Colobus guereza occidentalis*). Largely due to a very high local densities of red colobus monkeys, Kibale has one of the highest recorded primate biomasses in the world, with an estimated anthropoid biomass of 2317–3579 kg/km^2 in the Kanyawara area (Struhsaker, 1975).

Long-term systematic investigation of colobine ecology commenced in the Kibale Forest in 1970, with studies of red colobus by Struhsaker that were continued by Struhsaker, Leland, and Isbell up to 1988 (Struhsaker, 1974, 1975, 1978; Struhsaker and Leland 1979, 1985, 1987; Isbell, 1984), and of black-and-white colobus by Oates (1974, 1977*a,b,c*, 1978) (see also Struhsaker & Oates, 1975). Previous to this, there had been short studies of black-and-white colobus abundance by C. Koford (unpublished), red colobus vocalizations (Marler, 1970), and a comparison of red and black-and-white colobus behaviour (Clutton-Brock, 1972, 1974*b*, 1975*a*). Colobus monkeys were not hunted at the time these studies began, and had probably faced only light hunting pressure over several previous decades. In the first half of this century, humans occupied some of the hills in the Kibale Forest Reserve, but there had probably been no major disturbance to forested areas for 200–300 years (Struhsaker, 1975). However, when primate studies began in Kibale, commercial forestry had been in progress for some years, and many northern parts of the forest had lost most of their large trees. Skorupa (1988) has shown that this logging has severely depleted the populations of most anthropoids, including *P. badius*, but that it has apparently led to an increase in *C. guereza* populations. My *C. guereza* study area included part of a lightly-logged forest compartment (K14), while Struhsaker's *P. badius* study area (entirely within K30) had not been cut. These areas now constitute a Research Plot, which is connected by a research corridor to the 60 km^2 Kibale Nature Reserve, established in 1975.

The Kibale Forest supports a high density of crowned hawk-eagles (*Stephanoaetus coronatus*), which have been found to be major predators on the forest's monkeys, including colobus (Skorupa, 1989; Struhsaker & Leakey, 1990). Leopards may be present, but if so, their density is very low. Until recently, elephants were an important component of the forest ecosystem, but their numbers have been greatly reduced in recent years (Wing & Buss, 1970; Struhsaker, personal communication).

Tana

The Tana River rises on the eastern slopes of the Kenya Highlands and flows on a circuitous course to the Indian Ocean. On its lower flood-plain, groundwater supports patches of evergreen riverine forest (Figure 4.7), in an area that is otherwise thorn woodland with a low annual rainfall (400–500 mm) (Medley, 1990; Kinnaird, 1992). Some of the forest patches are inhabited by the Tana River red colobus monkey (*Procolobus badius rufomitratus*). Other monkeys in the Tana forests are the Tana River crested mangabey

Figure 4.7. Aerial view of gallery forest along the Tana River near Mchelelo (photograph by P. Kahumbu).

(*Cercocebus galeritus galeritus*), the yellow baboon (*Papio cynocephalus*), and a form of Sykes' monkey (*Cercopithecus albogularis albotorquatus*). The first thorough study of the Tana red colobus was conducted by Marsh in 1973–75 (Marsh, 1978*a,b,c*, 1979*a,b*, 1981*a,b,c*). Subsequently, there have been studies by Decker (1989) and Ochiago (1991). Marsh worked in what is now the Tana River National Primate Reserve, near Wenje, observing red colobus in the forest patches of Mchelelo (17 ha) and Congolani West (80 ha).

The Tana forests are notable for both their patchiness and their instability. These features are the result of both agricultural disturbance and relatively frequent changes in the course of the river channel. When the river course changes, there is a degeneration of forest that has grown up on raised sandy banks close to the river (Marsh, 1978*a,b*). Marsh found only 29 tree species in his main study area (Mchelelo), where a broken upper canopy reached to between 15 and 25 m (Marsh, 1978*b*). In the Primate Reserve as a whole, 86 tree species have been recorded, with the canopy dominated by *Pachystela msolo* (Sapotaceae), *Ficus sycomorus* (Moraceae), and *Diospyros mespiliformis* (Ebenaceae) (Medley, 1990).

Field study findings

The black-and-white colobus monkeys, subgenus Colobus

Schwarz (1929) regarded all populations of black and black-and-white colobus monkeys as one species (*Colobus polykomos*), but they are now widely regarded to be a diverse group of four or five species: *C. polykomos* (from which *C. vellerosus* is sometimes separated), *C. satanas*, *C. angolensis* and *C. guereza* (see Chapter 3). The ecology of this group is becoming relatively well known, especially through publications dealing with *C. guereza* and *C. satanas*. However, these publications give a somewhat distorted view of the group's ecology. *Colobus angolensis*, *C. polykomos* and *C. vellerosus* occupy a huge area from Guinea-Bissau in the west to the coastal forests of Kenya, and probably include the majority of lowland moist tropical forest populations of black-and-white colobus; although one long-term study has been completed on each of these forms, very little of the data from these studies has been published. *C. guereza*, the form about which most has been written, is unusual for the very wide range of habitats it occupies, occurring in montane forests, savanna gallery forests, and medium-altitude and lowland moist forests from Ethiopia to northern Tanzania and west to the Nigeria–Cameroon border.

Field studies

The first attempt to study the behaviour of a wild colobine seems to have been that of Wolfgang Ullrich, who observed *C. guereza* in montane forest on Mt. Meru, Tanzania, in June–July 1955 and October 1956–January 1957 (Ullrich, 1961). Subsequently, there have been a large number of other guereza field studies in East Africa and Ethiopia, but very few in the western part of the species' range (Table 4.4).

The black colobus monkey, *Colobus satanas*, is largely restricted to lowland moist forests in western Equatorial Africa, but it also occurs in the mountains of Bioko (Fernando Póo). Habituated groups of black colobus have been studied in the Douala-Edéa Reserve in coastal Cameroon by McKey (McKey, 1978*a*,*b*; McKey *et al.*, 1981; McKey & Waterman, 1982) and in inland Gabon by Harrison (Harrison, 1986; Harrison & Hladik, 1986). Some observations of unhabituated black colobus in Rio Muni are also reported by Sabater-Pi (1973).

The Angolan black-and-white colobus, *Colobus angolensis*, has a geographical distribution about as extensive as the guereza's (Figure 3.1), but

Table 4.4. *Field studies of* Colobus guereza

Site	Study dates	Topics investigated	Investigator (reference)
Mt. Meru, Tanzania	Jun–Jul 1955, Oct 1956–Jan 1957	General natural history	von Ullrich (1961)
Budongo Forest and Ishasha, Uganda	Oct 1964–May 1965	Group structure and vocalizations	Marler (1969, 1972)
Limuru, Kenya	Mar–Aug 1965	Social organization	Schenkel & Schenkel-Hulliger (1967)
Kibale Forest, Uganda	1965 (3 months)	Natural history and population census	C. B. Koford (personal communication)
Murchison Falls National Park, Uganda	1967 (5 weeks)	Social organization	Leskes & Acheson (1971)
Budongo Forest, Uganda	1967–68, 1970–71, 1972–73	Group structure and population distribution	Suzuki (1979)
Kibale Forest, Uganda	Aug–Oct 1970	Feeding behaviour	Clutton-Brock (1974*b*, 1975*a*)
Kibale Forest, Uganda	Oct 1970–Mar 1972, 1973, 1974	Ecology and social behaviour; food selection	Oates (1974, 1977*a*, *b*, *c*)
Murchison Falls National Park, Uganda	Jun and Oct 1971	Ecology and social behaviour	Oates (1974, 1977*a*, *b*)
Bole Valley, Ethiopia	May 1971, May–Sep 1972, June–Oct 1974	Social structure, ecology and population dynamics	Dunbar & Dunbar (1974, 1976); Dunbar (1987)
Lake Shalla, Ethiopia	Jul 1974	Social structure	Dunbar (1987)
Limuru, Kenya	Jan, Oct–Nov 1974	Positional behaviour	Morbeck (1977)
Lake Naivasha, Kenya	Sep 1974–Aug 1975	Positional behaviour, habitat structure and use	Rose (1977, 1978)
Arusha National Park, Tanzania	Aug–Sep 1972	Positional behaviour	Mittermeier & Fleagle (1976)

much less has been published on its natural history. It inhabits moist lowland forests, montane forests (up to at least 2700 m on the Ruwenzoris), and dry coastal forest. It appears to occur rarely in the gallery forests of the savanna zone, in which many guereza populations live. Groves (1973) conducted a short survey of *C. angolensis* populations in north-eastern Tanzania, and Moreno-Black (Moreno-Black & Maples, 1977; Moreno-Black & Bent, 1982) observed groups at Diani Beach, Kenya, over a 6-month period in 1972–73. During 8 months of 1991, Maisels *et al.* (1994) collected feeding observations of unhabituated *C. angolensis* in the Salonga National Park in the central Congo Basin. Observations on this species in the Nyungwe Forest of Rwanda by A. Vedder and in the Ituri Forest of Zaire by C. Bocian have not yet been published.

The West African, or ursine, black-and-white colobus monkey, *Colobus polykomos*, is an inhabitant of moist lowland forest from southern Sierra Leone to western Côte d'Ivoire, but it also occurs north of the moist forest zone in gallery forests and forest fragments in the Guinea savanna zone. The only long-term study of a habituated group is that by Dasilva (1989, 1992) at Tiwai in Sierra Leone, but some behavioural observations have also been made at Kilimi in northern Sierra Leone (Harding, 1984) and in the Tai Forest, Côte d'Ivoire (Galat & Galat-Luong, 1985).

Colobus vellerosus occurs immediately to the east of its close relative, *C. polykomos*. It inhabits moist and dry lowland forests, as well as forest outliers and gallery forest far into the savanna zone (Booth, 1956; Geerling & Bokdam, 1973; Sayer & Green, 1984). Like the guereza, *C. vellerosus* frequently travels on the ground between forest patches in the savanna zone (Booth, 1956). A long-term study of *C. vellerosus* was made in 1975–76 by Olson in the Bia National Park, Ghana; results from that study have been presented at several meetings (Olson, 1980; Curtin & Olson, 1984; Olson & Curtin, 1984), but only partially published (Olson, 1986).

In the remainder of this chapter, when these five forms (including the all-black *C. satanas*) are referred to collectively, they will be called 'black-and-white colobus'.

Social organization

With the exception of some populations of *C. angolensis*, most black-and-white colobus have been found to live in relatively small social groups (16 or fewer individuals). These groups typically contain 2–6 adult females and 1–3 adult males (Table 4.5). A single fully adult male appears to be the norm

Table 4.5. *Social organization of black-and-white colobus monkeys. Where investigators' study groups changed through time, data are used from the date when size and composition was first assessed*

		Total group size			Group composition					
					Adult males		Adult females		Immature/ indeterminate	
Species	Study site	mean	range	*N*	mean	range	mean	range	mean	Reference(s)
C. guereza	Budongo, Uganda	8.4	2–13	14	1.2	1–3	3.6	2–6	3.5	Marler (1969)
C. guereza	Budongo, Uganda	6.9	2–13	25	1.1	1–2	2.8	1–5	3.0	Suzuki (1979)
C. guereza	Ishasha, Uganda	6.3	3–9	4	1.3	1–2	2.8	1–4	2.3	Marler (1969)
C. guereza	Chobe, Uganda	12		1	2		3		7	Leskes & Acheson (1971)
C. guereza	Bigodi (Kibale), Uganda	6		1	1		3		2	Clutton-Brock (1975*a*)
C. guereza	Kanyawara (Kibale), Uganda	11.4	9–15	7	1.4	1–4	3.4	3–4	6.5	Oates (1977*b*)
C. guereza	Chobe, Uganda	6.0	5–7	2	1.0	1	3.0	2–4	2.0	Oates (1977*b*)
C. guereza	Lake Naivasha, Kenya	19		1	2		3		14	Rose (1977)
C. guereza	Arusha, Tanzania	5.4[a]	4–7[a]	5	1.4	1–2	1.8	1–3	2.2	Groves (1973)
C. guereza	Bole, Ethiopia	7.4	4–12	5	1.6	1–3	2.6	2–5	3.2	Dunbar & Dunbar (1974)
C. guereza	Lake Shalla, Ethiopia	7.8	6–10	6	1.0	1	2.0	1–3	4.7	Dunbar (1987)
C. angolensis	N.E. Tanzania	4.9	2–9[b]	10	1.2	1–2	1.6	0–3	2.1	Groves (1973)
C. angolensis	Diani Beach, Kenya	8		1	2		3		3	Moreno-Black and Maples (1977)
C. angolensis	Ituri, Zaire	19		1	4		7		8	C. Bocian (personal communication)
C. satanas	Douala-Edéa, Cameroon	12		1	3		7		2	McKey (1978*a*)
C. satanas	Lopé, Gabon	9		1	2		5		2	M. J. S. Harrison (personal communication)
C. vellerosus	Bia, Ghana	16	16	2	3.0	2–4	6.5	6–7	6.5	Olson (1980)[c]
C. polykomos	Tai, Côte d'Ivoire	11.5	11–12	2	2.5		4		5	Galat & Galat-Luong (1985, Table 2)
C. polykomos	Tiwai, Sierra Leone	11		1	3	4	4			Dasilva (1989)

[a]Mean size was 8.3, range 5–18, if four extra groups are included whose composition was not reported.
[b]Range was 2–16, if 27 other groups are included whose composition was not reported.
[c]Before one of the groups split.

in *C. guereza*, the species whose social behaviour is best known, but groups with two or more adult males appear to be commonplace in the other species.

Although it displays great ecological flexibility, *C. guereza* appears to show little variability in social organization throughout its wide range. There are commonly just two or three adult females in a group; these females form a cohesive group core within which there is a high frequency of allogrooming (Oates, 1977*b*). Many groups contain only a single fully adult male, but groups with more than one adult male do occur. Multi-male groups are typically larger than others, with both more males and more females. Such multi-male groups appear to result from both the maturation of animals within a group, and immigration; they tend to revert to smaller, uni-male structures through group fission or male competition and emigration (Oates, 1977*b*; Dunbar, 1987). Although small groups have been found in all intensive guereza studies, it is possible that larger groups occur in some areas. For instance, although Fay (1985) reported groups of 5–10 individuals on the lower Gounda River in northern Central African Republic, he said that large groups of up to 30–40 occurred in gallery forests in the south of the country.

All studied groups of *C. polykomos* have been small, with no more than 11 individuals, and those whose age–sex composition has been assessed (at Tiwai and Tai) have contained three or four adult females and one to three adult males. Over a 23-month period, Dasilva (1989) monitored a *polykomos* group at Tiwai which initially (April 1984) contained three adult males and four adult females. The females, as in a guereza group, maintained close spatial relationships with one another and groomed each other more frequently than they groomed males. Adult males rarely interacted with one another and displayed a clear dominance hierarchy; less pronounced dominance relationships were detected among adult females. No male takeovers or group fissions occurred during the study, but one adult male and two adult females disappeared from the group; one of these females is suspected to have died. A young male growing to adolescence in the group was increasingly harassed by adult males and eventually left. On another visit to Tiwai in April 1987, Dasilva (1989) observed a new adult female travelling with the study group; the female appeared to have bloodstains on her tail and was harassed by other females. Dasilva interpreted these observations as evidence of between-group migration by both males and females.

The two groups of *C. vellerosus* studied by Olson in the Bia National Park, Ghana, were larger than typical *guereza* or *polykomos* groups. One initially contained 16 individuals, including four adult males and six adult females; this group grew to 18 through the birth of two infants before splitting into two new groups, one of 13 and one of five, each with two adult males and three females. The other group started with 16 members (including two adult males and seven females), and grew to 21 through births.

McKey's study group of *Colobus satanas* at Douala-Edéa contained six adult females, which were accompanied by between one and three adult males (McKey & Waterman, 1982) over an 18-month period. At the beginning of a 9-month study, Harrison's Lopé group of *satanas* contained five adult females and two adult males. Later, one adult male and one adult female (with a young infant) left the group, and two adult females entered the group; the female who left the group was seen in a neighbouring group (M. J. S. Harrison, personal communication).

Groups of Angolan colobus in north-eastern Tanzania are reported to be small (2–9 individuals), with one or two adult males (Groves, 1973). Fitting this pattern are groups seen in the Ituri Forest, Zaire, with an average estimated size of 6.6 (Thomas, 1991); in the Salonga National Park, Zaire, with groups of 3–7 individuals (Maisels *et al.*, 1994); and a group studied in the Diani Beach forest on the Kenya coast, with two adult males, three adult females and three immature animals (Moreno-Black & Maples, 1977). On two-thirds of study days, the Diani group was seen to join a larger multi-male group (Moreno-Black & Bent, 1982). In this 'super-troop', individuals associated peacefully for several hours. Such super-troop formation may not be uncommon in *C. angolensis*, at least in the eastern part of its range. In 1972, in the Sango Bay forests on the shores of Lake Victoria, Uganda, I observed two large groups, one of at least 30 and the other of at least 51 individuals; these large groups each contained several adult males, and the largest group was judged to consist of three smaller groups associating closely together (Oates, 1974). Very large aggregations of Angolan colobus (more than 300 individuals) have been observed in the Nyungwe Forest, Rwanda (A. Vedder, personal communication).

Diet

An influential early paper by Booth (1956) on the synecology of Ghanaian forest primates referred to the colobus monkeys as 'purely leaf-eating', based on casual feeding observations and a study of stomach contents. A more concentrated observational study of a black-and-white colobus population, by Ullrich (1961), described guerezas eating both leaves and fruit (including seeds), but this account seems to have done little to change the generally-held view of these animals as leaf-eaters. This view was supported by the results of the first studies to focus on the diet of the black-and-white colobus, studies which found leaves to be the predominant food item. These studies were in the Kibale Forest where, in August and September 1970, J. Brooke and T. Clutton-Brock found leaves and shoots to comprise > 67% of feeding observations of *C. guereza* at Bigodi, and > 92% of observations at Kanyawara (Clutton-Brock, 1975*a*). A longer study at Kanyawara reported an annual diet containing 81% foliage (Oates, 1977*a*). As studies have broadened to include other areas and other species, it has become evident that black-and-white colobus are by no means obligate folivores. Great variation in diet has been found, and fruits or parts of fruits are the predominant food items in some places or in some seasons.

Table 4.6 summarizes annual diets quantified from observational sampling of habituated groups. At Kibale, young leaves were the guereza's major food item, but mature leaves and fruits were eaten in large quantities in months when preferred young leaves were scarce (e.g. the wet months of August and September 1971) (Oates, 1974). Similarly, in a mid-wet-season (June) sample in gallery forest at Chobe, Uganda, there was much heavier feeding on mature leaves and fruits than in a late wet-season sample (October) (Oates, 1977*a*). In the Bole Valley, Ethiopia, Dunbar & Dunbar (1974) found that fruit (especially figs) comprised more than 53% of guereza feeding observations during the wet-season months of May and June 1972. It seems that young foliage is the preferred food of guerezas, and that mature foliage and fruits are eaten when young leaves are in short supply (which is often in the middle of the rainy season). When fruits are consumed, these are often figs; studies so far do not indicate a strong preference for unripe versus ripe figs. Kibale guerezas also feed heavily on the foliage of climbing and scrambling plants in low second growth, when preferred young leaves are scarce (Oates, 1977*a*).

The seed-eating habits of *Colobus satanas* at Douala-Edéa are now well known. In McKey's study group, 53% of the annual diet was made up of seeds, which, with young leaves, were preferred dietary items. Seeds were eaten from a wide variety of plants, including some of the most abundant forest trees; leaf feeding was concentrated on lianas and relatively uncommon deciduous trees (McKey *et al.*, 1981). When both seeds and young leaves were scarce in the forest, mature leaves were heavily eaten and the diet as a whole became more diverse; examples of these times are the late rainy season (September–November) in 1974, and the dry season (January–February) in 1975. At these times, the black colobus group was more widely dispersed, moved further each day, and used a larger area of its range over a 5-day period (McKey & Waterman, 1982). In months when mature leaves constituted a major part of the diet, the black colobus frequently displayed 'treefall foraging', coming low in canopy gaps (and sometimes to the ground) and feeding on climbing plants and tree saplings (McKey, 1978*b*).

Table 4.6. *Annual diets of black-and-white colobus monkeys*

Species	Study site	Per cent composition of diet by food part							Number of species in diet	Reference	Notes
		YL	ML	UL	FL	FR	(SD)	OT			
C. guereza	Kibale Forest	65.2	13.1	2.7	2.2	14.5		2.4	43	Oates (1977*a*)	Table VII, excluding undetermined items
C. satanas	Douala-Edéa	18.1	20.5		3.3	53.2	(53.2[a])	4.8	84	McKey *et al.* (1981)	Average contribution of items/month
C. satanas	Lopé	23	3		5	64	(60)	4	65	M. Harrison (personal communication)	
C. polykomos	Tiwai	29.7	26.4	1.6	2.7	35.0	(31.8)	3.1	56	Dasilva (1989)	Table 6.2; total <100%

YL, immature foliage, including leaf buds; ML, mature foliage; UL, foliage of undetermined age; FL, flowers and floral buds; FR, fruits, including seeds; (SD), % seeds alone, where this stated; OT, other, including stem and bark.
[a]included seeds and pericarps.

The black colobus studied by Harrison at Lopé included even more seeds in their diet than did those at Douala-Edéa. Seeds comprised 60% of all feeding records during a 9-month study (Harrison, 1986, and personal communication) and mature leaves formed only 3% of records (see Table 4.6). Young leaves (23% of records) were eaten most when seeds were scarce, suggesting that seeds are the preferred item. When the black colobus fed heavily on seeds, their diet became more diverse in terms of species and the monkeys used more of their range (the opposite of the Douala-Edéa situation).

In Dasilva's Tiwai study of *C. polykomos*, young leaves, mature leaves and seeds contributed about 30% each to the total of feeding scores over one year (Dasilva, 1989, 1992, 1994). The monkeys foraged frequently in liane tangles, with the result that about 50% of feeding records came from lianes. In a parallel with the foraging of black colobus at Douala-Edéa, foliage food items (and mature foliage, in particular) were predominantly from lianes, while most seeds were harvested from trees, especially leguminous species. Seed-eating was concentrated in the September–March period (late wet season and throughout the dry season), when seeds were most abundantly available in the forest. Feeding on the mature foliage of lianes peaked in the middle of the wet season (July–early September), a time when both preferred young leaves and seeds were scarce; at this time the group used a smaller part of its range and made shorter day-journeys, a parallel with the behaviour of black colobus at Lopé.

Limited information from Olson's study of *Colobus vellerosus* at Bia in Ghana suggests that young leaves, mature leaves, fruit pulp and seeds are all important dietary items, with fruit and seeds most important in the dry season. A study group used a smaller part of its range in the dry season than during the rains (Curtin & Olson, 1984).

The majority of food items seen to be eaten by *C. angolensis* in Moreno-Black's Diani study were leaves, but some seed-eating was also observed (Moreno-Black & Maples, 1977). A. Vedder (personal communication) reports that leaves and seeds were eaten by *C. angolensis* at Nyungwe, where abundant trunk-encrusting lichen was also a common food. Like black-and-white colobus elsewhere, Angolan colobus came to the ground at Nyungwe to eat herbaceous climbers. In the 8-month study of this species in Salonga, Zaire, Maisels *et al.* (1994) found that seeds and young leaves were the most commonly eaten food items, but that the diet shifted opportunistically; unripe leguminous seeds appeared to be a preferred item when they were available.

Feeding Sites

Not only have black-and-white colobus been reported as feeding often at or near the ground (especially on climbing plants in canopy gaps), their feeding as a whole tends to be in the lower layers of the forest canopy. Booth (1956) described black-and-white colobus in Ghana as feeding predominantly in the middle and lower canopy of the forest, while travelling and resting in the upper and middle canopy. Systematic sampling in the Kibale Forest confirmed this view; 70% of guereza feeding observations came from below a height of 18 m, while over 50% of inactivity was above this height; feeding was concentrated on small and medium-sized trees in the middle storey of the canopy and at the forest edge (Oates, 1977*a*).

Dasilva (1989) found that feeding by *C. polykomos* at Tiwai occurred significantly lower in the canopy than did travelling or resting, and although M. J. S. Harrison found (personal communication) that *C. satanas* at Lopé spent 85% of their time in the highest part of the forest canopy, he did note that the monkeys would descend to lower levels to feed.

Activity patterns

Black-and-white colobus are noted for their inactivity. Scan samples taken over 1 year on the *C. guereza* study group at Kanyawara in the Kibale Forest found that group members were inactive or 'resting' (not moving, feeding, or engaged in overt social interactions) on 57% of scans, whereas they were feeding on 20% of scans and moving on 5% (Oates, 1977*a*). Similarly, 59% of scans of *C. polykomos* at Tiwai scored resting animals (Dasilva, 1989, 1992).

At Kanyawara, a typical guereza group's day consisted of about five periods of movement and feeding, punctuated by long periods of rest. The day often began with sunbathing on exposed branches, and ended with an intense bout of feeding (Oates, 1977*a*). This activity pattern is thought to be related to the guereza's high-fibre (and relatively low-energy) foliage diet at Kanyawara and to the relatively high altitude of this site (1500 m), where mornings are cool. Food energy is subsidized by direct sun energy to maintain body temperature at a time when temperatures are lowest and the stomach is empty, and movement is minimized during long fermentation periods after stomach fills. Similar evidence of energetic economy comes from the Tiwai study of *C. polykomos*: in the wet season, when fewest young leaves or seeds were available and most mature leaves were eaten, feeding, moving and social behaviour were reduced and resting was increased; at this time, when the black-and-white colobus may have had difficulty meeting their energy needs, their use of energy-conserving postures increased (Dasilva, 1992, 1993).

Range size and defence

The ability of *Colobus guereza* to subsist on mature leaves and its habit of living in small groups are probably directly related to its occupancy, at some sites, of very small home ranges. At several sites, guereza groups have been found to have long-term residence in areas of between 1 and 2 ha (Schenkel & Schenkel-Hulliger, 1967; Fay, 1985; Dunbar, 1987). These small ranges occur at sites where guerezas are occupying small forest patches; where more than one group is present in a small patch, they defend territories, and defence is by males (Schenkel & Schenkel-Hulliger, 1967). In larger forest blocks, home ranges are larger, and may overlap extensively (Oates, 1977*b*).

Ranging information from long-term studies of habituated groups is summarized in Table 4.7. In each of these studies, quantitative analyses of ranging behaviour were based on the use by study groups of 0.25 ha (50 m × 50 m) grid cells, or quadrats. At Kibale, the guereza study group had a range of 29 ha, which it shared with five other groups; within this range was an intensively-used core area of 2.8 ha, which was defended by adult males (Oates, 1977*b*). Core areas in Kibale seem to be essentially a dispersed set of the same small territories occupied by guerezas in small forest patches (Dunbar, 1987).

The Tiwai study group of *C. polykomos* had a total annual range of 22 ha. Dasilva (1989) describes a pattern of considerable range overlap, but hostile (though infrequent) encounters between groups when they came into proximity. As at Limuru and Kibale, males were the individuals involved in aggressive interactions (especially chasing) during group encounters.

McKey's Douala-Edéa group of *C. satanas* had a home range of 60 ha (McKey & Waterman, 1982), while Harrison's Lopé study group used an even larger area: 84 ha during nine day samples, and 184 ha if lacunae and cells entered outside samples were taken into account (M. J. S. Harrison, personal communication). These Lopé black colobus, whose diet consisted predominantly of seeds, did not focus their activities strongly in one part of their range. Harrison suggests that the large range and dispersed ranging pattern of the Lopé colobus might be the result of foraging for the highest-quality available seeds (with extra travel costs compensated for by high nutritional returns when seeds were found), or might be the result of efforts to

Table 4.7. *Ranging behaviour of black-and-white colobus monkeys*

Species	Study site	Study group size	Home range size (ha) based on 50×50 m cells entered	Per cent of cells used 50% of time	Daily travel distance (m)		Reference
					mean	range	
C. guereza	Kibale Forest	11–14	28	12.6	535	288–1004	Oates (1974, 1977*a*)
C. satanas	Douala-Edéa	13–16	60	12.6	459	<100–>800	McKey (1978*a*); McKey & Waterman (1982)
C. satanas	Lopé	9–14	84	23.8	510	40–1100	M. Harrison (personal communication)
C. polykomos	Tiwai	9–11	24	18.6	832	350–1410	Dasilva (1989)
C. vellerosus	Bia	13–18	48[a]	?	307[b]	75–752	Olson & Curtin (1984); Olson (1986)

[a]Based on 361 days' data; 31 ha used during 5 days/month, 12-month sample.
[b]Average of data for wet and dry season.

increase the diversity of the seed diet and thus minimize the intake of toxins from any one kind of seed. These hypotheses are not mutually exclusive, but sufficient data are not yet available to test them fully.

Territoriality and loud calling

In addition to their handsome coats, one of the most striking features of black-and-white colobus is the distinctive high-volume, low-frequency, resonating loud call of adult males, termed a 'roar' by Schenkel & Schenkel-Hulliger (1967), and rendered by Hill & Booth (1957) as 'rurr rurr rurr rurr'. The roar varies across the black-and-white colobus species in pitch and tempo; pulses within each 'rurr' are very low-pitched and relatively widely spaced in *guereza* and *vellerosus*, more closely spaced and higher-pitched in *angolensis* and *polykomos*, and highest-pitched in *satanas* (Oates & Trocco, 1983).

Among guerezas, roaring typically occurs at low intensity during the day, often evoked by disturbances such as the proximity of a potential predator; close to dawn, and at night, higher-intensity contagious roaring is common (Marler, 1972; Oates, 1977*b*). Roaring is not used directly in territorial encounters, however, but Marler (1969) has argued that it functions to maintain group spacing. In *C. polykomos* and other species, several adult males typically roar together when more than one is present in a group. Dasilva (1989) cited evidence suggesting that the number of fully adult males in a group affects the outcome of aggressive group interactions, with a reduction in male numbers leading to a group having a greater probability of being displaced by another. Dasilva pointed out that only fully mature males roar. Because roaring probably provides clues to male strength and condition, it may be best understood as a component of male–male competitive behaviour, one outcome of which may be the exclusion of other males from an area in which females are present. Tending to support this view is evidence from Lopé, where the population density of *C. satanas* is low, and groups have much larger home ranges than in Kibale or at Tiwai (Table 4.7). At Lopé both intergroup encounters and male roaring are rare. However, like most displays, roaring probably has more than one function; for instance, it is certainly used by Kibale guerezas to intimidate predators.

Reproduction

Like most Asian colobines, but unlike the African red and olive colobus, black-and-white colobus females do not exhibit pronounced perineal swellings (see Chapter 3). In south-western Uganda, guerezas produced young throughout the year, with no strong evidence of seasonality (Oates, 1974), but all births of *C. polykomos* recorded at Tiwai in Sierra Leone occurred in the dry-season months of December–February (Dasilva, 1989). This difference in pattern may be related to the more strongly seasonal environment of Sierra Leone compared with that of western Uganda. Sparse data from the wild suggest that females have interbirth intervals of around 24 months (Struhsaker & Leland, 1987; Dasilva, 1989).

Infants of *C. guereza, vellerosus, polykomos* and *angolensis* are born with pink skin and completely white coats, and over 3–4 months the coat and skin darken to the adult pattern. When they are very young and conspicuous, the infants attract considerable attention from females other than their mothers, and will often be taken and held by these other females for long periods (Leskes & Acheson, 1971; Wooldridge, 1971; Oates, 1977*b*; Dasilva, 1989). Infants of *C. satanas* are less conspicuously different from their parents, and 'allomothering' appears to be relatively uncommon in this species (M. J. S. Harrison, personal communication). Allomothering is discussed further in Chapter 11.

Summary of black-and-white colobus natural history

Although many of the differences between the various populations of black-and-white and black colobus may be seen as variations on one basic theme, some of the variations may have species-specific, genetically based components. For instance, the preferred foods of all populations are young leaves and/or seeds. When preferred foods are in short supply, all populations increase their feeding on mature leaves. Often, much of this feeding is from climbing plants – either in treefall gaps in the canopy, or in liane tangles growing on canopy trees. During climber foraging, the colobus will come to the ground. At these times, when preferred food is scarce, ranging may increase (*C. satanas*, Douala-Edéa) or decrease (*C. satanas* at Lopé, *C. polykomos* at Tiwai), the nature of the response presumably being contingent on the distribution of palatable food and the costs and benefits associated with its exploitation at any one site.

However, although all the black-and-white colobus will fall back on mature leaves during a resource bottleneck, there seems to be a major difference between populations in the extent to which they can rely on mature foliage and in the degree to which seeds are a favoured food item. *C. guereza*, for instance, does seem to be more truly folivorous than *C. satanas*, whose emphasis on seeds appears to be more than a proximate response to a lack of palatable foliage. Comparing only Kibale and Douala-Edéa, it could be argued that the differences in the diets of the two populations are the result of an abundance of digestible foliage at Kibale (and perhaps a relative scarcity of suitable seeds), compared with a scarcity of high-quality foliage at Douala-Edéa (and an abundance of suitable seeds). Indeed, the foliage of common trees at Kibale, compared with that at Douala-Edéa, has been shown to be of higher quality for a colobine monkey: it has a lower fibre content, a higher protein content, and a lower tannin content (Gartlan *et al.*, 1980; Oates *et al.*, 1990; and Chapter 9, this volume). However, foliage at the Lopé black colobus study site is of higher quality than that at the Douala-Edéa site, yet the colobus at Lopé eat even more seeds than they do at Douala-Edéa, and select seeds in preference to young leaves. M. J. S. Harrison (personal communication) has suggested that *C. satanas* is morphologically adapted to a different diet, citing as evidence the studies by Janis (1984), which compared the molars of *C. guereza* and *C. satanas*; *guereza* molars have more extensive buccal shearing surfaces, while *satanas* molars have larger buccal tip crushing areas (used for cracking hard, brittle items such as seeds). Similarly, Dasilva (1989) has drawn attention to the studies of Hull (1979) which show that the teeth, palate and jaws of *C. guereza* differ considerably from those of the other black-and-white colobus species, and that the *C. polykomos* dentition closely resembles that of *C. satanas*. At Tiwai, *C. polykomos* gnaw through large, woody legume pods and eat the seeds; these pods are not opened by the sympatric red colobus.

The ability to subsist for a time on mature foliage has been cited as one feature that allows *C. guereza* to thrive in highly seasonal gallery forests (Oates, 1977*a*); another is its willingness to travel on the ground between forest patches (Dunbar & Dunbar, 1974; Oates, 1977*a*; Fay, 1985). Some insight on the guereza's ability to deal with mature foliage is provided by the study of Watkins *et al.* (1985) on captive monkeys. This study showed that guerezas have an impressive ability to digest fibre, exceeding the efficiency predicted for their body size from models based on ruminant digestion. Guerezas also have a low metabolic rate (Müller *et al.*, 1983) and display energy-conserving behaviour such as early-morning sunbathing (Oates, 1977*a*). All these features suggest an *evolved* strategy that allows subsistence, at least on a seasonal basis, on a low-quality diet. *C. guereza* is not alone in flourishing in gallery forest habitats in the savanna zone; for instance, several populations of *polykomos* and *vellerosus* occupy similar habitats in West Africa. But *guereza* appears to be more typically associated with this vegetation than are other species in the group, and its social behaviour may contain special elements of adaptation to the niche it commonly occupies.

Although social groups of all black-and-white colobus species are typically small, *C. guereza* does appear to be different from the other species, in that groups with only a single adult male predominate. Groups with more than one male do occur in *C. guereza*, but they tend to be unstable and to revert to uni-male structures. In the other four species, multi-male groups seem to be the norm rather than the exception. The basis of this difference is not clear. Proximally it may depend on patterns of male–male aggression, which might themselves be responses to aspects of the behaviour of females, such as their social cohesion and/or the synchrony of their reproduction; either or both of these factors could influence the ability of males to monopolize access to females in reproductive condition. These factors might in turn be related to patterns of food availability in time and space. However, it is not evident that food distribution acts on black-and-white colobus social behaviour in a direct, proximate fashion. *C. guereza* occupies a great range of habitats, and varies greatly in its population density, but varies little in its patterns of social organization. Thus, its social characteristics may arise from an evolved genetic substrate that is a response to historic (very long-term) patterns of food availability and other environmental features, rather than to the precise conditions of modern habitats.

Some populations of *C. angolensis* seem to be unique among black-and-white colobus in the habit of forming large aggregations from smaller groups. The functional significance of these aggregations remains unclear. They too appear to be long-standing features of the Angolan colobus adaptive repertoire, expressed under a variety of present conditions, for they have been observed in widely different habitats. In these features, the black-and-white colobus provide some evidence in support of the idea that aspects of social organization can be among the least plastic elements of primate behaviour.

The hypothesis that differences in observed behaviour between black-and-white colobus species result in part from different evolved strategies could be tested by studies on sympatric populations. *Colobus guereza* and *C. angolensis* occur together in the Ituri Forest of eastern Zaire (Thomas, 1991), and there is probably sympatry between *C. guereza* and *C. satanas* in south-east Cameroon and north-west Congo Republic (Mitani, 1990). In the Ituri Forest, *C. guereza* is reported to be most common in roadside secondary forest and *C. angolensis* in primary forest (Thomas, 1991). A study of niche separation between these two species in the Ituri Forest is being conducted by C. Bocian.

The red colobus monkeys, subgenus **Piliocolobus**

Field studies

Only five of the 14–17 different forms of red colobus have been the subject of relatively long-term field studies involving habituated groups: *P. b. temminckii* has been studied at Abuko in The Gambia, and at Fathala, Senegal; *P. b. badius* has been studied at Tiwai in Sierra Leone; *P. b. tephrosceles* has been studied at Gombe, Tanzania, and in the Kibale Forest, Uganda; *P. b. rufomitratus* has been studied on the Tana River, Kenya; and *P. b. kirkii* has been studied on Zanzibar (Table 4.8). Although red colobus are often considered to be essentially moist-forest animals, the Abuko, Fathala, Tana and Zanzibar populations all live in habitats in which there is a prolonged dry season and an annual rainfall of less than 1500 mm.

Social organization

One of the very first reports on the natural history of red colobus monkeys was by Peter Marler. Marler made brief visits to the Kibale Forest, Uganda, in 1965, and noted that red colobus there had a much larger group size (30–50 individuals) than the *C. guereza* groups he was familiar with in Budongo Forest to the north. Marler (1970) observed that red colobus had a graded vocal communication system, whereas black-and-white colobus apparently used a set of more discrete calls; he ascribed this difference to an emphasis on intragroup communication by red colobus living in larger and more complex groups.

Marler's observation that red colobus live in large social groups has been confirmed by longer studies of the Kibale population, and of many other populations. The data in Table 4.8 suggest that although red colobus groups may range in size from 12 to ≥ 80, typical groups number between 25 and 40 monkeys. Groups usually contain at least three adult males and many adult females (with females commonly outnumbering males 2 : 1).

Not all red colobus live in large multi-male groups, however. In particular, groups on the Tana River in Kenya are markedly smaller than those at most other sites. Marsh (1979*b*) found that, although Tana groups contained a large number of females (mean 9.6), most groups had only a single fully adult male.

Although average group size at the Tana River is smaller than at other sites, the range of group sizes found at Tana overlaps that found in the Kibale Forest and elsewhere, and the average group size is still higher than that in

Table 4.8. *Social organization of red colobus monkeys.*

Subpecies	Study site	Total group size			Group composition					Reference
					Adult males		Adult females		Immature/ indeterminate	
		mean	range	*N*	mean	range	mean	range	mean	
P. b. temminckii	Senegal	28.5	21–30	4	–	–	–	–	–	Struhsaker (1975)[a]
P. b. temminckii	Fathala, Senegal	25.2	14–62	6	5.5	4–13	10.7	5–27	9.0	Gatinot (1977)
P. b. temminckii	E. Gambia	16.5	12–21	2	3.0	2–4	6.5	3–10	7.0	Struhsaker (1975)[a]
P. b. temminckii	Abuko, Gambia	26.0	26	2	2.5	2–3	10.5	8–13	13.0	Starin (1991)
P. b. badius	Tai, Côte d'Ivoire	51	–	1	–	–	–	–	–	Struhsaker (1975)[a]
P. b. badius	Tai, Côte d'Ivoire	32	–	1	3	–	13	–	16	Galat & Galat-Luong (1985)[a]
P. b. badius	Tiwai, Sierra Leone	33	–	1	7	–	13	–	13	A. G. Davies (unpublished data)
P. b. preussi	Korup, Cameroon	40	–	1	–	–	–	–	–	Struhsaker (1975)[a]
P. b. oustaleti	Badane, C.A.R.	16.0	14–18	2	–	–	–	–	–	Galat-Luong & Galat (1979)[a]
P. b. tholloni	Salonga, Zaire	>50	–	1	–	–	–	–	–	Maisels *et al.* (1994)
P. b. tephrosceles	Kibale, Uganda (Kanyawara)	34	9–68	?	–	3–>10	–	8–>21	–	Struhsaker (1975); Struhsaker & Leland (1987)
P. b. tephrosceles	Gombe, Tanzania	82	–	1	11	–	24	–	47	Clutton-Brock (1972)
P. b. rufomitratus	Tana River	18.0	12–30	13	1.5	1–2	9.6	5–18	6.9	Marsh (1979*b*)
P. b. gordonorum	Magombero, Tanzania	24.8	21–33	4	2.8	1–5	9.3	7–11	12.8	T. T. Struhsaker (personal communication)[a]
P. b. kirkii	Jozani, Zanzibar	33.6	25–46	7	4.9	3–9	8.9	6–16	19.9	Silkiluwasha (1981)[b]

[a]Counts apparently based on one or a few encounters only.
[b]Means of observations from 1977 to 1979.

the great majority of black-and-white colobus populations (Table 4.5). With the exception of the *C. angolensis* 'super-groups', it is fair to say that red colobus live in larger social groups than do black-and-white colobus.

Relationships within red colobus groups, and the dynamics of these groups, also appear to differ fundamentally from those of black-and-white colobus. Both adult and juvenile females have a much greater tendency to transfer between groups than do males, so that the female membership of groups is fluid (Marsh, 1979*a*,*b*; Struhsaker & Leland, 1979; Starin, 1981). Within groups, adult females exchange relatively little grooming, while adult males groom each other frequently; males are more stable elements in groups than are females, and adult males within a group (probably close relatives) often cooperate with each other in aggressive interactions with other groups (Struhsaker & Leland, 1979). In the small, isolated population studied by Starin (1991) at Abuko, The Gambia, females also took an active, aggressive role in intertroop encounters. Unlike in black-and-white colobus, infant transfer between females appeared to be very uncommon in red colobus (Struhsaker & Oates, 1975).

Diet

Table 4.9 compares the annual diets of red colobus populations at Fathala, Abuko, Tiwai, Salonga, Kibale, Gombe, Tana and Zanzibar. In terms of the relative proportions of mature leaves, young leaves, and fruits in the diet, these red colobus populations were as varied as those of black-and-white colobus. As in black-and-white colobus, certain young foliage and fruit items (as well as flowers) have been found to be eaten in preference to most mature leaves. Preferred items, such as leaf buds, flowers, or immature seeds, are typically available on a seasonal basis and their consumption therefore fluctuates with season. As for black-and-white colobus, mature foliage seems to function as a standby item that becomes most important when preferred items are scarce. For instance, on the Tana River, which is the most strongly seasonal of the study sites and the one with the lowest tree-species diversity, mature leaves reached their highest frequency in the red colobus diet in the months (June and July) when young leaves reached their lowest level of abundance in the forest (Marsh, 1981*b*). Mature-foliage-feeding by red colobus is often concentrated on leaf petioles or on the raches of large compound leaves (Clutton-Brock, 1975*a*; Struhsaker, 1975; A. G. Davies, personal communication).

If the general kinds of food eaten by red colobus are similar to those eaten by the black-and-white species, how do the two groups differ? In the same

Table 4.9. *Annual diets of red colobus monkeys*

Study site	Per cent composition of diet by food part							Number of species in diet	Reference
	YL	ML	UL	FL	FR	(SD)	OT		
Fathala, Senegal	41.5	6.5		8.7	35.9	(18.5)	7.4	39	Gatinot (1977)[a]
Abuko, Gambia	34.9	11.8	0.1	8.7	41.6	(2.9)	2.9	89	Starin (1991)
Tiwai, Sierra Leone	31.7	20.2		16.1	31.2	(25.3)	0.8		A. G. Davies (personal communication)
Salonga, Zaire	54.3	6.4		1.4	37.8	(25.7)		84	Maisels *et al.* (1994)
Kibale, Uganda	50.6	23.1	10.4	11.9	5.7	(0.8)	1.2	≥55	Struhsaker (1975)[b]
Gombe, Tanzania	34.8[c]	44.1		6.8	11.4		2.9	>58	Clutton-Brock (1975*a*)
Tana River, Kenya	52.4	11.5	0.9	6.2	25.0	(0.9)	4.0	22	Marsh (1981*b*)
Jozani, Zanzibar	46.7	7.3	7.3[d]	10.6	31.7		2.3	63[e]	Mturi (1993)
	53.4	6.3	5.6[d]	5.4	31.2		1.3	62[f]	

YL, immature foliage, including leaf buds; ML, mature foliage; UL, foliage or undetermined age; FL, flowers and floral buds; FR, fruits, including seeds; (SD), % seeds alone, where this stated; OT, other, including stem and bark.

[a]Data from 1975–76

[b]From Table 35; total exceeds 100%

[c]Includes 'new leaves' plus 'shoots'.

[d]'Leaf stalk'.

[e]Group I, from Table 12.4; total exceeds 100%.

[f]Group II, from Table 12.4; total exceeds 100%.

habitat, red colobus have been found to eat a more varied diet than have black-and-white colobus. For instance, at Kibale the monthly Shannon index of food-species diversity averaged 1.72 for guereza, and 2.61 for red colobus (Struhsaker & Oates, 1975). At Salonga, Zaire, red colobus were seen to eat 80 different plant species in an 8-month period when *C. angolensis* ate 46 species (Maisels *et al.*, 1994). Such diversity may in part result from a greater emphasis by red colobus on ephemerally available items such as flowers, floral buds and leaf buds. At Kibale, 11.9% of one year's red colobus feeding observations were of flowers and floral buds and 11.1% were of leaf buds, compared with 2.2% and 4.1% of guereza observations for the same items (Struhsaker, 1975: Table 35; Oates, 1977*a*: Table 7).

Feeding sites

In addition to emphasizing different items in their diet, red colobus obtain much of their food at different sites in the forest canopy from those exploited by black-and-white colobus. Much red colobus food is taken from the largest, most common trees in their habitat (for instance, *Newtonia buchanani* at Gombe and Kibale, an emergent that is not eaten by *C. guereza*). Such trees often have clumped distributions in the forest (Clutton-Brock, 1975*a*). In the forests of Ghana, Booth (1956) reported that the red colobus 'sleeps in, escapes to, and largely feeds in the tall emergent trees, rarely visiting the lower levels of the forest'. At Kanyawara in the Kibale Forest, whereas more than 70% of guereza feeding records came from below 13 m, less than 35% of the feeding records of red colobus were from below this height (Struhsaker, 1975; Figure 41). At Tiwai, red colobus obtain most of their mature leaves from trees, while black-and-white colobus feed heavily on the mature foliage of climbers.

Activity patterns

When studied by the same methods at the same site, red colobus have been found to spend less time resting and much more time feeding than black-and-white colobus. At Kanyawara in the Kibale Forest, for instance, red colobus were found to allocate 45% of their time to feeding and 35% to resting, while black-and-white colobus allocated 20% and 57% of their time, respectively, to these same activities (Struhsaker & Oates, 1975). At Tiwai, samples of the activities of red colobus gave scores of 37% for feeding and 55% for resting, compared with 28% and 61% for these same activities in black-and-white colobus (Dasilva, 1989). Such differences are likely to be related to the higher quality and perhaps smaller average size (and therefore greater per-item handling time) of the items that figure prominently in the diet of the red colobus.

A comparison of red colobus activity budgets measured at different sites (Table 4.10) shows some variation from site to site in the percentage of time spent feeding, with an especially high percentage in Struhsaker's CW study group at Kanyawara. Clutton-Brock (1974*a*) and Struhsaker (1975) have suggested that the differences in their Kanyawara results are due to the use of different sampling procedures. Although different sampling techniques might be partly responsible for differences in activity scores, a comparative study by Marsh (1978*c*) has shown that the differences between the Kanyawara CW and Tana study groups are probably real and not artefacts of sampling. Because Clutton-Brock and Struhsaker observed different groups at Kanyawara, it may be that the two groups had significant behavioural differences, as Whitesides (1991) has found in two groups of Diana guenons at Tiwai. If the Kanyawara CW group is not considered, sampled groups of red colobus are relatively similar in their activity budgets, especially in time spent resting (51–61% of observations).

Range size and defence

Ranging information collected from habituated red colobus groups is presented in Table 4.11. Overall, annual home range sizes are of a similar order of magnitude to those found in black-and-white colobus (Table 4.7). However, where both kinds of colobus have been studied in the same habitat using identical methods (i.e. at Kibale and Tiwai), black-and-white colobus ranges have been found to be around 50% smaller than those of red colobus. In each of these cases the red colobus study group was at least 50% larger than the black-and-white colobus group. In Kibale, a guereza group not only used a smaller area than a red colobus group, it also travelled slightly less far each day on average and regularly supplanted other groups encountered in the core of its range. The focal red colobus group at Kibale had complete overlap in its range with two other groups; although these groups were sometimes very tolerant of each other's close proximity, they also displayed a dominance hierarchy and were able to supplant one another on the basis of the number of adult males in the group, their physical condition and their fighting abilities (Struhsaker & Leland, 1979). As in black-and-white colobus, females and immature animals avoided these conflicts.

In some habitats, however, red colobus have been found to use very small ranges, while black-and-white colobus may use large ones. Marsh's Tana River red colobus study group at Mchelelo had a mean size of 21.5 and

Table 4.10. *Activity budgets (percentage of samples) of red colobus monkeys*

Study site	Feed	Rest	Move	Other[a]	Reference	Notes
Gombe, Tanzania	25	54	8	9	Clutton-Brock (1974*a*)	Median percentage scores for 32 sample days over 8 months
Bigodi, Kibale Forest	25	60	7	9	Clutton-Brock (1974*a*)	Median of 12 days in 1 month
Kanyawara, Kibale Forest	30	61	4	6	Clutton-Brock (1974*a*)	Median of 10 days in 2 months
Kanyawara, Kibale Forest	47	35	9	8	Struhsaker (1975)	Samples across 12 months; scores for clinging infants excluded
Kanyawara, Kibale Forest	46	31	11	12	Marsh (1978*c*)	One 3-day sample, clinging infant scores excluded
Tana River, Kenya	32	51	8	9	Marsh (1978*c*)	Samples across 12 months, clinging infant scores excluded
Abuko, Gambia	25	51	13	10	Starin (1991)	Samples across 12 months, excluding animals one year old or less
Tiwai, Sierra Leone	37	55	5	3	A. G. Davies (personal communication)	Samples across 12 months

[a]Includes social interactions and play.

Table 4.11. *Ranging behaviour of red colobus monkeys*

Subspecies	Study site	Study group size	Annual home range size (ha)	Percentage of cells used 50% of time	Daily travel distance (m)		Reference
					mean	range	
P. b. temminckii	Abuko, Gambia	24–30	34[a]	–	–	–	Starin (1991)
P. b. badius	Tiwai, Sierra Leone	29–34	55[b]	–	–	–	A. G. Davies (personal communication)
P. b. tephrosceles	Gombe, Tanzania	82	114[c]	–	–	–	Clutton-Brock (1975*b*)
P. b. tephrosceles	Kibale, Uganda	19–22	65	17.3	649	223–1185	Struhsaker (1975)[d]
P. b. rufomitratus	Tana River, Kenya	16–33	9	13.9	603	<200–>1000	Marsh (1981*a*)

[a] Range size July 1978–September 1982.
[b] Based on cells of 50×50 m.
[c] Based on cells of 100×100 yd (range based on cells of 50×50 m would probably be 50–75% of this figure).
[d] Data from November 1970 to October 1971 (percentage of cells used 50% of time calculated from data in Struhsaker (1975), Fig. 46).

used only 9 ha during 12 5-day samples (Marsh, 1981*a*); Harrison's Lopé black colobus group, with a mean size of 11.8, used 84 ha in only 9 months (M. Harrison, personal communication). In contrast to the situation in Kibale, the Mchelelo group had little range overlap with a neighbouring group, and most shared areas were entered only rarely; occasional aggressive interactions occurred between the two groups (Marsh, 1979*b*). The Mchelelo group, with a very small home range, travelled a similar distance each day to that of the guereza study group at Kibale, and distributed its activities across its range in a very similar way to the guerezas (compare Tables 4.7 and 4.11). In contrast, the Lopé black colobus group had a large range, but six other groups were seen within this range and very few aggressive encounters between groups were witnessed (M. Harrison, personal communication).

At Abuko, the focal red colobus group shared 60% of its range with two other groups in the small (<1 km^2) study area (Starin, 1991). Aggressive encounters with these other groups were observed once during every 47 observation hours, on average. As in Kibale, the outcomes of these encounters were not based on location. Rather, they appeared to be based on dynamic dominance relationships between groups; one group would consistently supplant another wherever the two groups met, until a change in their relationship occurred, and the other group would consistently win. Both males and females took part in intergroup encounters, each sex directing its aggression towards males in opposing groups. The size of a group and the number of males it contained did not affect the outcome of encounters.

It appears that there are few consistent differences between red colobus and black-and-white colobus in their patterns of range use or in the ways in which groups interact. Instead, these behaviours appear to vary greatly from place to place and are strongly influenced by the size and configuration of the wooded area inhabited by a population, and by the distribution of preferred food items in space and time within this habitat. Although *Colobus guereza* may have a greater tendency to use and defend small areas than the other taxa, even this species shows considerable flexibility in both range size and group interaction behaviour.

Adult male red colobus monkeys do not possess a low-frequency, high-amplitude 'loud call' equivalent to the roar of black-and-white colobus. However, adult males in adjacent groups often use a 'nyow' vocalization in countercalling, and this may have an influence on group spacing (Struhsaker & Leland, 1979; A. G. Davies and J. F. Oates, unpublished observations).

Reproduction

Female red colobus often display distinct swelling of the perineal tissues. The swelling is periodic, and both Struhsaker (1975) and Starin (1991) have provided some evidence of a positive correlation between swelling and the frequency with which females copulate. This suggests that the swelling occurs around the time of ovulation, but physiological studies are lacking.

The size of the perineal swelling varies considerably among red colobus populations. The swelling is relatively small in *P. b. tephrosceles* (maximum length 7.5 cm), large in *P. b. temminckii* (length up to 10 cm) and in *P. b. badius* (length up to 13 cm), and extremely large in *P. b. preussi* (estimated to equal 25–33% of body volume) (Struhsaker, 1975; Starin, 1991).

Both Struhsaker (1975) and Starin (1991) describe dominant adult males as performing most of the copulations within red colobus groups, apparently as a result of both male–male competition and female choice. Starin (1991) describes swollen females at Abuko giving conspicuous courtship displays, combining leaping about with a quavering call. Females of *temminckii* and *badius* also often produced quaver calls during copulation (Struhsaker, 1975; Starin, 1991; A. G. Davies and J. F. Oates, unpublished observations), while *preussi* females have been noted to quaver both during copulation, and immediately before and after copulation (Struhsaker, 1975). In contrast, *tephrosceles* males in Kibale gave rapid quaver calls when harassing other copulating pairs of monkeys (Struhsaker, 1975).

A. G. Davies (personal communication) has found a seasonal pattern in the distribution of female red colobus copulation quavers at Tiwai, with the highest frequency of quavering in the February–May period, and most births occurring in the early dry season (October–December). In the Abuko and Fathala *temminckii* populations, which inhabit a strongly seasonal environment, the great majority of births also appeared to be concentrated in the dry season (Starin, 1991). In the Kibale Forest, which has a less seasonal environment, the *tephrosceles* population exhibited no distinct breeding season, but births were clustered in peaks that tended to coincide with rainy months (Struhsaker & Leland, 1987). Struhsaker & Leland (1987) report an interbirth interval for Kibale red colobus of 25.5 ± 5.1 months; Starin reports a similar range at Abuko (27.8–32 months).

Newborn infant red colobus have different coat coloration from adults. The coloration varies from one subspecies to another (Struhsaker, 1975), but is never as conspicuous as the all-white coat of the black-and-white colobus or the bright colours of some Asian colobine neonates. Mothers very rarely allow other females to handle or carry their young infants, a distinct difference both from the black-and-white species and from most Asian colobines.

Summary of red colobus natural history

Red colobus typically concentrate their feeding in some of the largest and commonest tree species in their habitat, even when this requires terrestrial locomotion. In the Casamance, for instance, Temminck's red colobus frequently travel on the ground between clumps of trees (Gatinot, 1976; Struhsaker, 1975), in parallel with the behaviour of black-and-white colobus in gallery forest habitats.

The trees used for food by red colobus are often clumped in their distribution. Since the monkeys' preferred foods are seasonally available items such as young leaves, buds, unripe fruits and seeds, and flowers, red colobus foraging involves the exploitation of a set of large patches that are clumped both spatially and temporally, and often widely dispersed. This pattern of food distribution may explain both the extensive range overlap and the large social group size typical of many red colobus populations. Except where constrained by habitat (such as the fragmented forest on the Tana River), red colobus live in larger groups than do most black-and-white colobus, or indeed most of the arboreal monkeys of the African forests. Red colobus groups typically have several fully adult males and at some sites these males have been observed to cooperate in aggressive interactions with other groups. Female red colobus apparently transfer between groups more frequently than do males, leading to a description of red colobus society as patrilineal (Struhsaker & Leland, 1985).

Where red colobus groups contain several males, these males appear to cooperate in aggressive interactions with other groups, and this aggression is not site-specific. This could be related to the exploitation of a dispersed and possibly unpredictable food supply, so that males gain more in fitness by cooperating to defend both females and ephemerally occupied sites than they would by defending a fixed area of their home range. With males tending to remain with their male relatives and cooperate with them, females transfer between groups. As a result, female members of a group are often not closely related. The low degree of relatedness among females may be the ultimate explanation for the reluctance of mothers to allow their young infants to be handled by other females.

In a situation where males benefit from cooperation and large multi-male groups are common, females may be able to increase the fitness of their offspring by encouraging male–male competition, resulting in the evolution of conspicuous sexual swellings and, in some cases, of copulatory calls. Although the function of these copulatory calls has yet to be rigorously investigated, it is tempting to speculate that the calls (and pre-copulation displays at Abuko) serve to recruit additional males to the scene and incite competition among them, the function proposed for chimpanzee copulatory calls by Hauser (1990). A. G. Davies (personal communication) has observed serious fights between red colobus males at Tiwai at times of intense female quavering; at other times of year, single males tend to monopolize individual swollen females.

There may be a relationship between the seasonality of red colobus habitats, the seasonality of breeding, the size of female perineal swellings and the prevalence of copulation quaver calls in red colobus populations. Female strategies that tend to incite male–male competition may be of most value to females in situations where several females tend to be in breeding condition simultaneously. In such situations, males may achieve greatest reproductive success through a scramble competition to inseminate as many of the females as possible.

The olive colobus monkey, subgenus **Procolobus**

The olive colobus shares several morphological features with members of the red colobus group (Kuhn, 1967), but at 4–5 kg its body weight is only half the average weight of a red colobus (Chapter 3). This species, the only member of its subgenus, has the most limited geographical distribution of the three African colobine subgenera, occurring only in the moist coastal forest zone from southern Sierra Leone to eastern Nigeria (Oates, 1981).

Field studies

The observations by Booth of the olive colobus and other monkeys in Ghana in the early 1950s (Booth, 1956, 1957) were one of the earliest attempts to study the natural history of African forest primates. Although Booth did not watch habituated groups or collect quantitative behavioural data, he did gather a good deal of accurate information on the natural history of the olive colobus. He described the species as a thicket-haunter that prefers to feed in dense foliage at low levels, and is most common where the forest has been disturbed by treefalls or cultivation, in swampy areas, and along river banks. Booth also drew attention to the species' habit of associating with *Cercopithecus* monkeys, and to the unique behaviour of mothers, who carry their young infants in their mouths.

Following Booth's work, the olive colobus remained neglected during the burgeoning of primate field studies in the 1960s and early 1970s, when various forms of black-and-white and red colobus monkeys were receiving concentrated attention. Kuhn (1964) made a few observations in the course of collecting specimens for anatomical study, but little more was learned until 1976, when Galat and Galat-Luong began studying the primate community of the Tai National Park in Côte d'Ivoire. Between 1977 and 1983 the Galats made behavioural observations of several Tai primates, including the olive colobus (Galat-Luong, 1983; Galat & Galat-Luong, 1985), but they do not appear to have worked with habituated groups. In 1979, I began looking for a site at which to observe olive colobus, and subsequently began a study of the species at Tiwai Island in Sierra Leone in 1982; in this study a group was habituated, and its behaviour was sampled at intervals over a 32-month period from May 1983 to December 1985. Some results from that study are given in Oates (1988*b*) and Oates & Whitesides (1990); other observations are reported here and have not been published previously.

Association with other species

The tendency of olive colobus monkeys to associate with groups of other forest monkeys, particularly *Cercopithecus* species, is one of the best-documented features of the species' natural history. The studies in Sierra Leone have shown that, while association between a group of olive colobus and a group of another species commonly extends for several hours, the association can become more-or-less permanent, extending over several years. At Tiwai, one group of *Procolobus verus* was seen associating with one particular group of *Cercopithecus diana* from at least February 1983 to December 1988, and members of the olive colobus group were within 50 m of a member of the Diana monkey group on 83% of scan samples of behaviour (Oates & Whitesides, 1990). The interaction between the two species was essentially one-way, with the olive colobus orienting positively to the Diana monkeys, and the Dianas largely ignoring the olive colobus (Oates & Whitesides, 1990).

Social organization

Hill (1952) stated that *Procolobus verus* was generally found only in ones and twos. Hill was apparently relying on reports given to him by the naturalists G. S. Cansdale (from Ghana) and J. W. Lester (from Sierra Leone). Booth (1957) contradicted these reports, saying that he had never seen an undoubtedly solitary individual and that groups most commonly ranged in size from 10 to 15 individuals. Because of the secretive habits of the olive colobus and the lack of conspicuous sexual dimorphism, Booth had difficulty in assessing the sex-composition of groups. However, in one group of 11 monkeys, he shot two adult males.

More recent observations suggest that olive colobus groups are usually smaller than those reported by Booth. Galat-Luong (1983) counted 17 groups in Tai in 1976–82, and reports a range of 3–14 individuals per group, with a mean of 7.1. However, Galat-Luong acknowledges the difficulty of counting all group members. Two Tai groups whose composition was accurately censused contained six and seven individuals, each with a single adult male and two adult females (Galat & Galat-Luong, 1985). The habituated Tiwai group initially contained 11 monkeys, with two adult males and five adult females, but by the end of the study had declined to three individuals (one adult male, one adult female and a juvenile); a second Tiwai group had one adult male, one adult female and an infant (Oates, 1988*b*).

Both male and female composition of the Tiwai study group appeared to be unstable, but females were a much more dynamic element than males (unpublished observations). Between August 1983 and January 1984, one adult female disappeared from the group, and between April and June 1984 two more adult females disappeared. Although it is possible that all these females died, this is thought to be unlikely. Between July 1984 and February 1985, a new adult female joined the two that remained in the group, and in March 1985 another new female was seen briefly. By October 1985 the two original adult males were still present, but only one adult female remained. In November and December, one of the males was no longer seen. These observations are suggestive of a female transfer social system like that of red colobus, but more evidence is needed before any firm conclusions can be reached.

Social interactions within olive colobus groups are relatively infrequent. At Tiwai, most grooming was seen in the early morning or late evening (when the group was congregated at a sleeping site), or around the time of copulation: males and females were often seen to groom each other before and, especially, after a copulation. Adult males directed most of their grooming at adult females and rarely groomed each other or juveniles. Adult females distributed their grooming more evenly, but they groomed adult males more frequently than they groomed other classes of animals (unpublished observations).

Diet

The only large sample of feeding observations is from Tiwai, where 587 records collected in 11 months between June 1983 and October 1985 indicated a diet dominated by young foliage (59% of records) (Oates, 1988*b*). Fruits and fruit parts made up 19% of records (14.4% seeds; 4.2% whole fruits or other fruit parts), mature foliage accounted for 11% of records (9% petioles; 2% whole leaves or blades), and flowers and floral buds accounted for 7%. There were large seasonal changes in the diet, with seeds eaten much more frequently in the dry season than in the rains. The young foliage of climbing plants appeared to be a dietary staple.

Feeding sites

Booth's description of the feeding sites preferred by olive colobus was confirmed by the observations at Tiwai. Over 50% of feeding observations were collected from below 15 m. The monkeys avoided feeding in the crowns of tall trees and instead fed most frequently in the middle canopy, in liane tangles, and in dense low growth in canopy gaps (Oates, 1988*b*). In Tai National Park, Galat & Galat-Luong (1985) also recorded olive colobus more frequently in lower forest strata and much less frequently in emergents than red or black-and-white colobus in the same habitat; for instance, observations in emergents were 42% for red colobus but only 15% for olive colobus.

Activity patterns

The activities of the Tiwai study group were influenced to a considerable extent by the group of Diana monkeys with which they travelled. If the Diana group began a major travel from one part of their range to the other, the olive colobus would usually interrupt their feeding or resting and join their movement. Perhaps because of this, activities scored as travelling occupied a much greater proportion (25%) of scan samples than has been recorded for other colobus monkeys; feeding comprised 27% of records and resting 40%.

Range size and defence

Observations at Tiwai showed a strong link between the home range of the olive colobus study group and the home range of the *Cercopithecus diana* group with which it habitually associated. The olive colobus study group used 113 0.25-ha grid cells (28.25 ha) on 33 sample days between June 1983 and October 1985; 107 (95%) of these cells were among the 123 cells in which the Diana group was seen. The Diana group defended a territory against other Dianas, with intergroup calling and chases occurring frequently at range boundaries. Sometimes during these territorial encounters, adult males in the olive colobus study group exchanged 'laughing' calls with adult males associating with the neighbouring Diana group, but there was little evidence of active territoriality on the part of the olive colobus. The olive colobus did not appear to 'use' Diana monkeys as assistants in territorial defence, and the outcome of the Diana encounters was not evidently influenced by the behaviour of the olive colobus (Oates & Whitesides, 1990).

Reproduction

The Tiwai observations reported in this section have not been published previously.

Adult female olive colobus sometimes display large perineal swellings that are similar in size, relative to their overall body size, to those of *P. b. badius* (Figure 4.8). In Sierra Leone I recorded swellings as pink or greyish-pink in colour, with a maximum width of about 6 cm and a maximum length of about 5 cm. Swellings on individual females persisted for at least 8 weeks and possibly longer, increasing and decreasing in size over this period. Adult males showed particular interest in females with large swellings, but copulations with females having small swellings or no obvious swelling also occurred. Copulations followed presentations by females to males.

On 49 days when the behaviour of the Tiwai study group was sampled for a full day, one of the two adult males (RM) was seen to copulate 46 times, and the other (PF) copulated 8 times. These were clearly observed copulations only; many more probably occurred unobserved. Four different females were mated by the males, but on any one day most copulations were generally with a single female. In two months, male RM was observed in a consort relationship with an obviously swollen female, following her closely for several days and mating frequently. For instance, in June 1984 RM consorted for three days with one female who had a large swelling; he was seen to copulate 18 times with this female. Male PF was seen to copulate with the same female only once during these three days; within 2 min of PF's copulation, the female copulated with RM. No overt aggression between the two males was seen.

The great majority of copulations at Tiwai were observed between March and August and, with one exception, swollen females were only seen in April–August. Females with large swellings (4–6 cm in length) were only

Figure 4.8. An adult female olive colobus monkey (standing over a juvenile) presents her perineal swelling (arrowed) to an adult male at right (photographed at Tiwai by J. Oates).

seen in June–August (however, no observations were made in the month of September). Very small infants were only observed in November–April, the dry season, with a majority apparently being born in January. These observations suggest both a distinct breeding season and a typical colobine gestation period of about 6 months.

Newborn infants have a darker coat than adults, but are not conspicuously different in colour from their mother. As in red colobus, they have not been seen to be handled by animals other than their mother. Olive colobus are remarkable for the way in which young infants are carried. At Tiwai, the mother held the infant in her mouth, usually grasping the flank in such a way that the belly of the infant was tucked in to her neck, with the tail wrapped around the mother's neck and upper back. Infants were only carried during travel from place to place; when the female stopped moving she put the infant down.

Booth (1957), who first described this behaviour, said that older infants clung to the mother's belly. Galat-Luong & Galat (1978) reported that, following the early stages of oral transport, and before being carried ventrally, the infant was carried around the mother's neck. At Tiwai, infants were never seen either clinging to the mother's venter, or being carried around her neck, without being held in the mouth. Perhaps different populations behave differently. As Tiwai infants got older they were carried less frequently. Six-month-old infants were carried only occasionally, for instance when the mother jumped across a large gap in the canopy. The largest infant seen to be carried in the mouth was estimated to be at least 10 months old.

Summary of olive colobus natural history

In its mating system the olive colobus resembles red more than black-and-white colobus. Female sexual swellings occur in both olive and red colobus, and in each form females appear more inclined to disperse than do males (although evidence on this point is still sparse for olive colobus). Many features of the natural history of olive colobus appear to be related to small body size. Olive colobus are highly selective feeders, consuming only items of high digestibility and feeding preferentially in climber tangles and thickets (Oates, 1988*b*). Such a diet may limit the number of individuals that can feed together on a year-round basis, and this in turn may constrain group size. The combination of small body size and small group size seems likely to put olive colobus at greater risk to predators than the larger colobus species, thus favouring the strategy of association with other forest monkeys. The universal occurrence of this association behaviour among olive colobus populations in a variety of habitats has led us to the conclusion that it is part of a genetically-based evolved strategy that reduces predation risk (Oates & Whitesides, 1990). Other parts of this strategy are the species' drab coloration, low vocalization rate and habit of spending much time inactive in patches of thick vegetation (see also Galat & Galat-Luong, 1985).

Because olive colobus groups tend to feed in a dispersed pattern, frequently in thick growth, it may be difficult for males to monitor constantly the behaviour of females, or to detect the close approach of other males. The potential therefore exists for females to be inseminated by adult males outside their normal social group, and it should pay females even in uni-male groups to advertise their reproductive state with perineal swellings. A viable male counter-strategy in this situation is to consort closely with a swollen female and to mate with her frequently, and this is the behaviour observed at Tiwai. Such male behaviour would select for high sperm production and therefore explain the large testes noted in this species by Hill (1952), and visible in the male in Figure 4.8.

Booth (1957) suggested that the remarkable mouth-carrying of infant olive colobus was the result of a combination of several factors: the absence of a thumb, the shortness of the adult coat, and the density of the vegetation in which the monkeys frequently move. Booth surmized that this combination would make normal ventral clinging by infants unusually precarious.

Competitors and predators of African colobines

Feeding competition between African colobines and other animals has not been closely studied. Apart from other primates, only a few arboreal mammals approach colobines in size, and most of these (e.g. pangolins and viverrids) have very different diets. As folivores, tree hyraxes (*Dendrohyrax* spp.) and some anomalurid flying squirrels are potential competitors (Emmons *et al.*, 1983), but their diets are poorly known, and these animals usually occur at much lower densities than do monkeys (personal observation). Sciurid squirrels (especially the large, canopy-foraging *Protoxerus stangeri*) may compete with some colobus populations for seeds, but it appears that the feeding behaviour of squirrels and colobus monkeys has not yet been studied at the same site.

Apart from other colobines, colobus monkeys' closest dietary competitors are probably cercopithecine monkeys. At least two cercopithecine species commonly co-occur with any colobus population, but although there are often broad overlaps in the classes of food eaten by sympatric colobines and cercopithecines (e.g. the fruits and leaves of forest trees), overlap between specific food items is generally low. This low overlap is presumed to result largely from their differing digestive strategies and detoxifying capabilities.

In addition to humans, the crowned hawk-eagle (*Stephanoaetus coronatus*) is probably the predator that, historically, has posed the greatest threat to African colobines. This large raptor (around 80 cm in length) often has monkeys as a major prey item, and is able to kill even the largest colobus (Oates, 1977*a*; Rucks, 1978; Struhsaker, 1981*b*; Skorupa, 1989; Struhsaker & Leakey, 1990). The crowned eagle was once found throughout the African forest zone, and present range gaps may be due in part to the reduction of the eagle's prey by human hunters (Skorupa, 1989).

Chimpanzees are also significant colobus predators. Busse (1977) estimated that chimpanzees killed a minimum of 8–13% of the population of red colobus at Gombe, Tanzania, during 1973 and 1974, while Boesch & Boesch (1989) found that red colobus were the focus of the great majority (81%) of

135 chimpanzee hunting attempts observed in Tai in 1979–86. Busse has argued that such predation pressure could be a factor selecting for large group size in red colobus. However, one reason for the high level of chimpanzee predation on red colobus might be that the large size and noisy behaviour of this colobus make them readily detectable.

Leopards are widely sympatric with colobus monkeys and may be a threat on occasions when the monkeys come close to the ground. Of 215 leopard scats from the Tai Forest in Côte d'Ivoire examined by Hoppe-Dominik (1984), eight contained hair of red colobus monkeys and five contained black-and-white colobus hair; this might reflect leopard scavenging of eagle kills, rather than direct leopard predation (T. T. Struhsaker, personal communication).

In the Kibale Forest, Struhsaker & Leakey (1990) found black-and-white colobus to be captured more frequently than expected by crowned hawk-eagles, while red colobus were captured less frequently than expected. They attribute this to the aggressiveness of male red colobus which, they report, will attack crowned hawk-eagles in a closely-spaced coalition of four or more individuals. Struhsaker & Leakey also comment on the ability of red colobus males to supplant adult male chimpanzees from trees. Although Boesch & Boesch (1989) report that mobbing of hunting chimpanzees by adult red colobus in Tai is only rarely successful (two out of ten mobbing attempts in 68 hunts), even a low level of success might result in significant fitness benefits to males engaging in the behaviour (A. G. Davies, personal communication).

The olive colobus potentially faces an especially high risk of predation by chimpanzees, given its small body size, small group size and tendency to feed near the ground. Oates & Whitesides (1990) have suggested that chimpanzee predation has been the major factor selecting for the predator-avoidance strategy of the olive colobus.

Conclusions

The olive colobus has been shown to have a natural history distinctly different from that of the other subgenera of African colobines, the red and black-and-white colobus, which have about twice the body weight of the olive colobus. Within each of the other two subgenera there is very considerable variation in diet and patterns of social behaviour, and niche differences are relatively subtle. Early on in the history of ecologically-oriented primate field studies, Clutton-Brock (1974*b*) drew a distinction between large, multi-male groups of red colobus exploiting clumped and relatively unpredictable food supplies, and small (often uni-male) groups of black-and-white colobus, able to subsist for part of the year on a diet consisting largely of mature foliage. Since that time, a red colobus population has been described (at Tana River) in which individuals live in relatively small groups with usually a single adult male, and at times eat substantial quantities of mature foliage, while many populations within the black-and-white group have been found to have several males in a group. Indeed, some black-and-white colobus populations (e.g. *Colobus angolensis* in Rwanda) exhibit some of the largest social groups among primates as a whole.

Although there may not be radical differences between black-and-white and red colobus on an Africa-wide basis, some generalizations do hold (at least in the face of existing evidence). Red colobus generally get a large part of their food from some of the largest trees in their habitat, while members of the black-and-white group tend to feed lower in the canopy and more frequently exploit gap and edge vegetation, including lianes. Where they co-occur, red colobus select a more diverse diet than do black-and-white colobus and use larger home ranges. All the forms of colobus will eat seeds, but only in some of the black-and-white colobus do seeds dominate the diet. These behavioural differences probably have anatomical and physiological correlates, both in the food acquisition and processing systems, and in the locomotor system. Such aspects of structure and function have not yet been adequately explored, but several lines of evidence point to a greater ability of black-and-white colobus to cope with a diet particularly high in fibre. This is suggested not only by the differences in the diet of red and black-and-white colobus at Kibale, but also by the captive study of Watkins *et al.* (1985) and by a difference in the average fibre-content of fruit parts eaten by red colobus and black-and-white colobus at Tiwai: fruit foods averaged 21% fibre for red colobus compared with 28% for black-and-white colobus (Dasilva, 1994; G. Davies, personal communication). The greater energetic economy displayed by black-and-white colobus in habitats which they share with red colobus is probably another part of a difference in evolved niches.

Subtle, but probably long-standing, differences in the feeding niches of the different forms of colobus may have been an important ultimate determinant of differences in mating systems. Both red colobus and olive colobus, unlike members of the black-and-white group, display prominent female sexual swellings. Whereas both males and females move between groups in all kinds of colobus, the females of red and olive colobus appear to be more likely to transfer than the males, but in the black-and-white group, male-biased transfer seems to be more typical. As a correlate of this, individual female red and olive colobus tend to be less cooperative with one another than are female black-and-white colobus, but males tend to be more cooperative. In black-

and-white colobus, greater female–female cooperation is reflected in a willingness by mothers to allow infants to be handled by other females, whereas the existence of low-frequency loud calls in the males and the mechanisms to produce such calls (absent in red and olive colobus) may signify a greater phylogenetic history of competition among individual males.

Today, the feeding niche, mating system and social system of each population forms an adaptive complex within which no one facet can be understood in isolation. For instance, although the mating system of an ancestral red colobus population may have evolved as a response to a particular food-harvesting strategy (along the lines of the model proposed by Clutton-Brock & Harvey, 1978), this mating system is now grounded in what may be a relatively inflexible physiological substrate. Such a 'phylogenetic' constraint will tend to bias the kind of social system that can occur (e.g. a group of males cooperating with one another in competition with other males). As a result, the social system itself may have as much proximate influence on diet as diet has on social behaviour.

Acknowledgements

I am grateful to Liz Bennett, Glyn Davies and Tom Struhsaker for their helpful comments on earlier versions of this chapter. Research at Tiwai Island was supported by grants from the National Science Foundation and the Research Foundation of CUNY.

5

The ecology of Asian colobines

ELIZABETH L. BENNETT and A. GLYN DAVIES

Introduction

Asia contains a wide variety of forest types, from equable tropical evergreen rain-forests to montane, temperate and highly seasonal forests (Figure 5.1). Together, these support the six genera and 24 species of colobine monkeys which occur in Asia (Chapter 3). Unlike the African colobines, the Asian colobines show distinct geographical distributions in relation to separate taxonomic groupings. This precludes analysis of a single taxon over the whole Asian colobine range, as was done in the previous chapter (e.g. for red colobus), but lends itself to a geographical approach. Thus, we focus initially on *Semnopithecus* and *Trachypithecus* populations in India and Sri Lanka, and progress eastwards with the *Trachypithecus* and *Presbytis* species in South-east Asia, finishing with the 'odd-nosed' colobines (*Nasalis* and *Pygathrix*) in Borneo, Indo-China and China.

We give details from the main sites where colobine monkeys have been subjects of ecological studies (Figure 5.1), with the aim of describing socio-ecological characteristics of the different groups. Data from other sites are also included to show intraspecific variation and to give information on species that have been little studied.

Asian forest habitats

Vegetation formations are largely determined by climate, although soil, altitude and human interference all have additional influences, and climate varies enormously across the Indo-Malayan biogeographic realm (Figure 5.2), despite all sites being affected by at least one of two monsoons: from the south-west in June–July and from the north-east in December–January.

At Simla (3500 m above sea level (a.s.1.)), in the Himalayan foothills, the annual rainfall of 1700 mm is brought by the south-west monsoon in June

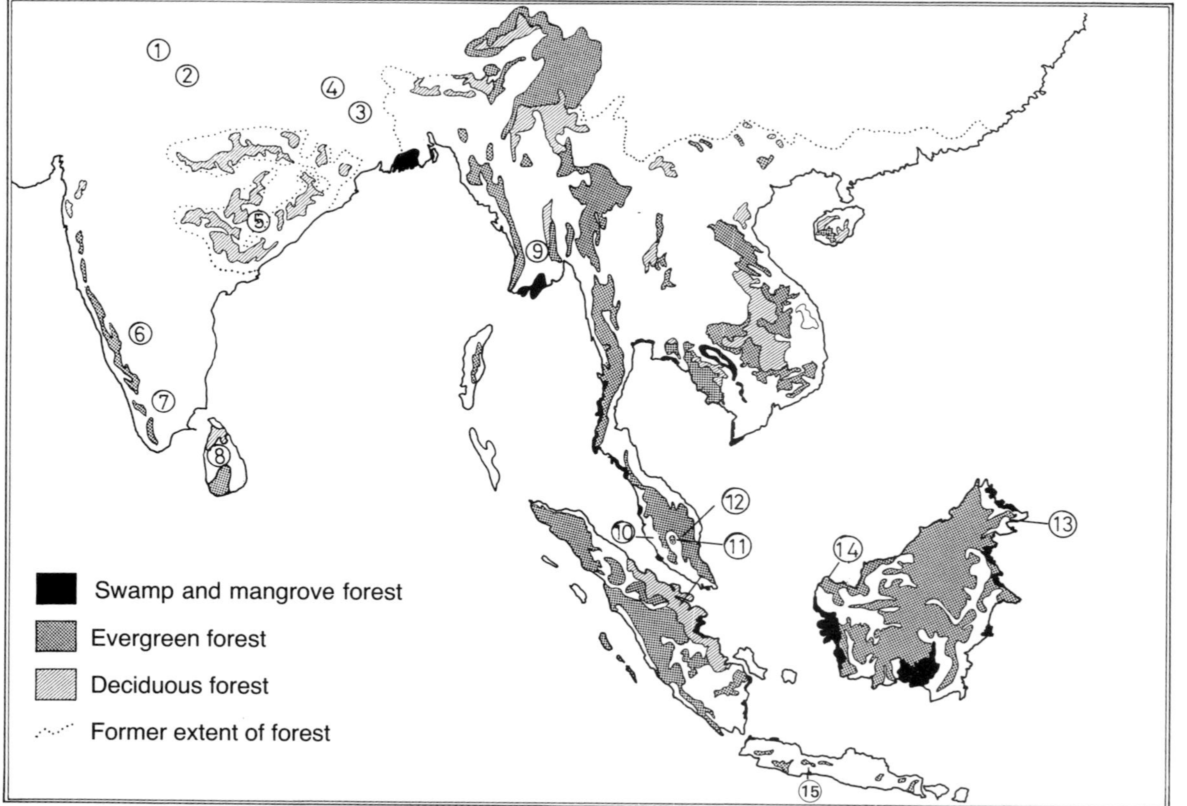

Figure 5.1. Distribution of evergreen, deciduous and swamp forests in Asia, and localitites of colobine socio-ecology study-sites (after Champion and Seth, 1968; Whitmore, 1984*a*; Collins *et al*., 1991). Sites: 1, Simla; 2, Bhimtal; 3, Junbesi; 4, Melemchi; 5, Kanha; 6, Periyar; 7, Kakachi; 8, Polonnaruwa; 9, Madhupur; 10, Kuala Selangor; 11, Kuala Lompat; 12, Sungai Tekam; 13, Sepilok; 14, Samunsam; 15, Pangandaran.

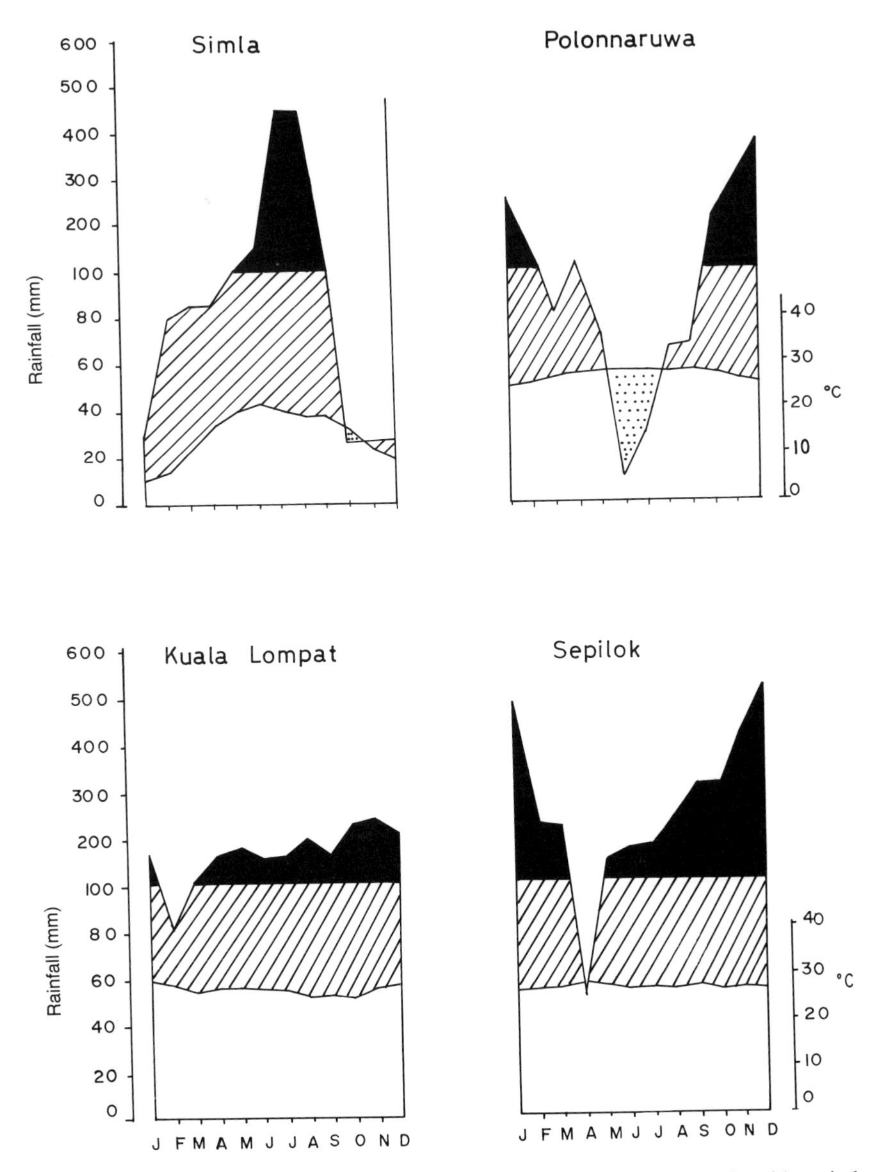

Figure 5.2. Climatic diagrams for four Asian study sites. Black, perhumid period (rainfall > 100 mm); hatched, humid period; dotted, arid period (after White, 1983).

(Sugiyama, 1976). Further south, the same monsoon brings most of the 1600 mm that falls on Kanha in the Indian central highlands (Newton, 1988*b*), but at the southern tip of India both monsoons give the high rainfall (3080 mm) at Kakachi (Oates *et al.*, 1980). In Polonnaruwa, in Sri Lanka, the south-west monsoon no longer brings much rain, because the uplands form a rain-shadow over the site, and most of the 1670 mm rainfall is brought by the north-east monsoon, blowing uninterrupted from the Bay of Bengal.

These patterns of rainfall are similar in South-east Asia, although precipitation tends to be higher, since there is an abundance of moisture supplied by the seas surrounding the main Sunda islands (Java, Sumatra and Borneo). Kuala Lompat in central Peninsular Malaysia is in the rain-shadow of two hill ranges, so rainfall (2000 mm) is low compared to other sites in the Peninsula, with a tendency to be heavier during the north-east monsoon (Raemaekers *et al.*, 1980). Sepilok, on the east coast of Borneo, receives most rain (3000 mm p.a.) from the north-east monsoon around the end of the year, but occasionally gets a secondary peak of rain in June, when the south-west monsoon strikes the other coast of Borneo.

This variation in rainfall, combined with altitudinal effects, gives a range of habitats for colobine monkeys: from the arid thorn forests of Rajasthan, and the oak–coniferous forests of the Himalayan foothills and southern China, to moist deciduous and dry deciduous forests of south Asia, and evergreen dipterocarp, mangrove and swamp forests of South-east Asia.

There are conspicuous differences in the structure of these different forest formations (Figure 5.3). In the semi-deciduous forest at Polonnaruwa, 5% of

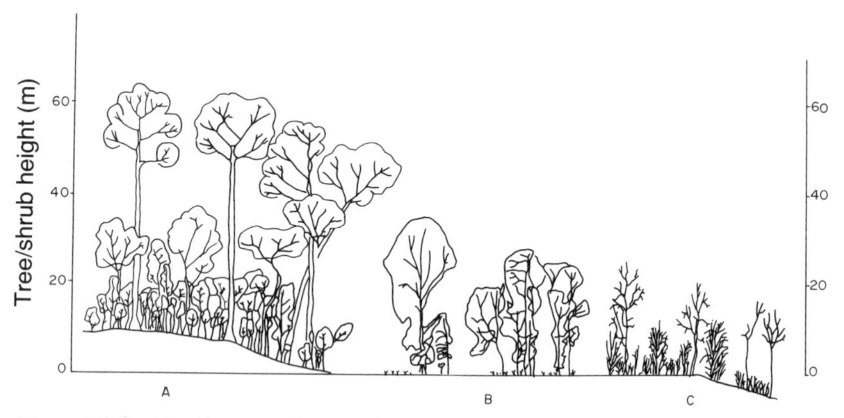

Figure 5.3. Profile diagrams of vegetation transects in three types of forest: A, lowland mixed dipterocarp forest at Sepilok; B, moist deciduous Sal forest at Kanha; C, dry deciduous forest at Kanha (Sources: Fox, 1973; Newton, 1985).

the tree stand reaches 40 m in height, but most trees (75%) are in the canopy, between 15 and 20 m tall, or the sub-canopy, between 5 and 15 m tall (Dittus, 1977). The dry deciduous and moist deciduous forests of the central Indian highlands differ markedly from one another (Figure 5.3), with thorny bushes dominating the former and the dipterocarp *Shorea robusta* dominating the latter (Newton, 1988*b*).

These deciduous formations are dwarfed by the majestic evergreen dipterocarp forests of South-east Asia, in which some emergent trees grow 80 m high, standing head and shoulders above the main canopy trees whose crowns are contiguous (Figure 5.3). Beneath, there is a dense stand of shade-tolerant understorey trees and saplings. For primates, arboreal pathways are offered in the canopy and, if the monkeys are small enough, in the understorey, where tree crowns interlink. Gaps between emergent trees are often too large to jump, so these trees are seldom used when travelling, although they are often used as night-sleeping sites. In open forest formations monkeys must travel along the ground to get between disconnected clumps of forest.

Differences in tree-species richness match this variation in structure (Figure 5.4). Tree-species diversity is greatest in the dipterocarp forests of South-east Asia (Whitmore, 1984*a*) where equable climatic conditions combine with forest fragmentation to encourage extensive species radiation, especially among the dipterocarps. There are roughly twice as many tree species in a 1-ha patch at Kuala Lompat or Sepilok, as in the evergreen forests of West Africa, such as Tai National Park in Côte d'Ivoire and Gola in Sierra Leone (Davies, 1987*a*).

In the semi-deciduous forest at Polonnaruwa and the moist deciduous forest at Kanha, tree-species richness is much lower: 61 species (41 excluding riverine species and cultivars) over 5 m tall, and 63 species greater than 2 m tall, respectively (Dittus, 1977; Newton, 1988*b*). At both sites, a single tree species dominates the flora: *Drypetes sepiara* in Polonnaruwa and *Shorea robusta* in Kanha. Importance Value Indexes (IVI)[1] are intermediate between those recorded in temperate forests, where only a few species predominate, and tropical evergreen forest, where no species predominate. Inevitably, this means that colobine monkeys in evergreen forests have more species from which to select food than do those in deciduous environments, with the greatest choice being in the south-east Asian dipterocarp forests.

Seasonal patterns of plant part production also influence primate food selection and, like species richness, they relate to climate (Figure 5.5). At

[1] Importance Value Index calculated as a combination of percentage tree frequency, percentage tree density and tree size.

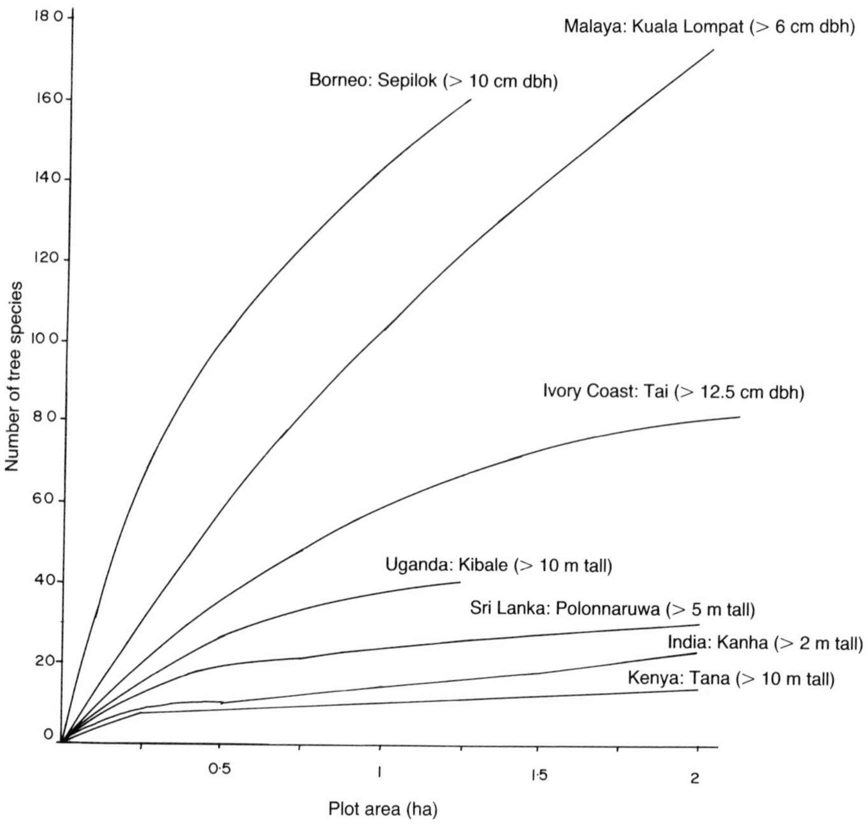

Figure 5.4. Comparison of the tree-species richness at seven sites where colobines have been studied, showing the great species richness of South-east Asian sites (sources: Davies, 1984; Raemaekers *et al.*, 1980; Bernard-Reversat *et al.*, 1978; Marsh, 1978*b*; Newton, 1988*b*; Dittus, 1977)

Kanha, 55 tree species are deciduous, one is semi-deciduous and only one is evergreen. The single peak of young-leaf production involves more than 96% of all trees. In contrast, double peaks of leaf production have been noted in South-east Asia, with little production during the driest or wettest months (Raemaekers *et al.*, 1980; Bennett, 1983; Davies, 1984). Production does not match the large-scale (gregarious) leaf production at Kanha, since half or less of the trees produce new leaves in peak periods and very few are deciduous, with notable exceptions such as *Intsia palembanica*.

There are single peaks of flower and fruit production at Kanha, Kuala Lompat and Sepilok, with Kanha again showing the greatest production; at Sepilok, only 10% of trees flowered or fruited during a one-year study period.

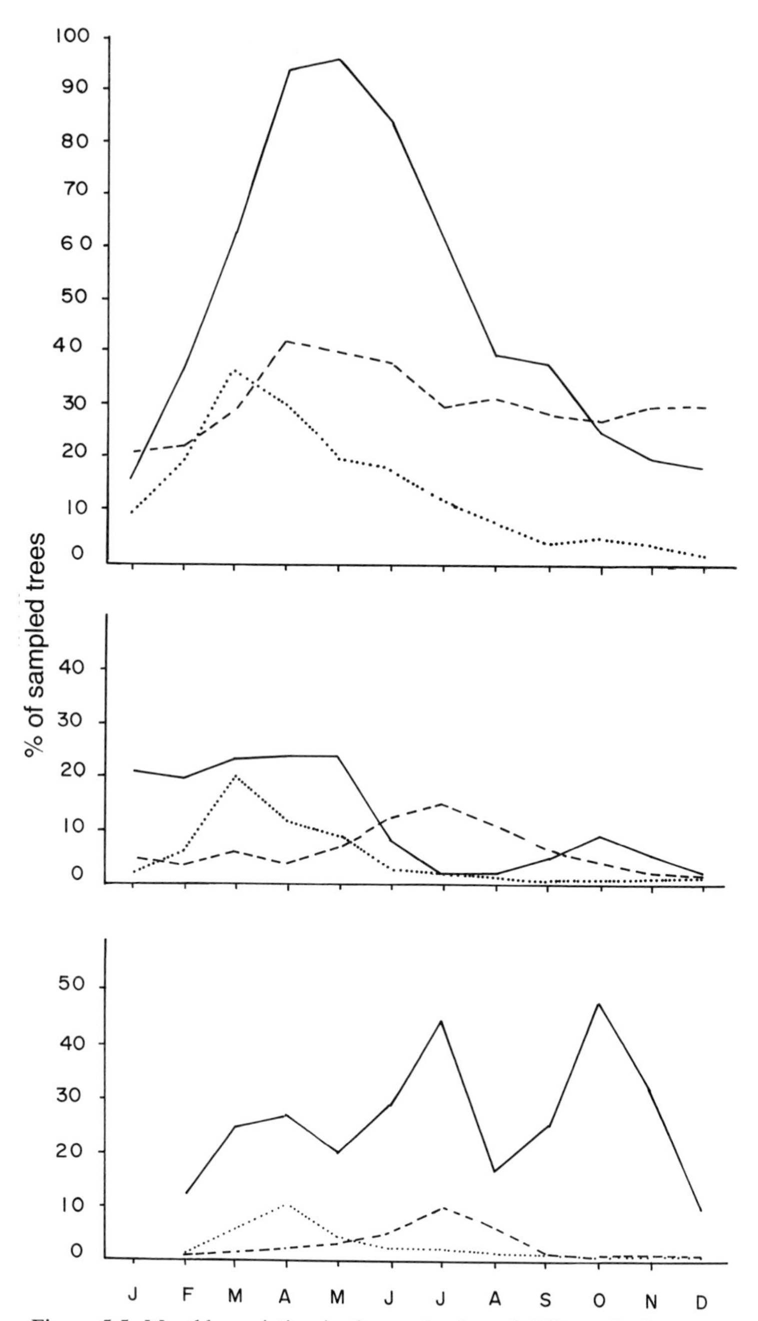

Figure 5.5. Monthly variation in the production of different leaf parts. Top, Kanha, 1981–82 (n = 49); middle, Kuala Lompat, 1976 (n = 900); bottom, Sepilok, 1980–81 (n = 627). Solid line, young-leaf parts; dotted line, floral parts; dashed line, fruit parts (after Newton, 1988*b*, Raemaekers *et al.*, 1980; Davies, 1984)

This, however, overlooks two important characteristics of South-east Asian dipterocarp forests. First, although many tree species produce their fruits synchronously, some species produce fruits in small quantities throughout the year (see e.g. Payne, 1980). Second, synchronized fruiting is variable with gregarious, or mast fruiting, occurring about every 15 years, when up to 50% of all trees produce fruits (Janzen, 1974).

Thus, leaf production is far more seasonal in the drier forests of South Asia than the wetter forests of South-east Asia, where there is always an abundance of mature leaves and substantial young-leaf production in most months. Fruit and flower production, however, can show erratic seasonal or annual patterns of production in both evergreen and deciduous forests.

Asian forest primate communities

The diurnal primate communities of the Indo-Malayan forests show a lower radiation of cercopithecine species than equivalent forests in Africa. In the dry forests of India and Sri Lanka, colobines are sympatric with a single macaque species: toque macaque (*Macaca sinica*) in Sri Lanka, bonnet macaque (*M. radiata*) in southern India and rhesus macaque (*M. mulatta*) in northern India and the Himalayas. In the Shola forest of south India, *T. johnii* is sympatric with its look-alike, the rare lion-tailed macaque (*M. silenus*).

In eastern India, the primate community is joined by gibbons (*Hylobates* spp.) which are distributed through most of Indo-China and South-east Asia; one and very rarely two species occur at a given site. In parts of Peninsular Malaysia and Sumatra, these are joined by the siamang (*H. syndactylus*), which is twice as large as the other gibbons, and in northern Sumatra and Borneo by the enormous orang-utan (*Pongo pygmaeus*) (> 50 kg). Only two macaques occur west of Wallace's Line in Peninsular Malaysia and insular South-east Asia: the long-tailed macaque (*M. fascicularis*) in riverine, coastal and farmland habitats and the pig-tailed macaque (*M. nemestrina*) in inland forests. However, in Thailand, Assam and Indo-China, there are several macaques which can occur alongside the Indo-Chinese colobines (Eudey, 1987).

Competition between colobine monkeys and simple-stomached primates like the macaques and gibbons has not been studied in detail, but comparisons of their diets where they are sympatric indicate that food requirements are very different (see e.g. Hladik, 1977*b*; Dittus, 1977; Curtin & Chivers, 1978; MacKinnon & MacKinnon, 1980), with correspondingly little conflict over food resources.

Predation of arboreal primates has hardly ever been reported in South-east

Asian forests (see Chapter 10); it appears to be much lower than that reported from African forests (see e.g. Aldrich-Blake, 1970; Busse, 1977). There are no raptors large enough to take anything but the smallest primates from the treetops, and although pythons are potential predators (Whitten, 1980), they appear to have minimal influence on populations. Species that live in more open forest formations and travel along the ground are more prone to predation (see e.g. Rudran, 1973*b*; Stanford, 1989). Furthermore, in areas where tribal people hunt primates, colobine populations may be reduced, as reported from Peninsular Malaysia (Chivers & Davies, 1979) and Siberut (Whitten, 1980), and even severely depleted as in Sarawak (Bennett, 1992).

Asian colobines

The western-most and most widespread colobine in Asia is the Hanuman langur, *Semnopithecus entellus*. It occurs throughout the Indian subcontinent, from the Himalayas in the north to Sri Lanka in the south. It features prominently in Indian mythology and has a long association with humans. Not surprisingly, it was the first colobine in Asia to be studied (see e.g. McCann, 1933) and it has now been investigated at more than 20 sites throughout India, Sri Lanka, Nepal and Bangladesh. Much of the work has focused on social behaviour rather than ecology (for reviews, see Oppenheimer, 1977; Vogel, 1971; Moore, 1985*a*; Newton, 1988*a*), and only recently has more information on the species' ecology become available.

S. entellus is an exceptionally adaptable species (Table 5.1). The 15 subspecies range in size from the Himalayan giants (*S. e. ajax*, *S. e. schistacea*), whose adult males weigh up to 20 kg, to the small Sri Lankan *S. e. thersites*, in which the adult male weighs 10.6 kg (Roonwal & Mohnot, 1977; Napier, 1985). The species occupies a wide range of habitats, from alpine scrub and sub-tropical pine forest to semi-evergreen deciduous and tropical thorn forests (Oppenheimer, 1977). It survives cold winters in the Himalayan foothills and the heat of the desert areas of Rajasthan. It also adapts well to man-made habitats, frequently raiding crops and entering village bazaars. Yet it is generally not found in evergreen forests, even where they occur within its geographical range, and in many of these respects it is unlike other Asian colobines, which tend to be arboreal, and confined to deciduous and evergreen forests.

Table 5.1. *Social organization of Hanuman langurs* (S. entellus) *in different sites in peninsular India*

	Abu	Dharwar	Dharwar	Jodhpur	Orcha	Rantham.	Singur
Habitat	ef	dd	gs	g/f	md	dd	v/f
Group size	21.3	15.5	15.3	35	19	45.7	12.8
% groups/bands that were all-male	1.7	1.5	1.0	1.0	3.7	7.7	1.0
Band size (mean)	8.2	15	8.7	15.7	1	16.6	–
Group home range (ha)	–	20	–	60–96	390	–	4
Population density (individuals/km^2)	31.6	85	16.6	18	4.4	14.6	12

Note: Rantham., Ranthambhore; ef, evergreen forest; dd, dry deciduous; gs, grasslands; g/f, grassland/forest; md, moist deciduous; v/f, villages and farms.
Sources: From Oppenheimer, 1977; Newton, 1987.

Kanha: Semnopithecus entellus entellus

Newton (1984, 1985, 1986, 1987, 1988*a*,*b*, 1992, 1994) monitored the Kanha meadows population of 360 *S. entellus* (1980–90) and conducted a detailed study for 3 years of one troop that occupied Kuloo chattan. More than 70% of his 75-ha study site was Sal (moist deciduous) forest, with meadows in the remaining area (a legacy of past slash-and-burn cultivation by the Baiga forest tribe), and small patches of dry deciduous forest covering boulder-strewn rocky outcrops. Apart from the cultivation (more than a hundred years ago), low-level hunting and some timber extraction (mostly 70 years ago), the habitat has been little influenced by man, and today the langur population experiences minimal human disturbance in an area similar to that in which the species is thought to have evolved.

The climate at Kanha was so seasonal that mature leaves were the only available food source for most of the year. Not surprisingly, therefore, leaves and leaf parts comprised half of the total annual diet, with mature leaves alone making up 35% (Figure 5.6). Young leaves contributed only 4% and buds 11% of the diet, flower parts 9.5% and fruit parts 24%. Minor items included gum, earth (most often from recently rebuilt termite mounds), stems, bark, herbs and invertebrates. Although the latter only comprised 3% of the diet, this is the highest recorded level of insectivory for any colobine (Newton, 1992).

The deciduous nature of the forest meant that the diet of the langurs varied greatly throughout the year. In the cold season when little else was available,

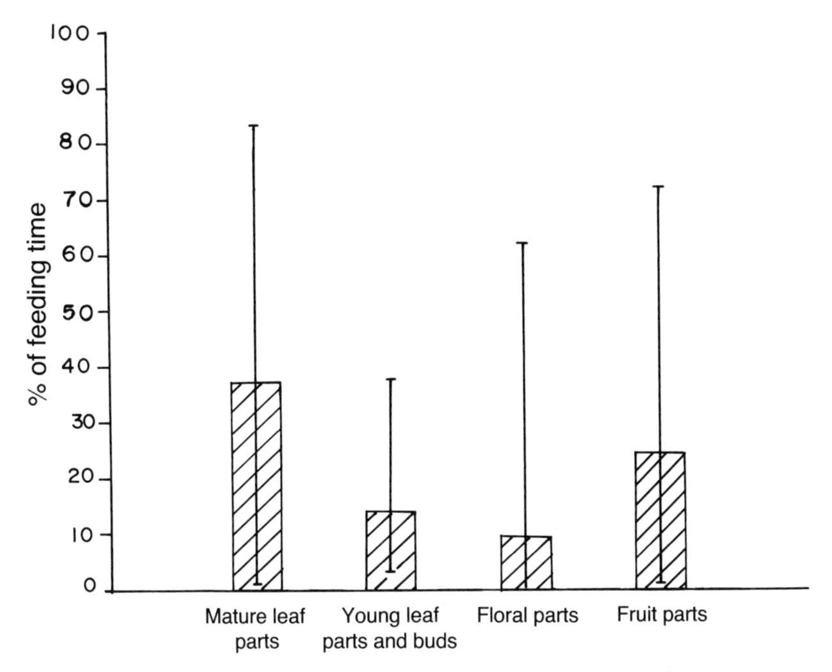

Figure 5.6. Annual diet of Hanuman langurs (*Semnopithecus entellus*) at Kanha (after Newton, 1992). Annual average and monthly maxima and minima are shown.

mature leaves were eaten in large quantities. By March, as soon as they sprouted, leaf buds, young leaves and flowers accounted for most of the diet. Fruit began to have more importance towards the end of the hot season and by June, flowers, fruits and invertebrates dominated the diet. In July, the eating of mature leaves recommenced in response to a scarcity of other items and, apart from a sudden and short upsurge in fruit intake in October, when *Flamingia semialata* shrubs were fruiting, mature leaves dominated the diet throughout the winter. No single species predominated over the whole year, and foods were selected from about 53 species of trees and climbers (Newton, 1992).

Although food selection is dictated to a large extent by seasonal availability, preferences could still be discerned. Young leaves were eaten in preference to mature leaves when both were available, and leaf buds, flowers and fruits were generally preferred to young leaves. Mature leaves of *Pterocarpus marsupium* were eaten only in the cold season. The super-abundant mature leaves of *Shorea robusta* were never eaten, nor were the mature leaves of *Diospyros*. Presumably, these items were innutritious or toxic, as is the case for fruits of *Caesaria tomentosa*, which were not eaten by the langurs and were used as a fish poison by the Baiga people. The animals spent most of their time feeding on the ground or less than 5 m above it, and long-distance travel was done on the ground. Home ranges and the average length of day ranges were relatively large, about 75 ha and 1 km, respectively, which is related to temporal and spatial distribution of food supplies and to a lesser extent to interactions with neighbouring groups (Newton, 1992).

The social system of *S. entellus* at Kanha comprises uni-male groups and all-male bands. The uni-male groups consisted of one adult male, about ten adult females and their progeny, and all-male bands comprised sub-adult and adult males. Both females and males migrated between bisexual groups, but the males did so much more frequently, and at least one bisexual group experienced a multi-male phase as a result of male immigration.

Home ranges overlapped considerably: 50% of the home range of Newton's main study group was also used by neighbouring uni-male groups and 44% was used by all-male bands, some areas being overlapped by both bands and groups. Intergroup encounters occurred almost every day and were usu ally aggressive, involving 'whooping' calls, displays and grappling. They could be violent, especially if they preceded or succeeded changes in group composition. They often involved group members other than the rival males.

In April 1981, adult males of a 16-member all-male band attacked and killed three out of six infants in the study group during four attacks on consecutive days. These assaults were directed at infant-bearing adult females and only infants were injured; the resident male was ineffectual in defending them. Subsequently the band males consorted and mated with adult females whilst a sub-group comprising the resident male with the three surviving infants and their presumed mothers remained separate from the band. Seventeen days after the initial infanticide, the resident male was replaced by a band male, and was never seen again. In the five months after the takeover, two of the three remaining infants disappeared but without direct evidence of infanticide (Newton, 1986).

The new male probably sired the ten infants born in the subsequent birth season and, despite this system of violent takeovers, the adult male remained resident and in control of the main study group for 10 years, until he was 17–19 years old (Newton, 1994).

Himalayas: Semnopithecus entellus ajax *and* Semnopithecus entellus schistacea

Harsh climatic conditions are experienced by Himalayan langurs (*S. entellus ajax* and *S. entellus schistacea*), which occur between about 1000 and 3500 m a.s.l. (review by Bishop, 1979). At these altitudes, temperatures often drop below freezing on winter nights and snow storms occur from December to February. Cold nights are partly compensated for by generally cloudless days during which the langurs spend much time warming themselves in the sun. Presumably as an adaptation to such winters, Himalayan *S. entellus* are much larger and have much longer, thicker coats than do any other subspecies. During the summer (July to October), temperatures are more moderate and most of the high annual rain (1640 to 2500 mm) falls.

As at Kanha, in such a seasonal environment the distribution and abundance of preferred foods varied markedly throughout the year. At Hatto, near Simla, the main winter foods were cones of *Abies* and *Taxus* spp., along with their bark, cambium and twigs; some herbs were also eaten. During the summer, the diet was much more diverse, with the monkeys eating many fruits, including acorns and pine-cones (Sugiyama, 1976). This frugivory is conspicuously like the diet of *S. entellus* at Kanha, and highlights the fact that *S. entellus* is not consistently folivorous (Sugiyama, 1976). None the less, the ability to subsist on fibrous leaves and barks is essential for sustaining populations at harsh times of year, when preferred foods are unavailable.

Social organization of Himalayan *S. entellus* subspecies is somewhat flexible (Table 5.2). Groups varied markedly in size, from seven to almost 100 animals, with averages for any one site ranging between 12 and 47 (Bishop, 1979). Mating was seasonal, occurring only from May to September. Unlike at Kanha, outside this period groups generally contained more than one adult male. Prior to and during the mating season, however, one male drove the others out of the group and injuries to adult males were common (Boggess, 1979). Some of the males emigrated permanently, others temporarily, and when outside a bisexual group they usually occurred as solitaries, although pairs have also been reported; all-male bands were rare (Boggess, 1979, 1980).

There was a general movement of males into and out of groups during the breeding season, and resident adult males were sometimes replaced. This was less violent than at Kanha, largely because fighting was restricted to the males, but groups did sometimes split up in this period (Bishop, 1979). Group takeovers were preceded by a non-group male tracking a uni-male group for days or weeks, often being attacked by the resident male. If the incoming

Table 5.2. *Social organization of Himalayan Hanuman langurs in different sites*

	Melemchi	Junbesi	Bhimtal	Simla
Altitude (m)	2800	3000	1200	2000
Habitat	Temperate and meadow	Temperate and meadow	Upper monsoon	Moist temperate
Group size (mean and range)	32	12 (7–19)	23 (15–30)	47 (19–98)
Adult males/group	2.5	2	2	2.9
Home range group (ha)	210	760	20	190
Uni-male/multi-male groups	0/3	1/5	1/3	2/8
All-male band size	1, 2	1, 2	10	4

Sources: From Bishop, 1979; Curtin, 1975; Boggess, 1976; Vogel, 1971; Sugiyama, 1976.

male was successful, he entered the group, rapidly excluded the incumbent male and took up residence with females already familiar with him. Adult males in uni-male groups at Junbesi were generally replaced within 3.5 years (Boggess, 1979).

While foraging, large groups occasionally split into sub-groups but were still generally cohesive, with adult males co-ordinating travel (Sugiyama, 1976), using the stereotypic 'au' call which is unique to Himalayan *S. entellus* (Vogel, 1971, 1973) to keep group members in contact.

At 2–12 km^2, home ranges in the Himalayas were much larger than those at Kanha, due to the scarcity of food in the winter. At Junbesi, summer ranges were centred in broad-leaved forests and winter ranges in meadows. Food sources were scarcer and more scattered in the meadows than in the forests, so home-range size in the winter was greater than in the summer (Curtin, 1975). Home-range overlap between neighbouring groups was small to moderate and, unlike at Kanha, intergroup encounters were rare. This was partly a result of the large home ranges, and partly because groups apparently avoided each other, meeting less frequently than expected from models of random travel (Bishop, 1979). When encounters did occur, they were protracted with much calling and chasing.

Polonnaruwa: Semnopithecus entellus thersites *and* Trachypithecus vetulus philbricki

In Sri Lanka, at the opposite end of its geographical range, *S. entellus* is represented by a much smaller subspecies in the dry deciduous forests of the lowlands. In the wetter evergreen forests of southern Sri Lanka, *S. entellus* is replaced by the purple-faced langur (*T. vetulus*), and in the intermediate semi-evergreen forests of central Sri Lanka, *S. entellus* and *T. vetulus* are sympatric. These forests are characterized by an upper tree layer which is defoliated during the dry season and a lower layer which remains evergreen (Dittus, 1977). In the centre of the region lies Polonnaruwa, an archaeological reserve of about 3 km^2 which has been the site of primate studies since the early 1960s.

Polonnaruwa receives an average rainfall of just over 1000 mm, which largely falls in December during the monsoon (Dittus, 1977; Hladik, 1977*b*). This is preceded by a period when high winds and cyclones rip across the country, sometimes doing considerable damage to the forests (Dittus, 1985). The rains are followed by a prolonged dry season between May and September, with high temperatures (around 29° C) and desiccating wind. The temperature remains high until the rains return in September when it declines slowly to a minimum (about 22° C) in January (Dittus, 1977).

Although temperature variation throughout the year is much less extreme than in Kanha and the Himalayas, rainfall and plant part production is still very seasonal. With the exception of *Ficus*, each tree species has a single annual cycle of leaf flush, flowering, fruiting and leaf fall (Rudran, 1973*b*). Leaf loss is greatest during the dry season, when most flowers and fruits are produced. At the start of the rains there is a major increase in young leaf production, which gradually declines so that, for most of the rainy season, few plant parts are available other than mature leaves (Hladik & Hladik, 1972).

From the 61 plant species and the variety of plant parts available in different seasons, the two langurs selected different diets, which resulted in their having different patterns of home-range use and possibly social behaviour. *T. vetulus* exploited a small number of common species, while *S. entellus* fed from a larger number of widely-spaced food sources. Consequently, *T. vetulus* had a monotonous diet, with only three tree species accounting for 70% of the annual intake and 28 species being fed from. In contrast, ten species accounted for 70% of the diet of *S. entellus*, no single species dominated and a total of 43 species were recorded as eaten (Hladik, 1977*b*).

The two langurs also differed in the plant parts they ate. *T. vetulus* had a high foliage intake but more than half of the diet of *S. entellus* comprised fruits and flowers (Figure 5.7). During the rainy season, both relied heavily on mature leaves. On some days, these comprised 80% of the diet of *T. vetulus* (Hladik, 1977*b*). Fruit-eating by both langurs was most common in the fruiting season, between April and August. Early in the season, *S. entellus* ate ripe *Ficus* and *Drypetes* fruits, and later on they switched to ripe fruits of *Walsura* and *Schleichera*. They also ate the fruits of *Strychnos potatorum*, which contained high levels of alkaloids that were probably toxic to *Macaca sinica*, which did not eat them (Hladik, 1977*b*). *T. vetulus* selected different fruits, such as *Elaeodendron* fruits, which were often unripe, dry, fibrous and coriaceous compared with the ripe, fleshy fruits eaten by *S. entellus* (Hladik, 1977*b*).

S. entellus at Polonnaruwa lived in groups of 20 to 30 animals containing more than one adult male. Unlike their Himalayan relatives, however, they maintained this structure throughout the year, with groups often splitting into sub-groups of about five individuals to forage during the day, especially when feeding from fruiting trees (Ripley, 1970). Few other details on their social system are available. Non-group males have been observed trying to enter bisexual groups, adult males migrate between groups and violent takeovers have been noted; extra-group males live in all-male or predominantly-male bands (Ripley, 1967).

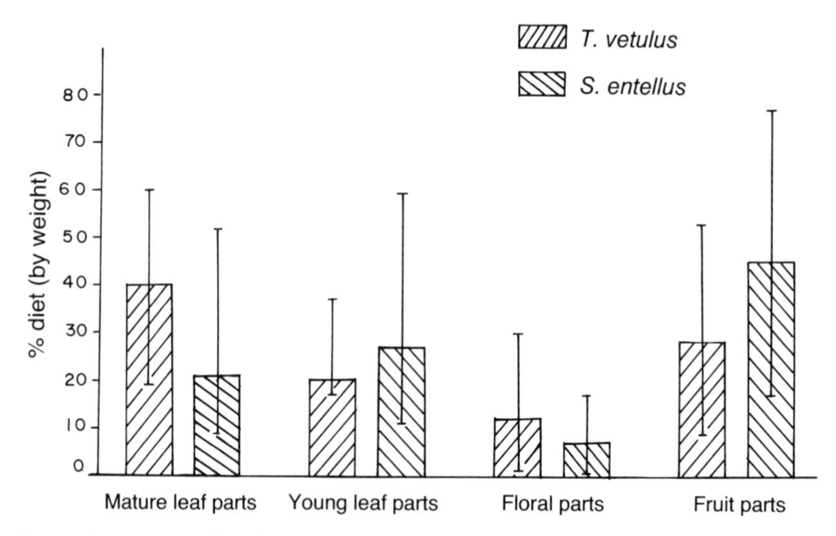

Figure 5.7. Annual diet of the Hanuman langur (*S. entellus*) and purple-faced langur (*T. vetulus*) at Polonnaruwa (after Hladik, 1977*b*).

T. vetulus lived in much smaller, more cohesive groups (mean 8.4 individuals), in which there was only one adult male. The uni-male social structure was maintained by takeovers, new males entering uni-male groups about once every 3 years (Rudran, 1973*a*). Takeovers were violent affairs, with non-group males fighting and eventually deposing the resident. All immature animals of both sexes were then expelled from the group by the new resident male, and joined predominantly male bands (Rudran, 1973*a*). As a result, more than half of the reproductive groups contained no subadult or juvenile animals.

Infants were missing, presumed dead, from groups of *T. vetulus* that had undergone a takeover. An adult female was once seen carrying a badly wounded infant during a takeover fight, so infanticide was assumed (Rudran, 1973*a*). Another female with a fairly old infant left her group when a new male took over. She stayed with the deposed male until the infant was fully independent, whereupon she rejoined a uni-male group (Rudran, 1973*a*).

Predominantly male bands of *T. vetulus* contained juvenile males and females, but no adult females, so they were reproductively inactive. They were less cohesive than uni-male groups, often breaking into smaller units that were submissive to uni-male groups. As a result, they occupied ecologically sub-optimal habitats, in which tree densities were low and canopies discontinuous. This forced animals to travel along the ground occasionally, making them susceptible to predation by dogs.

In spite of differences in the distribution of their food sources, both *S. entellus* and *T. vetulus* were territorial. *T. vetulus* lived in tiny home ranges (2 to 3 ha) and defended them fiercely against all conspecifics. Adult males even chastised members of their own group that ventured too close to neighbouring groups. *S. entellus* had much larger home ranges (17 to 18 ha), which were actively defended, and an average of two intergroup encounters occurred each day (Ripley, 1970; Hladik, 1977*b*). These two langurs had similar population densities; the highest recorded for any colobine in Asia (Chapter 10).

Western Ghats: **Trachypithecus johnii**

The closely related congener of *T. vetulus* is the Nilgiri langur, *T. johnii*, which is restricted to the Western Ghats of southern India. It occurs in the Shola forests of the southern hills above altitudes of 500 to 600 m, and also in some of the lush gallery forests which extend into the drier woodlands below (Poirier, 1970*b*). *T. johnii* has been studied in detail at four sites (Table 5.3): Ootacamund (Poirier, 1968, 1969*a*,*b*, 1970*a*,*b*), Periyar (Tanaka, 1965;

Table 5.3. *Socio-ecology of the Nilgiri langur* (T. johnii) *in southern India*

	Ootacamund	Periyar	Kakachi	Mundanthurai
Forest	Evergreen (Shola)	Deciduous and evergreen	Evergreen (Shola)	Gallery forest
Group size	8.9 (3–25)	7, 21, 27	17 (10–20)	7.6 (4–24)
Bisexual groups	Uni-male	Uni-male	Uni-male	Uni-male (some multi-male)
All-male bands	2.5	Solitary	Solitary	Small
Group change	Fission/fusion			Fission/fusion
Home range (ha)	50–260	5.6, 7.3, 8.3	24	?
Diet	Leaves, shoots, buds, flowers	Foliage, young leaves, *Tectona*	Mature and young leaves, flowers (115 plant species)	

Sources: From Poirier, 1968, 1969*a*,*b*, 1970*a*,*b*; Horwich, 1972; Oates, 1979; Oates *et al.*, 1980; Hohmann, 1989.

Horwich, 1972), Kakachi (Oates, 1979; Oates *et al.*, 1980) and Mundanthurai (Hohmann, 1989).

At Kakachi, rainfall was high (mean 3084 mm) and occurred during both monsoon periods (Oates *et al.*, 1980). In spite of the stony and occasionally rocky soils, this gave rise to tall closed-canopy forest. The forest was dominated by the upper-canopy species *Cullenia exarillata*, which accounted for almost half of the crown volume. Two other species (*Agrostistachys longifolia* and *Aglaia bourdilloni*) were also common but because they were smaller they contributed less to the overall biomass. The forest was evergreen; of 20 species enumerated, only one was deciduous (Oates *et al.*, 1980).

The diet of *T. johnii* was much more varied than that of *T. vetulus* at Polonnaruwa. The study group at Kakachi was recorded to eat from about 115 species of plant, including 53 trees, 32 climbers and 13 non-woody herbs (Oates *et al.*, 1980), and mature leaves formed a major part of the diet in spite of the forest apparently being less seasonal than others on the Indian subcontinent (Figure 5.8). In addition to eating of plant parts, soil-eating was seen once, and one case of insectivory was recorded (Oates *et al.*, 1980).

T. johnii showed strong food preferences; 45% of the diet came from only three tree species, and the two commonest trees in the forest contributed only 5% of the diet (Oates *et al.*, 1980). The small tree *Gomphandra coriacea* accounted for less than 2% of the tree crown area but contributed 21% of the diet. In addition to showing a preference for certain species, *T. johnii* favoured certain plant parts. For example, the mature leaf petioles of *Drypetes* and *Cullenia* were eaten but laminae were generally dropped. Similarly, the single large seeds of *Myristica* were eaten but the aril was usually discarded (J. F. Oates, personal communication).

Groups of *T. johnii* at Ootacamund averaged 8.9 individuals, and in Mundanthurai/Agastyamalai, 7.6 individuals (Hohmann, 1989), although groups of over 20 individuals have been recorded (Oates, 1979; Hohmann, 1989). At all sites, *T. johnii* usually lived in uni-male groups, with solitary males and predominantly-male bands being a small proportion of the population (Hohmann, 1989). Non-group males were recorded to oust resident males, and on one occasion when a predominantly male band invaded a uni-male group, the animals split into both uni-male and multi-male groups (Poirier, 1969*a*, *b*). In addition to this pattern of group change, four instances of group fission were noted by Hohmann (1989) with single young, subordinate males from within the group developing adult 'loud calls', having agonistic interactions with the dominant adult male, forming a stable sub-group (with both males and females) from the original group and occupying part of the original group's home range.

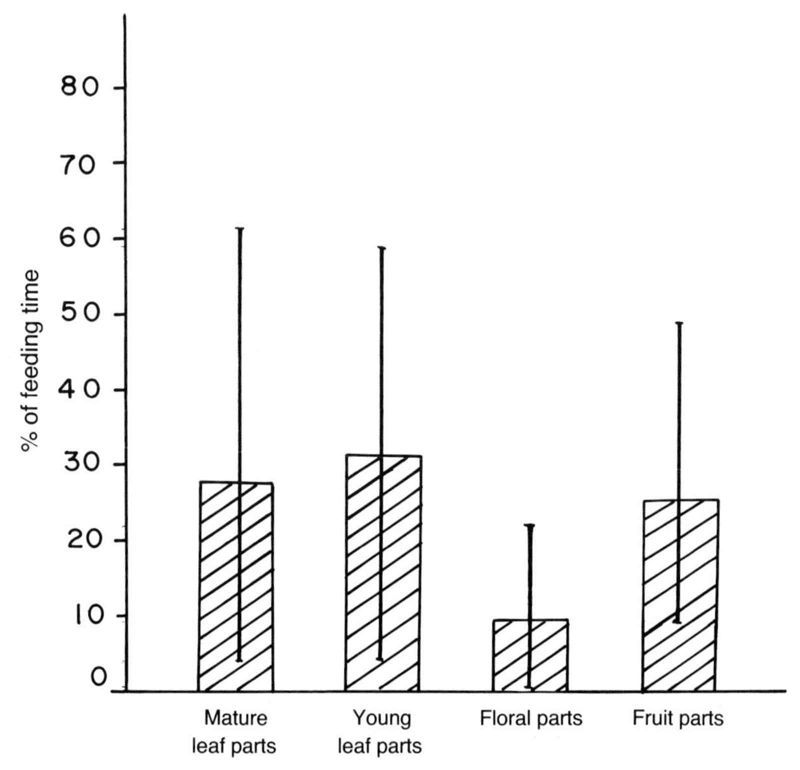

Figure 5.8. Annual diet (% of feeding time) of the Nilgiri langurs (*T. johnii*) at Kakachi (sources: Oates *et al.*, 1980; and J. F. Oates, personal communication).

This was a very different pattern of group formation to that recorded for other langurs so far described, involving far less aggression, and this even applies where takeovers occurred; sub-adult and juvenile animals were generally not excluded from the newly formed uni-male groups.

Uni-male groups lived in home ranges that were actively defended, with little home-range overlap (10% at Kakachi). Home-range size varied from 5.6 to 24 ha, and was influenced by the size of the group, its ability to displace its neighbours, and the vegetation of the home range (Poirier, 1969*b*).

Assam and Bangladesh: **Trachypithecus geei** *and* **Trachypithecus pileatus**

There is no suitable habitat for evergreen forest-dwelling colobines between the Western Ghats and Assam, although *S. entellus* occupies many deciduous forests in India. The Assamese forests mark the eastern limit of the distribution of *S. entellus* and the western boundary of the most diverse array of colobines in the world, spreading north-east as far as China and Vietnam and south-east into Thailand, the Malay Peninsula and Sunda Islands.

In western Assam and Bhutan, the golden langur (*Trachypithecus geei*) occurs but is replaced, east of the Manas River, by *T. pileatus*, which occurs in eastern Assam, Bangladesh and northern Burma (Roonwal & Mohnot, 1977; Gittins, 1980; Green, 1981; Farid Ahsan, 1984).

T. geei and *T. pileatus* are ecologically very similar, both occurring in Sal and other types of tall, moist deciduous forest (Mukherjee & Saha, 1974; Green, 1978, 1981; Mukherjee, 1978; Farid Ahsan, 1984). Both species live predominantly in uni-male groups (McCann, 1933; Mukherjee & Saha, 1974; Mukherjee, 1978; Gittins, 1980; Islam & Husain, 1982), although the average group size in *T. geei* may be slightly larger than in *T. pileatus* (Tables 5.4 and 5.5). Intergroup relations in both species are relatively relaxed, with groups approaching to within 15 m of each other, and even occasionally mixing, with little sign of aggression (Mukherjee & Saha, 1974; Islam & Husain, 1982; Stanford, 1991).

The Sal forest at Stanford's study site in Madhupur had low tree-species richness, only 28 tree species being recorded in 1.2 ha (5% of the study group's home range), and the diet of a study group of *T. pileatus* was correspondingly species-poor: 35 species of tree and liane in one year (Stanford, 1991). Mature leaves were often the only abundant food available and contributed a substantial portion of the diet (Figure 5.9). Young leaves and leaf buds were preferred to mature leaves but were available only briefly, so contributed only 16%. Fruit eating was divided into seed eating (9%), mostly from *Mallotus* and *Litsea*, and whole-fruit eating (24%) which included consumption of two acidic, pulpy whole-fruit species. Other items in the diet were flowers (7%), lepidopteran larvae (1.5%) and very small quantities of soil and sap. The seasonal variation in diet was manifested as rainy-season frugivory and dry-season folivory.

All of these features are reminiscent of the diet of *S. entellus* in the Sal forests at Kanha, but the two species' foraging strategies are very different, because the forest at Madhupur received 2280 mm rainfall p.a. compared with Kanha's 1600 mm. *T. pileatus* rarely travelled along the ground, had

Table 5.4. *Socio-ecology of capped langurs* (T. pileatus)

	Manas	Madhupur (1)	Madhupur (2)
Altitude (m)	70		20
Forest	Moist deciduous (Sal)	Moist deciduous (Sal)	Moist deciduous (Sal)
Bisexual groups	Uni-male	Uni-male	Uni-male
Group size	9.6 (7–13)	7.5 (2–14)	8.3 (5–13)
All-male bands	Yes	Yes and solitaries	Yes (4 individuals) and solitaries
Diet	Leaves and flowers	68% mature foliage	42% mature foliage
Home range (ha)	64		21.6 (14–24)
Territorial		No	No
Day range length (m)	800	450	324
Population density (individuals km^2)		92	52

Sources: From McCann, 1933; Pocock, 1939; Mukherjee, 1978; Gittins, 1980; Green, 1981; Islam & Husain, 1982; Farid Ahsan, 1984; Stanford, 1991.

Table 5.5. *Socio-ecology of* Trachypithecus *in north-east India, Bhutan, Bangladesh, and Malaysia*

	T. phayrei	*T. geei*	*T. obscurus*
Site	Various	Ripa and Manas, Assam	Kuala Lompat
Forest	Evergreen	Sal	Evergreen
Group size (range)	12.9 (3–30)	12.5 (3–40)	17
Bisexual groups	Uni- and multi-male	Uni-male	Uni-male (some multi-male)
All-male bands		Yes and solitaries	Yes and solitaries
Territorial		No	Semi-territorial
Home range (ha)	30	150–600	33
Diet	'Folivore'	'Folivore'	58% foliage
Population density (groups/km^2)	0.3[a], 1.2[b]		6.2

[a]Primary forest.
[b]Secondary forest.
Sources: T. phayrei: Farid Ahsan, 1984; Gittins, 1980; Fooden, 1971; Zhang *et al.*, 1981; Green, 1978
T. geei: Mukherjee and Saha, 1974; Gee, 1961, 1964; Mukherjee, 1978; Khajuria, 1977; Ghosh and Biswas, 1975; Oboussier and von Maydell, 1959
T. obscurus: Burton, 1981; Curtin and Chivers, 1978; Curtin, 1980; MacKinnon & MacKinnon, 1980; Marsh & Wilson, 1981.

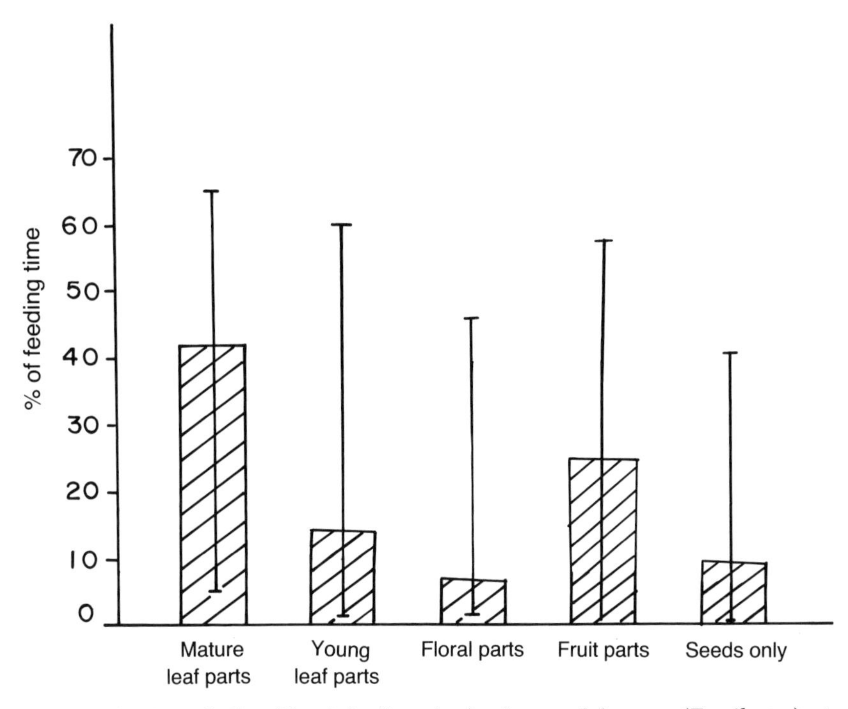

Figure 5.9. Annual diet (% of feeding time) of capped langurs (*T. pileatus*) at Madhupur (after Stanford, 1991).

shorter day ranges when mature foliage dominated the diet and longer ranges when fruits and flowers were sought. Diet also influenced activity patterns; more time was spent resting when mature leaves dominated the diet, and juveniles stopped playing during the same months (Stanford, 1991).

Stanford's main study group was uni-male and comprised 13 individuals, and other groups varied from 5 to 13 individuals. Overt social activities (grooming, copulations and affiliative behaviours) were rare, although females tended to affiliate more often after the birth season, especially through allomothering. This is not unusual for Asian colobines, but the intergroup relations were distinctive.

Both males and females migrated from their natal groups, and new groups were formed by females joining solitary or band males. Inevitably, adult males in bisexual, uni-male groups discouraged group members from affiliating with strange males, and displayed aggressively towards the latter. However, they did not attack intruding males even if they persistently followed the group. Similarly, there was complete overlap of home ranges between bisexual groups (home ranges varied between 16–24 ha) and neighbouring groups tolerated each other's proximity. Aggressive displays did occur when groups approached, but this did not prevent up to five groups feeding at the same time in a single *Ficus* tree! None of this compares with the aggressive intergroup interactions reported for uni-male groups of *T. vetulus* in semi-deciduous forests at Polonnaruwa.

Several more *Trachypithecus* forms occur in eastern Bangladesh, Burma, northern Thailand, Laos, Vietnam and southern China. Their taxonomy is somewhat confused and their ecology largely unstudied. They include *T. francoisi* and *T. phayrei* (see Chapter 3, pp. 68–69). The limited data (Table 5.5) on these species indicate that they too are largely folivorous; they live in predominantly uni-male groups, although some multi-male groups also occur (*T. phayrei*: Gittins, 1980; Farid Ahsan, 1984; *T. francoisi*: Tan, 1985).

Kuala Selangor and Pangandaran: **Trachypithecus cristatus** *and* **T. auratus**

Three other *Trachypithecus* species which have been studied intensively occur to the south: *T. obscurus, T. cristatus* and *T. auratus. T. cristatus* lives in inland, tall forests of Kampuchea, Vietnam and southern Thailand (Lekagul & McNeely, 1977), but throughout peninsular and insular South-east Asia (Sundaland) it is largely restricted to riverine and coastal swamp forests (Medway, 1977; Marsh & Wilson, 1981; Payne *et al.*, 1985). *T. auratus*, a close relative of *T. cristatus*, occurs in Java, and similarly is restricted to coastal and riverine habitats.

Due to the difficulties of following animals in mangroves, *T. cristatus* and *T. auratus* have mainly been studied in habitats at the fringe of their range; the two main study sites have been Kuala Selangor in Peninsular Malaysia and Pangandaran in western Java, respectively. The Kuala Selangor site is a small hill of parkland with a cleared understorey and many open spaces between patches of trees. It is contiguous with a large area of coastal mangrove, and groups move freely between the two habitats. The Pangandaran study area is also a mixture of habitats, with *T. auratus* using both indigenous riverine forest and exotic teak plantations (Kool, 1989).

T. cristatus and *T. auratus* live predominantly in uni-male groups of about 10–20 individuals (Bernstein, 1968; Wolf, 1978; Kool, 1989; Table 5.6). Extra-group males live either as solitaries or in all-male bands and, as in *T. vetulus*, violent male takeovers with subsequent infanticide have been recorded in *T. cristatus* (Wolf & Fleagle, 1977).

Like other *Trachypithecus* species, *T. auratus* is relatively folivorous, its diet comprising 55% leaves (Kool, 1989). Unlike for species in the more

Table 5.6. *Socio-ecology of* T. cristatus *and* T. auratus

	Kanchamburi, Thailand	Kuala Selangor, Malay Peninsula	Pangandaran, Java	Lombok
Species	*T. cristatus*	*T. cristatus*	*T. auratus*	*T. auratus*
Forest	Evergreen	Parkland	Mangrove/evergreen/teak plantation	Evergreen
Males in bisexual groups		Uni-male (mainly)	1–2/group	Uni-male
Group size	17 (9–30)	26.5 (11–38)	(6–23)	5–13
All-male bands		Yes and solitaries	Yes and solitaries	
Infanticide		Recorded		
Allo-mothering		Common	Common	
Diet		'Folivore'	55% foliage	
Home range (ha)			2.5–8	
Territorial		No	Yes	
Population density (individuals/km^2)			23–61	

Sources: From Bernstein, 1968; Kool, 1989; Wolf, 1978.

seasonal habitats further north, however, these are mostly young leaves, mature leaves comprising a mere 1% of the annual diet. In spite of mature leaves being such a small component of the diet, home ranges are small; at Pangandaran, ranges of *T. auratus* are only about 5 ha in area, and overlap very little (Kool, 1989). This implies that young leaves are relatively abundant throughout the year. At Kuala Selangor, the home ranges of *T. cristatus* are thought to be large, and considerable home-range overlap has been reported (Wolf, 1978).

Kuala Lompat: **Trachypithecus obscurus** *and* **Presbytis melalophos**

The inland forests of South-east Asia are dominated by trees in the Dipterocarpaceae, and their abundance in relation to other common families (e.g. Leguminosae, Euphorbiaceae, etc.) varies considerably between sites (Whitmore, 1984*b*; Waterman *et al.*, 1988). Dipterocarps provide little food for herbivorous mammals, including colobines (see e.g. Medway, 1972; Chivers, 1974; Marsh & Wilson, 1981; Whitmore, 1984*b*; Davies *et al.*, 1988), probably because of the relatively low ratio of protein to fibre and the abundance of terpenes in their leaves (Waterman *et al.*, 1988). Therefore, an abundance of these trees in a forest represents a corresponding scarcity of large trees from other families which might provide foods (Davies, 1984).

In the centre of Peninsular Malaysia lies Kuala Lompat, which has been the site of intensive primate studies since the late 1960s (Chivers, 1980). Dipterocarps only comprised 4–16% of the total tree basal area, whereas trees of the Leguminosae comprised 11–26% (Bennett, 1983). This makes it one of the most benign sites for colobines in the region (Waterman *et al.*, 1988), since seeds and foliage from this family are common foods.

The forest at Kuala Lompat was extremely diverse; the 1-km^2 colobine study area contained a minimum of 400 tree species and 110 liane species, with an average of 159 tree species per hectare (Raemaekers *et al.*, 1980; Bennett, 1983). In addition to legumes and dipterocarps, other common tree families included Annonaceae, Euphorbiaceae and Sapindaceae. Many of the common emergents had leaves which were relatively digestible, particularly young leaves (Waterman *et al.*, 1988). All plant parts were available from at least some species at all times of year, although there were marked peaks of flowering in March and fruiting from June to August (Raemaekers *et al.*, 1980; Bennett, 1983).

T. obscurus moved and fed predominantly in the upper canopy, where legumes and other colobine food families (e.g. Annonaceae, Myristicaceae) were relatively abundant (Curtin, 1980; MacKinnon & MacKinnon, 1980).

The anatomy and mode of locomotion of *T. obscurus* is suited for travel along boughs of these large trees (Fleagle, 1976, 1977, 1980). These monkeys were selective feeders; young leaves made up 36% of the diet and a further 11% comprised mature leaf parts (Figure 5.10). *T. obscurus* also ate the seeds, fruits and flowers of selected emergents when those items were in season (Curtin, 1980). Groups of *T. obscurus* were generally uni-male, and occupied home ranges of about 33 ha which were small enough to be defensible; groups were territorial (Curtin, 1980).

T. obscurus shared the forest at Kuala Lompat with *Presbytis melalophos*. The latter, like all *Presbytis* species, has a relatively smaller stomach and is more agile than *T. obscurus* (Chapters 3 and 7, this volume). *P. melalophos* obtained foods from all levels of the forest (Curtin, 1980; MacKinnon & MacKinnon, 1980); its anatomy and mode of locomotion are more suited to travelling on smaller supports than are those of *T. obscurus* (Fleagle, 1977, 1980). A total of 46% of the diet of *P. melalophos* comprised fruits and seeds, and a further 17% comprised flowers (Figure 5.10).

P. melalophos had a more diverse diet than *T. obscurus*, although it was still selective in the items taken: 45% of the diet came from only five species, and those only made up 9.6% of the trees in the forest (Bennett, 1983, 1986*b*). It ate a greater diversity of young leaves than seeds, and was particularly choosy about the fruits eaten; the majority of whole fruits in its diet came

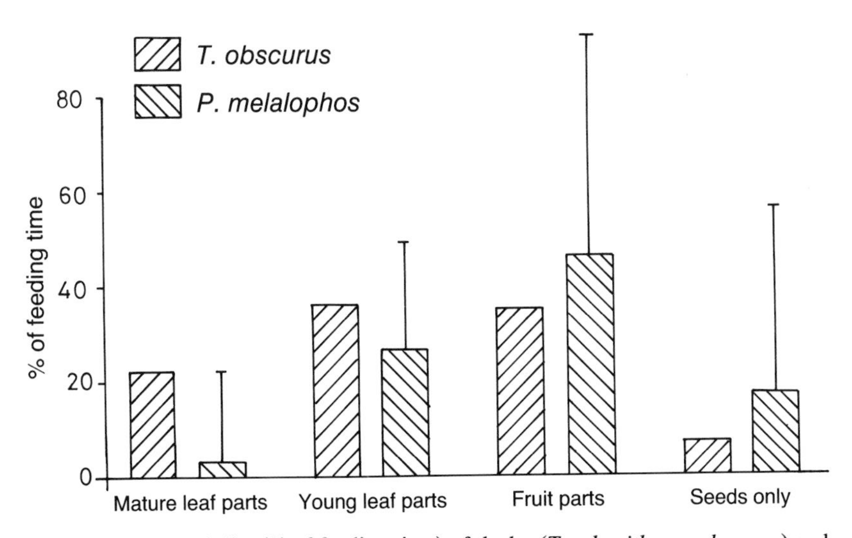

Figure 5.10. Annual diet (% of feeding time) of dusky (*Trachypithecus obscurus*) and banded (*Presbytis melalophos*) leaf-monkeys at Kuala Lompat (after Curtin, 1980; Bennett, 1983).

from only three species: *Mezzetia leptopoda, Monocarpia marginalis* and *Pterygota alata* (Bennett, 1983; Davies *et al.*, 1988). In the limited period around the turn of the year, when few favoured foods were available, the monkeys travelled less each day and fed on a greater variety of foods, mainly young leaves.

There were sufficient palatable mature leaves at Kuala Lompat to allow *P. melalophos* to subsist on them at such times, and up to 22% of the diet in any one month could comprise mature leaves (Bennett, 1986*b*). This allowed them to exist at relatively high biomasses (486 kg/km^2; Chapter 10) even though, overall, leaves did not form the major part of their diet.

The fruits, seeds and flowers favoured by *P. melalophos* were generally from large, common trees, with the exception of figs, which were eaten by *T. obscurus* but seldom by *P. melalophos* (Lambert, 1990). For most of the year, foods were abundant, and two or more groups commonly came together to feed in the large trees. Most food species of *P. melalophos* were trees and lianes with synchronized leafing, flowering and fruiting patterns, so foods were only available from any one species for a short time. The monkeys' food supply was curtailed by trees no longer producing preferred foods, rather than by other animals eating them. Thus, there was no advantage in defending food sources, and this is probably why the home ranges of *P. melalophos* groups overlapped almost entirely (Bennett, 1986*b*).

The behaviour of *P. melalophos* is relatively flexible, even within the same subspecies. Like Kuala Lompat, the Sungai Tekam area in Peninsular Malaysia comprised primary forest at the time of study. *P. melalophos* there lived in smaller home ranges than those at Kuala Lompat (14.5 ha compared to 29.5 ha) and were apparently territorial (Bennett, 1986*b*; A. D. Johns, 1986). This might be because food trees at Sungai Tekam were smaller on average, but more common in the forest, than those at Kuala Lompat (Bennett, 1986*b*). The cost of allowing other groups to feed in any one tree was great, but the number of such trees meant that a group could obtain all of its food from a relatively small area. Thus, the benefits of territoriality were greater than at Kuala Lompat, but the costs were less (Bennett, 1986*b*).

In spite of such behavioural flexibility, the social organization of *P. melalophos* was similar at both sites. The animals lived in uni-male groups of about 15 individuals (Bennett, 1983; Johns, 1983). The males of each group gave 'calling rounds' each night around dusk, at dawn and at intervals through the night, and it was the males which called and displayed when two groups approached each other in the forest (Curtin, 1980; Bennett, 1983). Males left their natal group when they were only about half the adult size. They then either led a solitary life or joined an all-male band. As with *T.*

vetulus, such bands were frequently chased by harem males, and at Sungai Tekam one all-male band was apparently confined to an area of scrubby, food-poor forest (Johns, 1983).

Sepilok: Presbytis rubicunda

In spite of their relatively benign climate, the forests at the extreme western and eastern sides of insular South-east Asia are particularly severe for colobines. In Borneo, the soils of inland areas are generally very poor, and low in nutrients (Davies & Baillie, 1988). There is a large number of dipterocarps and a greatly reduced abundance of food trees. At Sepilok in Sabah, for example, 63% of the basal area of forest trees comprised dipterocarps. The only common, large non-dipterocarp tree was the ironwood, *Eusideroxylon zwageri* (Lauraceae) whose mature leaves were also very indigestible (Davies, 1984). The favoured families for colobine food of Leguminosae and Annonaceae comprised a mere 0.7% and 0.8% of the basal area, respectively, despite the high species-richness of the forest (Waterman *et al.*, 1988).

This low digestibility of leaves of common trees means that there is no available niche for a *Trachypithecus* species, with its strategy of moving through the upper canopy and subsisting mainly on foliage. The only *Trachypithecus* species on Borneo, *T. cristatus*, is confined to mangrove, peat swamp and riverine forests of the coastal plain, as noted above. Both mangrove and riverine forests probably have higher primary productivity than do dipterocarp forests (mangroves in Sarawak: Chai, 1982; riverine forests in Kutai, Kalimantan: Wheatley, 1980), thus they are more likely to contain food for crown-dwelling folivores.

In inland forest, *P. rubicunda* at Sepilok fed on trees which are among the rarest in the forest. Dipterocarpaceae and Lauraceae contributed only 3.6% to the diet of *P. rubicunda*. The scarcity of trees with palatable food meant that *P. rubicunda* selected many of their foods from lianes, which accounted for 32% of the annual diet (Davies, 1991). Preferred foods of *P. rubicunda*, like those of *P. melalophos*, were young leaves and seeds (Figure 5.11). *P. rubicunda*, again like *P. melalophos*, increased the diversity of their diet when favoured foods were scarce but, unlike *P. melalophos*, they could not resort to eating a significant amount of mature leaves at such times. This is because there were almost no mature leaves digestible enough to be eaten; mature leaves comprised only 1% of the annual diet and never made up more than 7% of the diet in any single month (Davies *et al.*, 1988).

P. rubicunda lived in uni-male bisexual groups that were small: three to nine animals, mode about seven. Even these split into sub-groups to forage

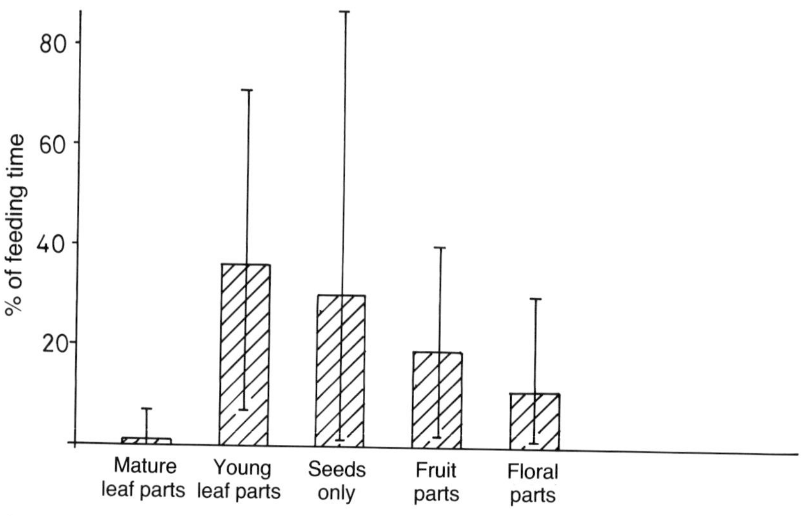

Figure 5.11. Annual diet (% of feeding time) of red leaf-monkeys (*P. rubicunda*) at Sepilok (after Davies *et al.*, 1988; Davies, 1991).

when young foliage was being taken from small understorey trees (Davies, 1987b). Sub-adult males travelled in uni-sexual bands and there is indirect evidence from Tanjung Puting (Kalimantan) to suggest that females also migrate between bisexual groups (Supriatna *et al.*, 1986).

Social change documented at Sepilok (Davies, 1987*b*) involved a violent fight between the resident male and a single invading male, followed by group fission. One resulting group comprised the invading male and two nulliparous females from the original group, and the other comprised the original adult male, two females with their infants and a juvenile male. This was reminiscent of the female *T. vetulus* that remained with a deposed male until her infant was independent. The new sub-group of the male and two nulliparous females eventually displaced the other sub-group to the forest edge, and took over the original group's home range.

Because of the scarcity of food, *P. rubicunda* foraged from widely scattered, often small food sources, and had a long average day-range length (890 m) and a very large home range (84 ha). In spite of the fact that their home ranges were so large, *P. rubicunda* vigorously chased away other groups whenever possible (Davies, 1984); presumably the cost of allowing others to share food sources was great.

P. rubicunda shared the forest at Sepilok with another *Presbytis* species, *P. hosei*. These were even rarer in the forest at Sepilok than were *P. rubi-*

cunda; the combined biomass of the two species was 64 kg/km^2 (Davies & Payne, 1982), which is the lowest biomass of colobines in any Old World rain-forest so far studied (Davies, 1984; Waterman *et al.*, 1988). This is because the number of colobines is limited by the lack of digestible foliage during periods when seeds and young leaves are scarce. At other sites in Borneo, *P. hosei* is more abundant and *P. rubicunda* far rarer (Rodman, 1978; Davies & Payne, 1982; Bennett, 1992), and it is unclear how they avoid direct competition, since they both seem to have similar anatomy, diet and social organization (see Chapter 10).

There is an underlying similarity between all *Presbytis* species, and differences in socio-ecological behaviour appear to be due to differences in habitat rather than to intrinsic differences between the animals (Table 5.7). *P. thomasi* in Sumatra apparently has a similar diet to *P. melalophos* at Kuala Lompat; more than 50% of the diet of *P. thomasi* comprised fruit and seeds, and only 36% was leaf matter (Gurmaya, 1986). They were studied in old rubber estates and fruit gardens, however, so their behaviour in primary forest has yet to be determined. *P. comata* in the submontane forests of Kamojong, western Java, appeared to be more folivorous than any other member of the genus, with 62% of the diet comprising young leaves and a further 6% comprising mature leaves (Ruhiyat, 1983). Furthermore, their diet is unusual in that 9% was branch-tips, fungi and pseudobulbs. This species also appears unusual in its social organization, with groups at Patenggang in west Java being reported as monogamous (Ruhiyat, 1983).

Samunsam: Nasalis larvatus

T. cristatus share their coastal swamp forests in Borneo with the bizarre-looking proboscis monkey, *Nasalis larvatus*. At Samunsam Wildlife Sanctuary in Sarawak, both species live in an area of mixed coastal forest adjoining the Samunsam River. Mangrove–nipa forest occurs along the lower reaches of the river, with *Avicennia* and *Sonneratia* close to the river-mouth giving way to *Rhizophora* further up-river. These are serially replaced by *Brugiera* and *Nypa fruticans*, which eventually give way to riverine forest up-river, where the water is no longer brackish. Away from the river, the low hills are covered by a mixture of heath and lowland dipterocarp forests (Bennett & Sebastian, 1988).

In spite of their large body size (Chapter 3) and relatively enormous stomach (Chivers & Hladik, 1980; Chapter 7), *N. larvatus* were extremely selective feeders, mainly eating fruits, seeds and young leaves. Mature leaves only comprised about 3% of the diet (Bennett & Sebastian, 1988) and preferred

Table 5.7. *Ecology of* Presbytis *species in different sites in South-east Asia*

	P. melalophos			*P. rubicunda*	
	Kuala Lompat, Malaya	Tekam, Malaya	Samunsam, Borneo	Sepilok, Borneo	Tanjung Puting, Borneo
Altitude (m a.s.l.)	50	100–500	0–300	0–150	
Forest	Evergreen lowland	Evergreen, hill	Riverine/heath/ dipterocarp	Evergreen	Evergreen mixed dipterocarp/ heath peatswamp
Group size	15 (12–18)	14	6	7 (3–12)	6.1 (3–8)
Males in bisexual groups	Uni-male	Uni-male	Uni-male	Uni-male	Uni-male
All-male bands	Yes (and solitary)			Band (2 individuals) and solitaries	
Allomothering	Rare			Rare	
Female emigration	Yes			Yes	Yes
Male emigration	Yes			Yes	Yes
Diet	Fruit and seeds (46%)	Fruit and seeds (60%)		Fruit and seeds (49%)	Fruit and seeds (52%)
Day range	717 (300–1360)	614 (450–935)		850 (225–1670)	
Home range (ha)	29.5	14.5		85	71
Territorial	79% overlap	Territorial		<10% overlap	25% overlap
Population density (group/km^2)	7.2	3.4		2.7	

foods were apparently scarce. Groups of *N. larvatus* had longer day-ranges (up to 2000 m) and larger home ranges (approximately 9 km^2) than any other forest colobine so far studied (Bennett & Sebastian, 1988). Moreover, the forest at Samunsam supported a very low biomass of *N. larvatus*. The animals remained within 1 km of the river, and even within this riverine forest, the biomass of *N. larvatus* was only 45.8 kg/km^2 (Bennett & Sebastian, 1988).

The taxonomic richness of trees in mangrove forest was extremely low; only ten genera (Anderson, 1980). This fact, combined with the apparent scarcity of foods, might explain why *N. larvatus* frequently used both mangrove and riverine forest: in Sarawak, they are generally found at the highest densities at ecotones. Exploiting two habitats is likely to give them a greater variety of food sources throughout the year. At Samunsam, a large proportion of groups of *N. larvatus* apparently spent more time feeding on young leaves in mangrove forest when fruit in the riverine forest was scarce (Bennett & Sebastian, 1988).

On a daily basis, ranging patterns of *N. larvatus* were greatly affected by

P. hosei		P. comata		P. thomasi	P. potenziani
Sepilok, Borneo	Kutai, Borneo	Patenggang, Java	Kamajong, Java	Bohorok, Sumatra	Paitan, Siberut
0–150		1700	1500		0–384
Evergreen	Evergreen	Submontane	Montane/ submontane	Rubber/gardens	Evergreen
7 (3–13)		6 (3–10)	7 (3–12)	6 (3–21)	3.7 (2–6)
Uni-male Band and solitaries		Monogamous?	Uni-male	Uni-male Bands (average 6, range 2–10)	Monogamous Solitaries
			Rare		
					?No
Yes				Yes	Yes
			Young leaves (61%)	Fruit and seeds (61–66%)	Fruit (50%)
			500 (250–900)	640 (150–1300)	
		14	38 (35–40)	14	11.5–30
		Much overlap	9% overlap		Much overlap
<1		5	3.5	3.4	4

Sources: From Bennett, 1983, 1986a, 1986b; Davies, 1984, 1987*b*, 1991; Davies and Payne, 1982; Gurmaya, 1986; Johns, 1983; Marsh and Wilson, 1981; Rodman, 1978; Ruhiyat, 1983; Supriatna *et al.*, 1986; Tilson and Tenaza, 1976; Watanabe, 1981; Whitten and Sadar, 1981.

their returning to sleep in trees next to rivers every night. They generally moved away from the river in the early morning, went somewhat further inland during the day and returned to trees adjacent to the river in the hour before dusk (Kern, 1964; Kawabe & Mano, 1972; Macdonald, 1982; Salter *et al.*, 1985; Bennett & Sebastian, 1988). The distance travelled away from the river is limited by the animals' need to return to it every evening. At Samunsam there was only one main river, which restricted the monkeys to a strip of forest not more than 1 km wide on either side of the river. At Sukau in Sabah there is a network of rivers and ox-bow lakes, and here *N. larvatus* range widely (R. Boonratana, personal communication). Travel away from the river is also limited by the animals' preferred habitats (mangrove and riverine forest) being along the river, where soils are richer and productivity higher (Bennett & Sebastian, 1988).

The social organization of *N. larvatus* is very flexible. Like many other colobines, it lives in uni-male groups. Home ranges overlap completely, however, and different groups frequently join together, particularly next to the river at night (Bennett, 1987; Bennett & Sebastian, 1988; Yeager, 1991). At Samunsam, each uni-male group spent the majority of its nights within 50 m of another, and more than 70% of its nights within 100 m of another. Occasionally, up to six groups (about 80 animals) congregated along one 200 m stretch of river (Bennett & Sebastian, 1988). In peat-swamp forest at Tanjung Puting, Kalimantan, particular harems apparently associated together more commonly than they did with others, indicating that there might be two levels of social organization (Yeager, 1991).

N. larvatus exhibit a further unusual degree of social flexibility. Not only did males leave their natal groups, but also females moved between groups relatively frequently (Bennett, 1988; Bennett & Sebastian, 1988). As a result, harems often changed composition. Females moved between groups both on their own and also while carrying infants (Bennett & Sebastian, 1988; Rajanathan & Bennett, 1990). In the latter case, they then either entered another harem, or joined an all-male group for at least a short period (Rajanathan & Bennett, 1990). This might have been to avoid infanticide, if there had been a recent male replacement (Bennett, 1987).

Paitan, Mentawai Islands: **Presbytis potenziani** *and* **Simias concolor**

On the opposite side of the Sunda shelf, lying off the west coast of Sumatra, are the Mentawai islands. The forests of this chain of four small islands also grow on poor soils, and contain a high proportion of dipterocarps. At Paitan on Siberut, the largest of the islands, 21% of the trees were dipterocarps (Whitten, 1982). This is similar to Sepilok, where dipterocarps comprised 27% of the trees (Waterman *et al.*, 1988). Similarly, legumes comprised 2.4% and 2.3% of stems at Paitan and Sepilok, respectively (Davies, 1984; Waterman *et al.*, 1988). The proportions of the main tree families at the two sites were similar, despite their being at opposite sides of the Sunda shelf.

The Mentawais have two colobines, one an endemic genus, *Simias concolor*, and the other an endemic species, *Presbytis potenziani*. They are unusual amongst Old World monkeys, because both species live in groups comprising a monogamous pair and their offspring (Tilson & Tenaza, 1976; Tilson, 1977; Watanabe, 1981; Table 5.8). Little is known about the ecology of either, except that mated pairs of *P. potenziani* duet in much the same way as do gibbons (Tilson & Tenaza, 1976). In the north of Siberut, *S. concolor* occur in harem groups with more than one adult female, whereas in the south of the island, they occur in monogamous groups (Watanabe, 1981). In the north, there has been more human disturbance than in the rest of Siberut, and

Table 5.8. *Socio-ecology of* N. larvatus *and* S. concolor

	N. larvatus		*S. concolor*			
	Samunsam	Tanjung Puting	Sarabua	Grukna	Paitan	Sirimuri
Altitude (m a.s.l.)	0–300				0–380	
Forest	Mangrove/riverine	Peat swamp	Evergreen	Logged forest	Evergreen	Evergreen
Group size	9 (4–20)	12 (3–23)	3 (2–5)	8 (3–20)	3.5	3.4
Bisexual groups	Uni-male	Uni-male	Monogamous	Uni-male	Monogamous	Monogamous
All-male bands	Yes (and solitary)	Yes		Solitary		Solitary
Male emigration	Yes	Yes				
Female emigration	Yes	Yes				
Allomothering	Yes	Yes				
Diet	50% fruit and seeds	40% fruit and seeds			'Folivore'	Leaves, fruit
Home range (ha)	900	130	6.5–20	3.5		30
Territorial	No	No				
Population density (groups/km^2)	0.52	5.2	2.5	30	Approx. 4	2.5

Sources: From Anon., 1980; Bennett and Sebastian, 1988; Tilson, 1977; Watanabe, 1981; Yeager, 1989, 1991

the variation in social organization was presumably related to a change from primary to secondary forest, or to hunting pressure.

Indo-China and China: **Pygathrix** *species*

The 'odd-nosed' species in the genus *Pygathrix* are some of the most endangered as well as some of the most unstudied primates in the Old World. These are the two forms of douc monkey, *P. nemaeus*, in Indo-China, and three species of snub-nosed monkey (*P.* (*Rhinopithecus*) *roxellana*, *P.* (*R.*) *brelichi* and *P.* (*R.*) *bieti*) in China. A fourth snub-nosed monkey, *P.* (*R.*) *avunculus*, occurs in northern Vietnam, and little is known about it except that it is apparently confined to forest over limestone, and more than 70% of its original habitat has been lost (MacKinnon, 1986).

The behaviour and ecology of wild populations of *Pygathrix nemaeus*, which occur in the forests of Vietnam, Kampuchea and Laos, are also almost totally unknown. On Mount Sontra, Vietnam, *P. nemaeus nemaeus* occurs in mixed, partly deciduous moist forests, where it subsists on leaves and fruits (Lippold, 1977). It lives in uni-male and multi-male groups of about eight animals, and there is evidence that, as in *N. larvatus*, these groups sometimes join to form larger groups (Gochfeld, 1974; Lippold, 1977; Table 5.9). Solitary males and solitary females have been recorded (Lippold, 1977), but it is unclear if this is due to both sexes moving between social groups, or to extreme disturbance of the animals and their habitat. The only other fact known about the species' behaviour is that it exhibits an unusually high degree of food-sharing between group members, both in the wild (Gochfeld, 1974) and in captivity (Kavanagh, 1972).

The three species of *Pygathrix* (*Rhinopithecus*) in China all live at high altitudes in steep, mountainous areas with very seasonal climates. *P. bieti* and *P. brelichi* have declined in numbers to only about 2000 and 800 animals, respectively (Schaller, 1985; Tan, 1985; MacKinnon, 1986; Long & Kirkpatrick, 1991). The Yunnan snub-nosed monkey, *P. bieti*, occurs in association with fir–larch forests within an altitudinal band between 3000 and 4300 m (Long & Kirkpatrick, 1991), and there is no evidence to show vertical migration between the summer and winter seasons. The staple foods are grass and lichens, supplemented seasonally with bark in winter and young leaves and fruits in summer. They have also been reported to eat emerging bamboo sprouts and young leaves of birch (*Betula* sp.) (Li *et al.*, 1982). Pine needles do not appear to be major food items (Long & Kirkpatrick, 1991), and it would be surprising for an animal with forestomach fermentation to ingest

Table 5.9. *Socio-ecology of odd-nosed monkeys of China and Indo-China*

	P. nemaeus	*P. roxellana*	*P. bieti*	*P. brelichi*
Site	Mount Sontra, Vietnam	Wolong, China	Yunling Mountains, China	Wuling Mountains, China
Altitude (m a.s.l.)	0–696	Winter: 1800–2800 Summer: 3000–3300	3000–4300	1500–2200
Forest	Semi-deciduous	Mixed conifer/broadleaf	Conifer forest, some broadleaf (<25 tree species)	Broad-leaf/conifer
Group size	8 (5–11)	–	23–200	6(3–10)
Congregations (individuals)	–	600 (smaller in winter)		400+
Bisexual groups	Uni-male/multi-male (fission–fusion)	?Uni-male (fission–fusion)	Multi-male	Uni-male
All-male bands	Solitary	Solitary		Yes
Male emigration	Yes			Yes
Female emigration	Yes (solitary)			
Allomothering	In captivity			
Seasonal births	?Feb–Jun	Yes	Jul/Aug	
Diet	Leaves and fruits	Broad-leaf foliage, fruits and bark	Conifer foliage buds	Broad-leaved foliage/ buds
Home range		10–30 km^2	40 km^2	>10 km^2
Population density	Low	Low	Low	Low

Sources: From: Bleisch *et al.*, 1993; Gochfeld, 1974; Happel & Check, 1986; Hu, 1981; Li *et al.*, 1982; Lippold, 1977; Long *et al.*, 1994; MacKinnon, 1986; Schaller, 1985; Wu, 1983; Zhao, 1988.

pine needles likely to contain bacteriostatic resins; however, deer in the temperate regions can feed on pines.

Group size of *P. bieti* is extremely variable, with reports of 23 (Li *et al.*, 1982), 100–269 (Wu, 1993), and <50–>200 animals (Long *et al.*, 1994). Inevitably, these groups are multi-male in composition. Between 35% and 90% of adults' time is spent on the ground, and home ranges could be over 40 km^2 (Wu, 1993; Long *et al.*, 1994). The species has suffered loss of habitat and intensive hunting and trapping, and is now restricted to a few forest patches (Zhao, 1988).

The Guizhou snub-nosed monkey, *P. brelichi*, is restricted to a single mountain range, where it occurs between 1500 and 2200 m in the mixed evergreen–deciduous broadleaf forests (Bleisch *et al.*, 1993). The dominant trees are oak (*Cyclobalanopsis* spp.) and beech (*Fagus*), in association with other species (e.g. *Acer, Camellia, Ilex*, etc.) and a dense understorey of dwarf bamboo. The main dietary items are the leaves and buds of broad-leaved trees throughout the year (Bleisch *et al.*, 1993).

As with all 'odd-nosed' monkeys, the social system of *P. brelichi* is flexible and based around a uni-male family unit of 3–10 animals. These units can form congregations of 400 animals, which may include all-male (adult and sub-adult) bands on the periphery (Bleisch *et al.*, 1993). Why this fission–fusion social system has arisen is unclear, but patchiness of food-tree distribution may be one factor, as seems to apply to *P. roxellana* (see below). *P. brelichi* seldom travels along the ground, but moves along large branches with quadrupedal walking, climbing, leaping, semi-brachiation and even full brachiation (Bleisch *et al.*, 1993).

P. roxellana occurs at higher latitudes, and experiences the longest winter and lowest average temperatures of any non-human primate in the world (Happel & Cheek, 1986). The climate is even more severe than that experienced by Himalayan *S. entellus*, with snow cover for up to six months of the year and frost on up to 280 days per year (Li *et al.*, 1982). As a result, the behaviour of *P. roxellana* varies greatly at different times of the year. In summer, it lives in mixed broadleaf–conifer and pure conifer forests between 3000 and 3300 m above sea level (Happel & Cheek, 1986; Schaller, 1985; Schaller *et al.*, 1985; MacKinnon, 1986). Here, it travels between 500 m and 2500 m each day (Hu, 1981) and its diet mainly comprises leaves of broadleaved trees such as *Sorbus yunnanensis* and *Euonymus porphyreus* (Happel & Cheek, 1986; Schaller, 1985). In winter, it descends to an altitude of about 1800 m to 2800 m and subsists to a large extent on several species of lichen, which, in the prevailing humid climate, thrive on tree trunks and branches (Schaller, 1985). In this season the monkeys travel much less each day than during the summer (Hu, 1981). The total area used by a group of *P. roxellana* during the course of a year is at least 30 km^2 (Schaller, 1985; Schaller *et al.*, 1985).

Trends in Asian colobine natural history

One question which arose during the early phases of socio-ecological research on forest primates was: How can two colobine species coexist, without one out-competing the other? In Asia the conclusion from early studies of two species at Polonnaruwa was that their socio-ecological strategies were sufficiently distinct that competition between them was minimized.

One species (*T. vetulus*) has a low-quality foliage diet, supplied by abundant evenly-distributed food sources, and lives in small uni-male bisexual groups that defend tiny home ranges. The other species (*S. entellus*) feeds from widely dispersed, large food sources (often fruit) and lives in large multi-male groups which range over much larger areas. This is analogous to the model proposed to explain how red colobus and black-and-white colobus can coexist in Kibale (Clutton-Brock, 1974*b*; Struhsaker & Oates, 1975), even though the dietary differences between the African species are less extreme than at the Asian site.

Results from subsequent comparative studies of the two leaf-monkeys at Kuala Lompat did not fully concur with this pattern, although there were some similarities. The two species differed in diet, with one highly folivorous and the other a folivore–semivore. The two species also used different arboreal supports and occupied different levels of the forest: *T. obscurus* on boughs in the middle and upper storeys, *P. melalophos* on smaller supports of both small and large trees. These differences in diet and locomotion, however, are not complemented by differences in social systems; both species lived in uni-male bisexual groups of about 15 animals, occupying similar-sized home ranges with only their degree of territoriality differing (Bennett and Caldecott, 1989).

When comparisons are extended to the inland forests of Borneo, even the major differences in the degree of folivory shown by sympatric species cease to apply. In northern Borneo two sympatric *Presbytis* species appear to have generally similar diets, social organization, population densities and, adjudged by their skeletal anatomy, similar locomotor behaviour. Only very detailed ecological studies over a long time will provide an understanding of the ecological separation between them, but it is clear that no single model can be used to describe ecological separation between sympatric colobines across Asia.

Adaptive strategies

Despite the lack of a simple hypothesis to explain ecological separation, it is still possible to highlight distinctive strategies adopted by the different genera of Asian colobines.

The studies of *Pygathrix*, *Nasalis* and *Simias* indicate a remarkably consistent adaptive strategy for this phylogenetically primitive group. Most species have a basic social unit of about 5–15 individuals, including a single adult male, adult females and their progeny. They are large-bodied (Chapter 3) and can subsist on low-quality diets at certain times of year, but exploit high-quality clumped food sources at others. They range over large distances, often along the ground, and their social system is sufficiently flexible to allow family units to join together, numbering tens if not hundreds of animals in single congregations. This is a successful strategy in the harsh montane forests of China, but is obviously not so successful in the more equable conditions that prevail over most of tropical Asia; it has not allowed *Nasalis*, nor probably *Simias*, to out-compete sympatric leaf-monkeys.

Presbytis and *Trachypithecus* show less flexible social organization and ranging patterns, but still have distinctive socio-ecological strategies (Figure 5.12). *Trachypithecus* species, whether in deciduous or evergreen forests, consistently have a high foliage intake, usually about 60% of the annual diet, with 20–40% mature foliage, even though mature leaf parts may not be preferred to seasonally available young leaves, buds, flowers or fruits. A high foliage intake correlates well with their large body size and/or large stomach relative to body size (Chapter 7), both features improving the opportunity for prolonged fermentation of high-fibre foodstuffs. It also correlates with their dentition, which is well-adapted for folivory (Chapter 6). Only in the exotic teak plantations at Pandangaran are *T. auratus* exceptional in having less than 20% mature leaf parts in the annual diet. At all sites, the basic social structure comprises uni-male bisexual groups and all-male bands, the latter including females in some species.

For *Presbytis* species, a distinctive feature of the feeding strategy is the high seed intake by species in dipterocarp forests, along with the fact that foliage, mostly young leaf parts, supplies less food than fruit parts (seeds, whole fruits and fruit flesh, combined). These species are small, gracile animals with relatively small stomachs in relation to body size, and teeth which are not especially adapted for folivory (Chapter 6). The uni-male bisexual group remains the basic social unit.

No generalizations can be made about the highly adaptable *Semnopithecus*. Its feeding strategy is intermediate between the folivorous *Trachypithecus*

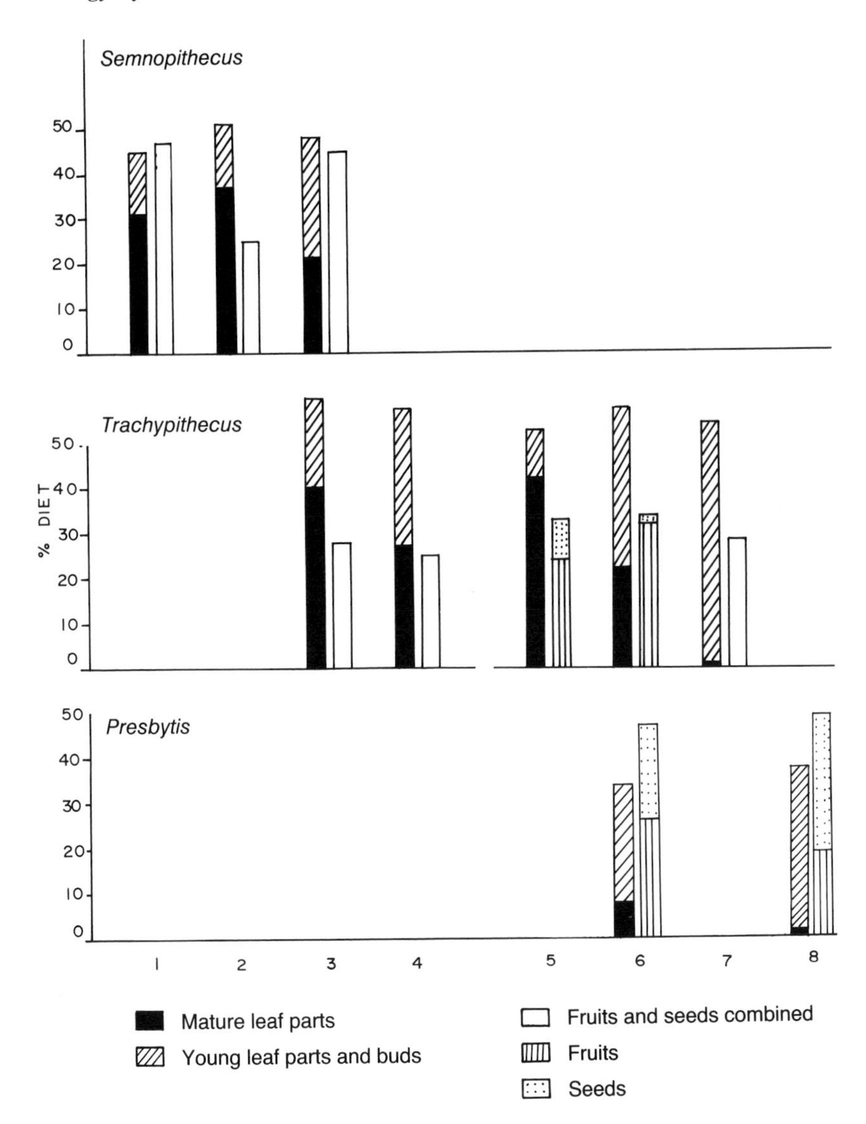

Figure 5.12. Summary diets of the different genera of Asian colobines, using information from studies that were conducted for a year on habituated study groups (sources in text). 1, Simla; 2, Kanha; 3, Polonnaruwa; 4, Kakachi; 5, Madhupur; 6, Kuala Lompat; 7, Pangandaran; 8, Sepilok.

and frugivorous *Presbytis*; foliage intake can be high, sometimes including a large proportion of mature leaves, but fruit parts also make up a large proportion of the diet at some sites. To this can be added three other characteristic features of the diet: the higher species diversity of the diet compared with that of *Trachypithecus* species in south Asia, the inclusion of a relatively large invertebrate intake (although still less than 5% of the annual diet), and a capacity to consume ripe fruit. As described in Chapter 11, the social organization can be either uni-male or multi-male bisexual groups.

The home range size of Asian colobines is variable. It may be in excess of 10 km^2 for the semi-migratory, semi-terrestrial *Pygathrix* spp. and *S. entellus*, but for the largely arboreal *Presbytis* and *Trachypithecus* it varies from 2.5 to 84 ha and is not simply correlated with the proportion of foliage or fruit parts in the diet. Similarly, territorial behaviour is displayed by species whether foliage or fruit parts dominate the diet.

There is therefore no simple correlation between diet and home-range size, and different socio-ecological strategies interplay with features of the habitat to produce different behavioural patterns. Among the most significant habitat features are:

1 The plant species diversity, and hence the range of plants and plant parts which can be selected as foods;
2 The seasonality, particularly the presence or absence of prolonged dry, cold or wet weather, when preferred food items (such as young-leaf parts and fruit parts) are unavailable; and
3 The chemical composition of plant parts, especially mature-leaf parts, which can sustain colobines at times that other items are unavailable.

For example, the seasonal nature of the south Asian deciduous and semi-deciduous forests largely excludes *Presbytis* species, which appear to be unable to subsist on mature leaves for long periods of time. On the other hand, no *Trachypithecus* species has successfully colonized the inland forests of Borneo, where the abundant mature leaves are of low quality and only a semivore/folivore strategy, which relies on a combined year-round availability of fruit and young-leaf parts, is viable.

After the highlighting of these threads holding together the fabric of ecological strategies displayed by different Asian colobine genera, it is quite apparent that they are very adaptable and can accommodate many environmental conditions. To explore the opportunities for adaptive strategies, the following chapters consider the flexibility of colobine anatomical and physiological systems, and describe in more detail the variation in population densities and social organization.

Acknowledgements

During the writing of this chapter, ELB was supported by the Wildlife Conservation Society (New York Zoological Society) and WWF Malaysia; additional facilities were provided by the University of Cambridge. AGD's work has been funded by WWF Malaysia, the National Science Foundation and the New York Zoological Society (to J. F. Oates), WWF-US, WWF-International and the Overseas Development Administration. Drs Kathy MacKinnon and Bill Bleisch provided information and papers on the Chinese primates and Drs J. M. Bennett, Mike Kavanagh, Paul Newton and John Oates commented on the draft manuscript. Milcah Karanja and Lydia Abiero assisted with typing and Dennis Milewa drew clear figures. Many thanks to all of these people and organizations.

6

Functional morphology of colobine teeth

PETER W. LUCAS and MARK F. TEAFORD

Introduction

As the reader will gather from other chapters in this book, most species of colobine monkeys eat large quantities of young leaves. Many also eat considerable amounts of large thinly-covered seeds. In other words, colobines may be fairly adept at processing both leaves and seeds. If that is the case, the question we wish to ask is the following: could colobine teeth exhibit a similar duality of function: i.e. could they be useful for both leaf-eating and seed-eating? To begin to answer this question, we first describe the form of the colobine dentition and emphasize those features which distinguish it from those of other primates and cercopithecines. Second, we attempt to explain these features. From our analysis, we suggest that a change of perspective is needed: colobine teeth may well be adapted for processing both leaves and certain kinds of seeds.

Colobine dental morphology

Colobines have the same number of teeth as other catarrhines, with a dental formula of 2–1–2–3. The maxillary and mandibular incisors are heteromorphic, so that those that are isolated and unworn can generally be correctly identified and sided in the jaw (Figure 6.1). The maxillary lateral incisor is mesiodistally narrower and more pointed than the central (Vogel, 1966). The mandibular incisors are fairly similar in size, but the lateral is again more pointed than the central, so that its high point usually lies close to the mesial corner and the incisal edge slopes downward (towards the base of the crown) as one moves distally. Unlike the mandibular incisors of Papionini, those of colobines do have enamel on their lingual surface (Delson, 1973). The maxillary incisors are generally mesiodistally larger than the mandibular (Delson,

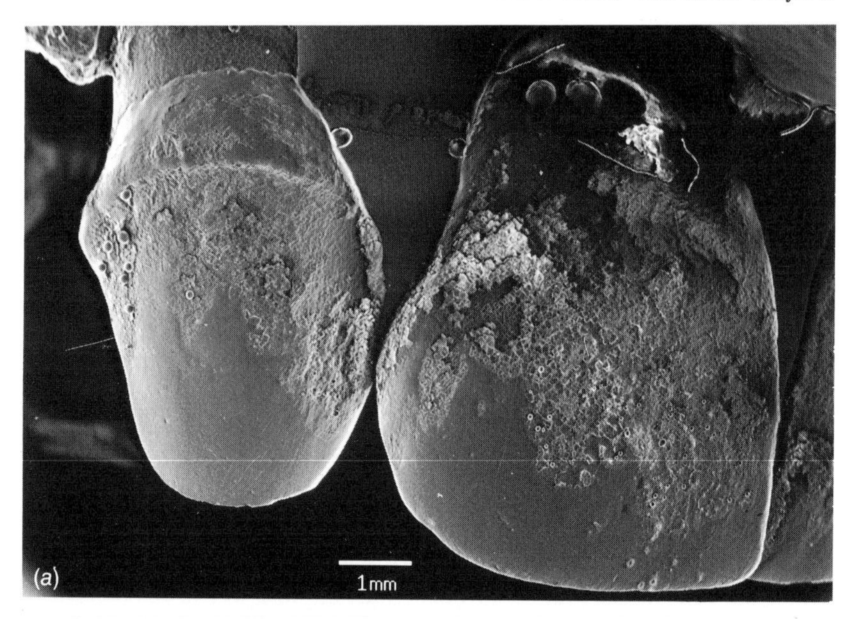

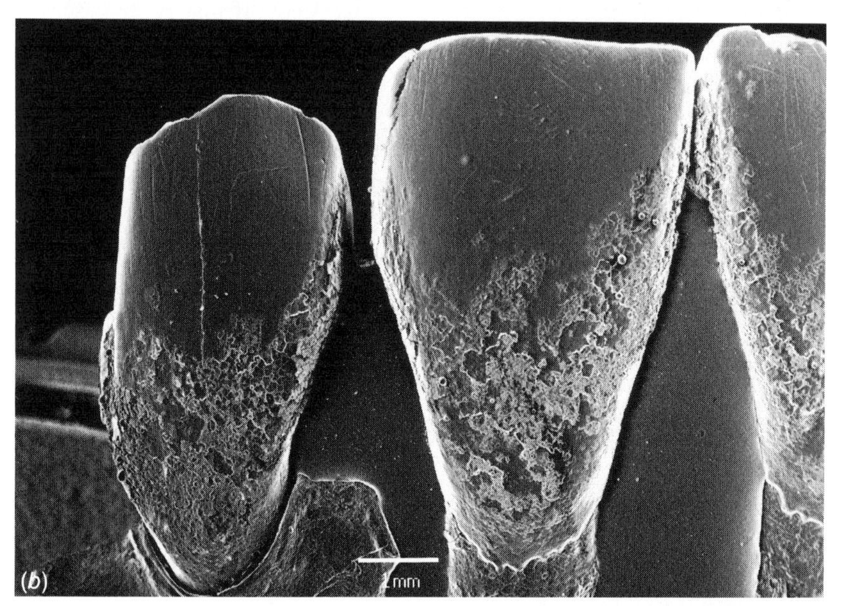

Figure 6.1. Scanning electron micrographs of colobine incisors (*Trachypithecus cristatus*, NMNH 25227, National Museum of Natural History, Smithsonian Institution, Washington, DC). In both cases, right = mesial. (*a*) Right maxillary incisors; (*b*) right mandibular incisors.

1973), but colobine incisors are still smaller than those of cercopithecines of similar body size (Hylander, 1975).

Probably the most unusual trait of colobine incisors is the high incidence of underbite in certain taxa – ranging from approximately 30% (in *Procolobus badius* and *Trachypithecus cristatus*) to nearly 100% (in *Presbytis rubicunda*) (see Figure 6.2). As might be expected, the incidence of underbite is negatively correlated with maxillary incisor size and palate length (Colyer, 1936; Sirianni, 1979; Swindler, 1979; Teaford, 1983*b*; Emel & Swindler, 1992).

Colobine canines are similar to those of other cercopithecids and project significantly beyond the other teeth around them (see Figure 6.2). However, the degree of projection of the canines of males is generally less than in larger cercopithecines and is also less in Asian colobines than in African forms (Lucas *et al.*, 1986*a*). The mandibular canines frequently have a distal heel, which often becomes more noticeable with wear (see Figure 6.3*c*), and the maxillary canines have a sharp distolingual edge and exhibit a characteristic groove running along the long axis of the mesial surface of the crown (Figure 6.3*a* and *b*). The sharpness of canines is critical for their role in

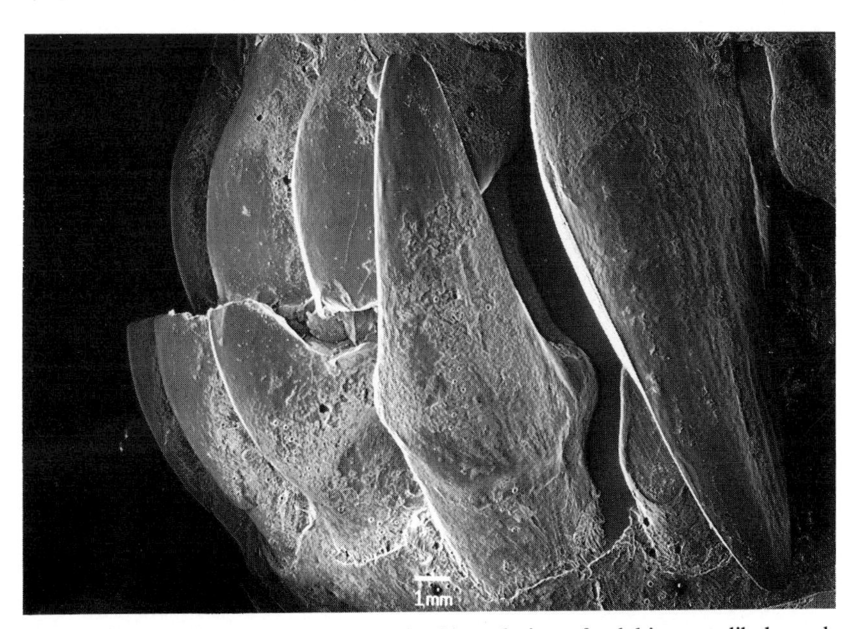

Figure 6.2. Scanning electron micrograph of lateral view of colobine mandibular and maxillary anterior teeth (*Presbytis comata*, NMNH 196809, National Museum of Natural History, Smithsonian Institution, Washington, DC). Left = anterior. Note: (1) underbite (tips of mandibular incisors are anterior to tips of maxillary incisors); and (2) projection of canines beyond plane of cusp-to-cusp occlusion of incisors.

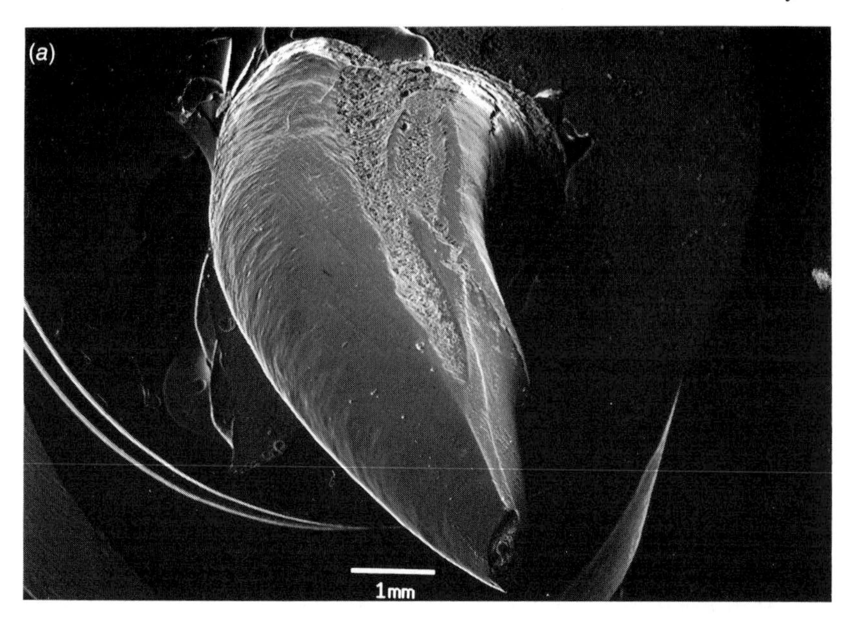
(a)
1mm

(b)
1mm

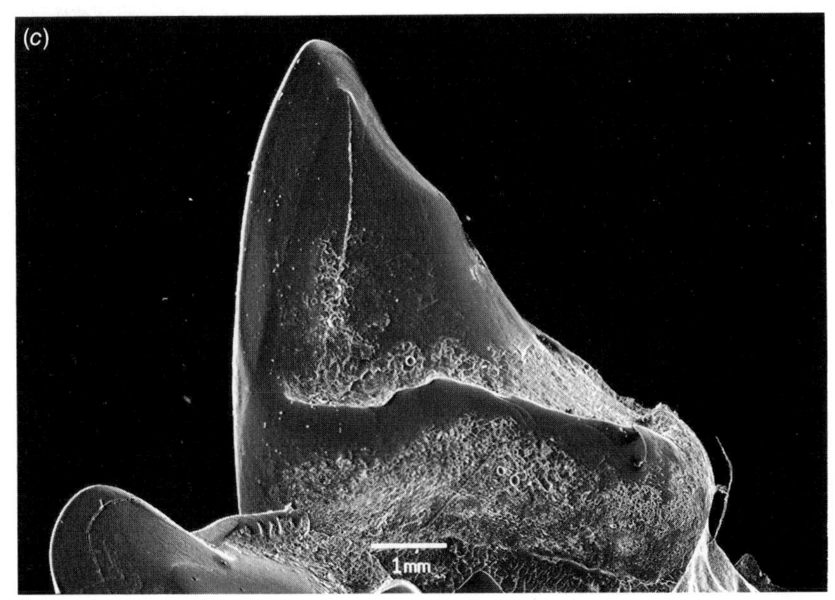

Figure 6.3. Scanning electron micrographs of colobine canines. (*a*) Right maxillary canine of *Presbytis comata* (NMNH 197649, National Museum of Natural History, Smithsonian Institution, Washington, DC) looking from canine tip, along mesial face of tooth, towards base. Note groove running along the middle of the mesial face with wear facets on either side. (*b*) The same right maxillary canine, looking at the distal surface. Note pronounced distolingual wear facet in bottom third of micrograph. (*c*) Right mandibular canine of *Trachypithecus cristatus* (ANSP 20212, Academy of Natural Sciences, Philadelphia, PA). Left = mesial. Note pronounced distal heel.

crack propagation (Freeman, 1992). Males generally have larger canines (as indicated by basal area) than do females (Swindler & Orlosky, 1974; Swindler, 1976) although the degree of dimorphism varies between species (Leutenegger, 1971; Hull, 1979) and canine height may not vary between the sexes in certain species. For example, in subspecies of *Colobus guereza*, the canines of females may be so enlarged as to be visually indistinguishable from those of males, just as in gibbons (Lucas *et al.*, 1986*a*).

As in other cercopithecids, colobine maxillary premolars are fairly similar to each other (Figure 6.4). Delson (1973, p. 299) has accurately described them as 'D-shaped, with a straight buccal face'. Each has a small mesial fossa or fovea, a buccal and a lingual cusp, and a large distal basin (or trigon). The most mesial of the maxillary premolars (P3)[1] is easily the most

[1] Primate premolars are traditionally numbered under the assumption that, during the course of their evolution, modern species have lost one or more premolars as compared with primitive mammals. Thus the first, or most mesial, premolar found in Old World monkey jaws is generally known as P3 and the second is known as P4 (Swindler, 1976).

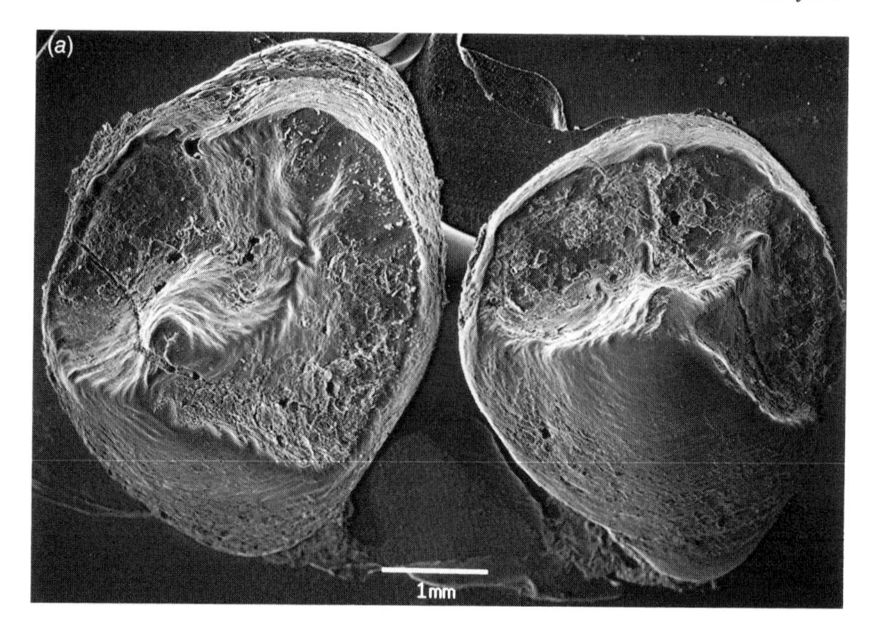
(a)
1mm

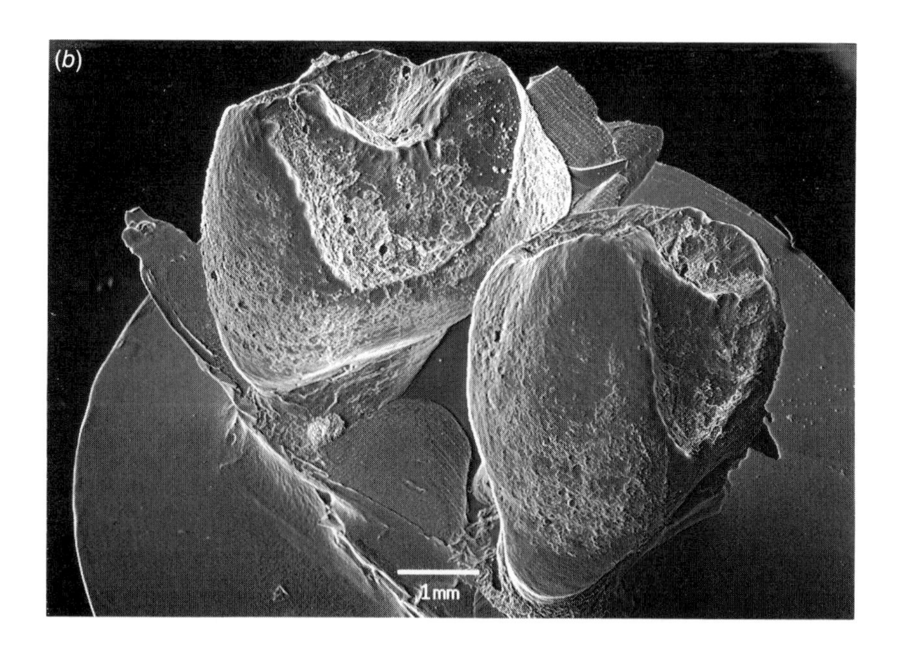
(b)
1mm

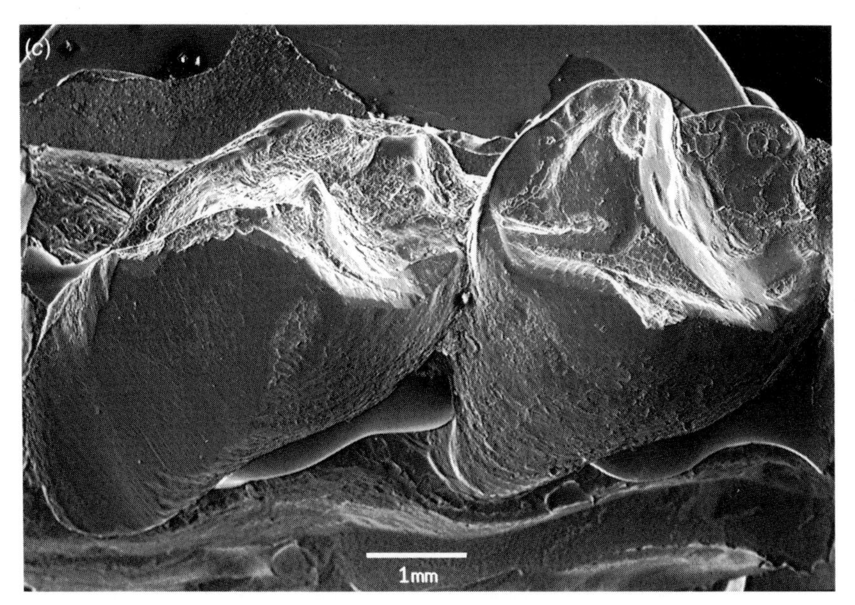

Figure 6.4. Scanning electron micrographs of colobine premolars. (*a*) Occlusal view of left maxillary premolars of *Presbytis comata* (NMNH 198280, National Museum of Natural History, Smithsonian Institution, Washington, DC). Right = mesial; bottom = buccal. (*b*) Oblique view of same maxillary premolars. Note extension of enamel down mesiobuccal corner of P_3. (*c*) Slightly oblique occlusal view of left mandibular premolars of *Presbytis comata* (NMNH 198281, National Museum of Natural History, Smithsonian Institution, Washington, DC). Left = mesial; bottom = buccal. Note pronounced mesiobuccal flange on P_3.

distinctive of the two, with a relatively large extension of enamel down the mesiobuccal root also known as the mesiobuccal flange. The lingual cusp of P3 is also reduced in most colobines (Delson, 1973; Swindler, 1976). The other maxillary premolar (P4) is larger in basal area than its mesial counterpart, although most of this is due to the increased buccolingual width of the tooth.

The mandibular premolars in all cercopithecids are very heteromorphic, and those of colobines are no exception. The more mesial mandibular premolar (P3) is dominated by a large cusp and a large mesiobuccal flange against which the upper canine is sharpened (Walker, 1984). Any basin distal to the main cusp is rudimentary at best. As might be expected from the differences in canine size between the sexes, males tend to have larger P3 flanges than do females. The mandibular fourth premolar is a more molariform tooth, with a mesial and a distal basin on either side of a buccal and a lingual cusp. However, the only significant difference between colobines and cercopithecines for this tooth is that colobines tend to have better-developed mesiobuccal flanges on their P4s (Zingeser, 1969, Delson, 1973).

Colobines have classic bilophodont molars, meaning that, as one moves mesiodistally along the tooth, there is first a mesial basin or fovea, then a mesial pair of cusps connected by a transverse crest or loph, a large central basin, a distal pair of cusps also connected by a transverse crest or loph, and finally a distal fovea (see Figure 6.5). In both the upper and lower molars, the cusps sit along the buccal and lingual margins of the crown, with the buccal pair of cusps and the lingual pair of cusps each connected by marginal ridges (see Figure 6.5). Third molars are slightly different from the first and second molars. The maxillary third molar is often smaller than the other upper molars, due to a reduction in the size of the distal loph. The mandibular M3 is usually, mesiodistally, the longest of the lower molars due to the presence of a fifth cusp along its distal margin (see Figure 6.5). However, the position of this cusp on M3 is subject to some variability in colobines (Swindler *et al.*, 1967). It may even be absent in some smaller species (e.g. *Presbytis*) (Delson, 1973).

Probably the most striking contrast between colobine and cercopithecine molars involves the amount of occlusal relief (Figure 6.6). Colobines have relatively taller cusps that are set closer to the margins of the tooth than in cercopithecines. As a result, colobines have longer molar crests than do cercopithecines (Kay, 1978; Kay & Hylander, 1978; Benefit, 1987). Variations in the lengths of these crests appear to be ecologically important, longer ridges being associated with a greater intake of leaves even within the Colobinae (Kay, 1978; Kay & Hylander, 1978). For example, *Trachypithecus cristatus* has longer crests than does *Presbytis rubicunda* (Teaford, 1983*a*).

The size of the molars of colobines is also greater, relative to body weight, than in cercopithecines (Kay, 1975). It has been found that the ratio of the first to the third molars of either jaw in cercopithecids is inversely related to the proportion of feeding time devoted to the consumption of leaves and seeds. Furthermore, the shape of the lower third molar is an important dietary indicator on its own, in that the breadth:length ratio is very highly correlated with the percentage of leaves in the diet (Lucas *et al.*, 1986*b*).

What teeth do

In order to understand the functional implications of colobine dental morphology, it is necessary to have a better grasp of the physical properties of food items, so that we can understand how they are broken down in the mouth. This is because the answer to the question 'what do teeth do?' is that 'they

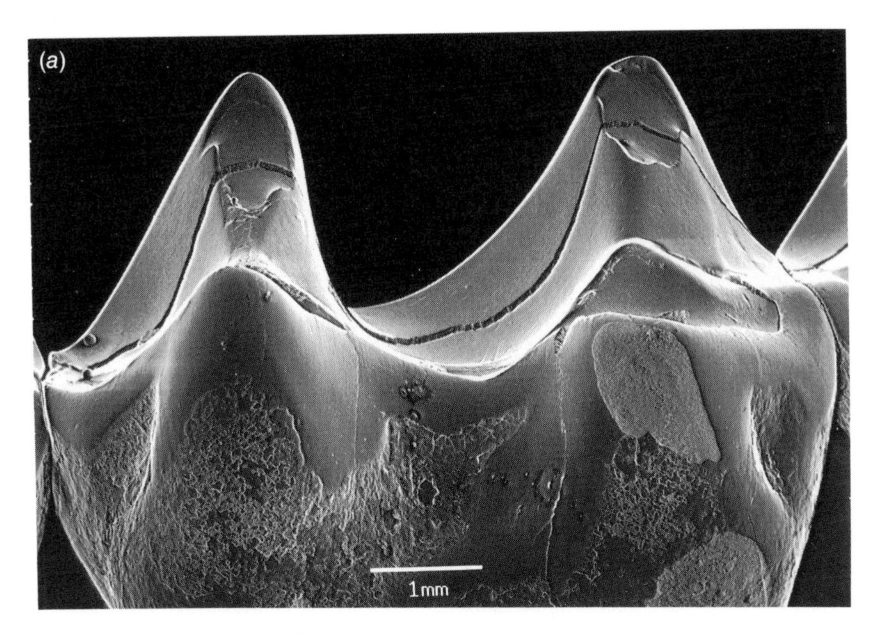

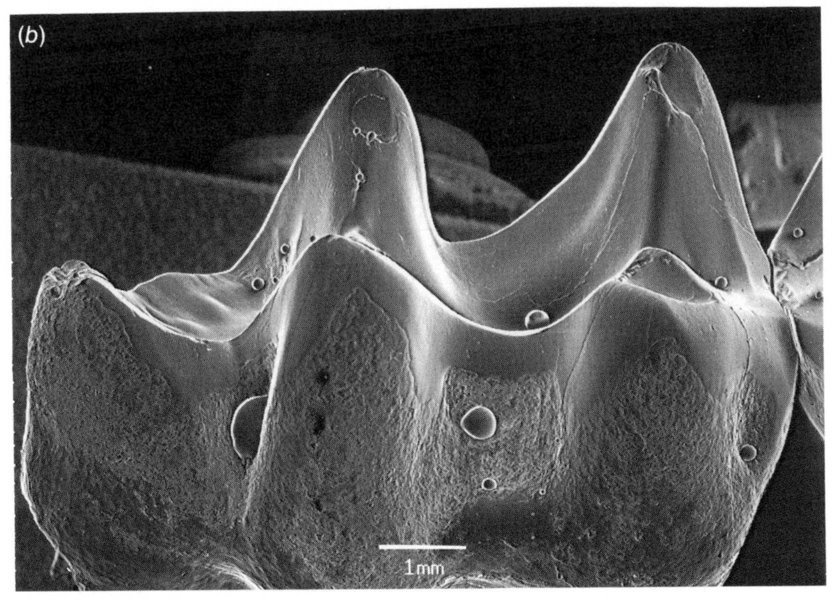

Figure 6.5. Scanning electron micrograph of buccal view of right mandibular M2 (*a*) and M3 (*b*) of *Procolobus badius* (NMNH 378647, National Museum of Natural History, Smithsonian Institution, Washington, DC). Right = mesial.

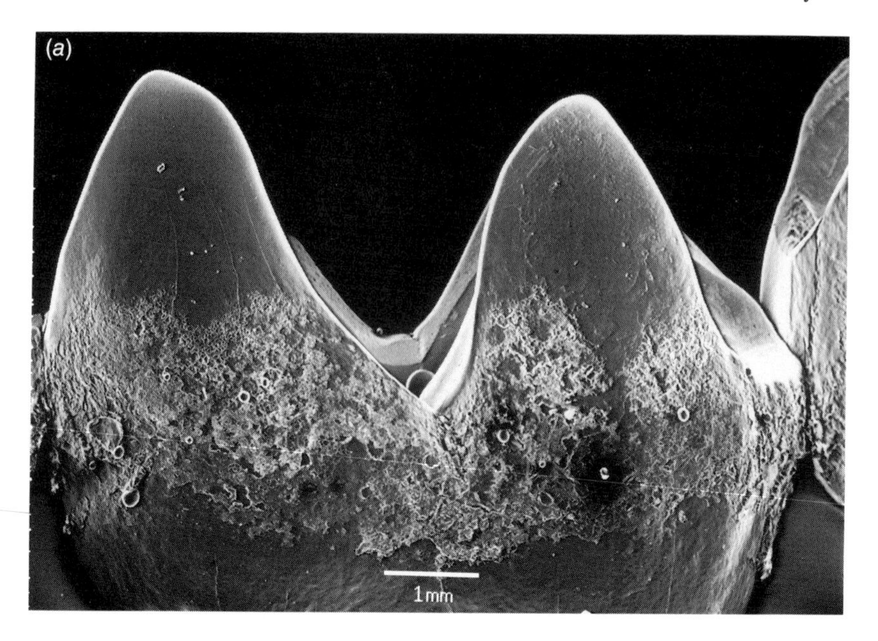

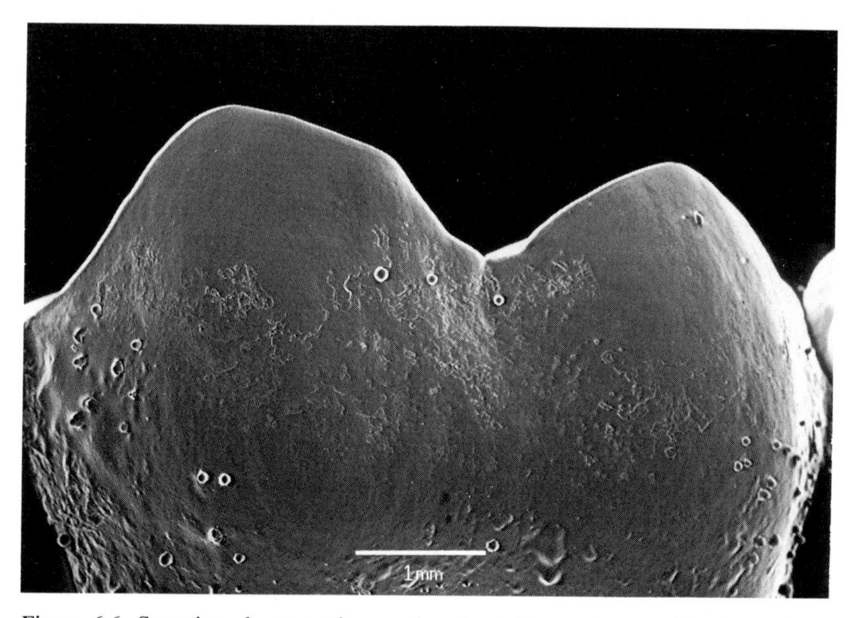

Figure 6.6. Scanning electron micrographs of colobine and cercopithecine molars. Lingual view of right M_2s of (*a*) *Procolobus badius* (NMNH 381458, National Museum of Natural History, Smithsonian Institution, Washington, DC) and (*b*) *Cercocebus albigena* (NMNH 164578). Note difference in cusp height.

fracture and fragment solid foods' and not, as is commonly supposed, that 'they crush, grind, tear, cut or shear'. Such terms are members of a large family of words that simply denote fracture and, by masquerading as explanation, have stunted the understanding of how teeth work. They can be used (e.g. by materials scientists) as descriptive shorthand, but they are not analytical – i.e. they are ambiguous descriptions of successful events and cannot be used for analysing the conditions necessary for the event to have taken place. Besides, no single term can explain what teeth do, because it cannot combine a description of tooth surfaces and tooth movement with a description of the fracture behaviour of food between the teeth. Thus the answer to 'why are there different shapes of teeth?' is 'because the fracture behaviour of foods is variable'. The first step to understanding dental–dietary adaptation is to consider the fracture properties of foods because it is to these that teeth are ultimately 'adapted'.

There are two major fracture-related properties. One is strength,[2] which can be measured in two ways. Fracture strength is the maximum stress that a structure can take before starting to fracture. It depends on the size of the structure. Yield strength is more fundamental and is the stress which causes a permanent change of shape. Both are very relevant to the fracture of rigid foods like hard-shelled seeds. However, in softer foods, like leaves and the inside of seeds, cracks can be started easily and the major problem for the dentition is getting these cracks to propagate. From this perspective, then, a general aim of the teeth is to *fragment* foods, not just to fracture them (Lucas, 1994). The resistance of a structure to propagating a crack is its fracture toughness – the work required to fracture a unit area of tissue.[3] By such a definition, something that requires relatively little work to fragment would be brittle. Many brittle solids are also rigid. However, rigidity or stiffness is a separate property from toughness and would be defined by the elastic modulus.[4]

The definition of toughness is both convenient and important. It allows objective comparisons with other materials (e.g. gels have a fracture toughness of 0.5–40 J/m^2, whereas plastics have a fracture toughness of 300–8000 J/m^2). It fits in much more easily with the ecological discussion of feeding strategies than does fracture strength, because it is energetically based (Choong *et al.*, 1992). In addition, there is evidence from humans that toughness can be perceived in the mouth (Sim *et al.*, 1993) which raises the possibility that mammals can detect toughness and respond to it.

[2] 1 Pa = 1 N/m^2 in SI units.
[3] J/m^2 in SI units.
[4] 1 Pa = 1 N/m^2 in SI units.

Modes of fracture

As noted previously, no single term can describe the general pattern of stress in food loaded by teeth. However, a basic understanding of fracture mechanics (the initiation and propagation of cracks within a material) leads to a much clearer picture of what teeth actually do. There are three modes of fracture which reflect three different states of stress in the material ahead of the crack tip. These different states are presumably what 'crushing' and 'cutting' attempt to describe. In any case, the stresses ahead of the crack are all-important. From this perspective, 'crushing' fracture is certainly not due to compressive stresses. Likewise, 'cutting' fracture may be due to tension or to one of two shearing modes.

This is shown schematically in Figure 6.7, where the modes of fracture are identified with Roman numerals. A notch is shown in a block of hypothetical material in which a crack will initiate and propagate in the same direction as the longest dimension of the notch, i.e. vertically down the block without deviating. In mode I fracture, the material ahead of the crack tip is subject purely to tension. To produce this, the material to the left and right of the notch have to be moved directly away from each other. In both modes II and III, material ahead of the crack is sheared, with the difference that in mode II, crack growth is parallel to the loading direction, whereas in mode III, it is perpendicular to it.

For example, closing a pair of scissor blades onto a thin piece of paper produces a mode III fracture, because the fractured surfaces are twisted (out-of-plane shear). However, if after a small notch is made in mode III, the

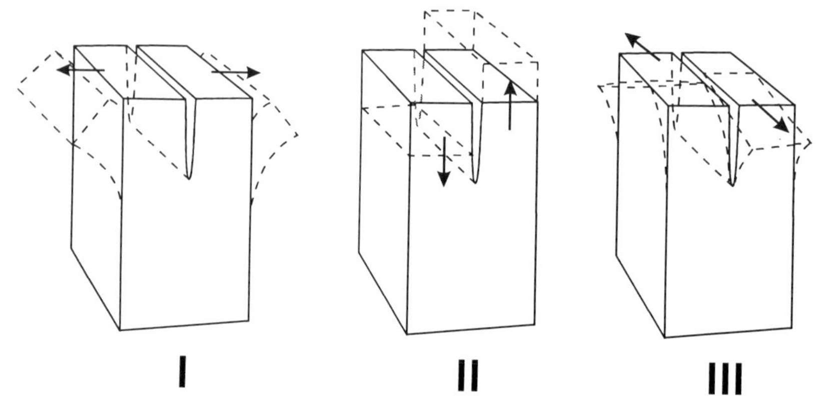

Figure 6.7. The three modes of fracture. Arrows indicate the direction in which surfaces are moved (see text).

scissors, with blades open but stationary, are advanced into the paper, then a mode-I fracture results. There is no twisting, no shear at all – the sides of the blades force the material apart in tension. In contrast, the sharp edges of a paper hole-puncher produce mode-II fracture (in-plane shear). All these alternatives could be termed 'cutting'! These modes of fracture are geometrically defined. Complex fractures of 'real' materials can mix two or three modes. However, once a crack is initiated in 'mixed mode', one mode predominates, and this is usually mode I if it is present (Cotterell & Mai, 1989).

In the above mode-III example, important changes in the mode of fracture with scissors take place as the thickness of the paper is increased (Lucas & Pereira, 1991). As the blades are closed, their sides become increasingly important in pushing the paper aside. This is now mode-I fracture, and each blade acts like a wedge. A blade that is used to fracture a thin sheet, in either modes II or III, converts the work needed to move the blade directly into work of fracture (Atkins & Mai, 1979), but a wedge stores energy in the thicker material by deforming it. This stored energy is expended when the crack starts to run (Vincent *et al.*, 1991). Instead of the crack tip being constrained by the blade, it now runs free, ahead of the blade. This has several advantages for comminution: cracking is faster; the blades do not have to close completely on the food, which saves on their wear; it is also often cheaper, because a free-running crack can run through the material along the cheapest path rather than have its path dictated by the blade.

As a result of the above, wedges would be predicted to have different design features from blades (Figure 6.8). Both should be sharp, sharpness being measured by the radius of curvature of the tip. The length is important, since this should be longer than the food object to be cracked, especially if cracks do not spread easily in the food material. In addition, a wedge requires an even height, or else the crack will tend to arrest and not fragment the object. Its included angle is also critical and should be distributed symmetrically about the direction of applied force, in order that the crack is directed in the path of wedge movement. The included angle of a blade is not critical for fracture. One edge should be directed parallel to the direction of applied force, with the possibility of a small relief angle behind the edge. Most bladed systems do not load the entire blade at once, so reducing the forces associated with crack initiation by point contacts. This point loading induces cracking by twisting (mode-III fracture). In other words, mode-III fracture is much more common than mode II fracture in bladed systems. Before we show how far the above theoretical framework is relevant to colobine dental morphology, we briefly describe the anatomy of the food on which it acts.

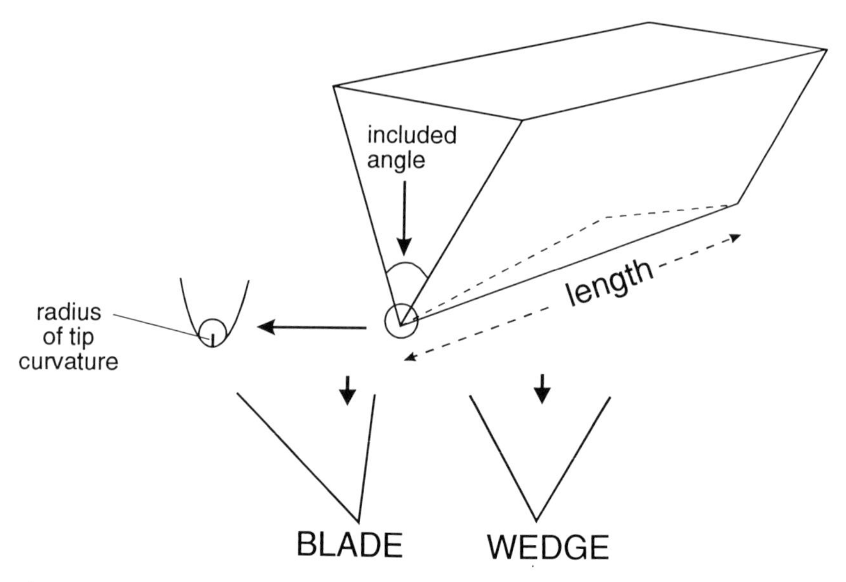

Figure 6.8. The features of a blade are compared to those of a wedge. Both share a need for sharpness, measured as the radius of curvature of the tip, and length. However, the included angle of a wedge is critical and is distributed symmetrically about the direction of movement (small arrow). A blade is often inflected (dashed line).

Colobine foods

It is best to use functional rather than botanical terms to describe plant tissues when discussing their mechanics. In any case, the embryological derivation of functionally equivalent plant tissues, on which the botanical name depends, is variable and not always known.

There are several difficulties with the measurement of the fracture properties of primate foods. First, these properties do not generally survive storage. Turgor pressure and moisture content may be critical and, unless a laboratory is near by, field tests are necessary. Second, current field tests often do not make fundamental measurements and depend greatly on the particular piece of equipment used. Though they provide quantification, their results need to be viewed with a great deal of caution. Third, most mechanical tests require the dimensions of the test specimen to be known accurately in relation to the direction of the load. Fourth, it may be necessary to measure toughness in different modes of fracture (Wright, 1992). Fifth, many primates, particularly colobines, eat a plant part before it matures, making it necessary to know how

Table 6.1. *Comparison of the mechanical properties of a thin seed coat (*Millettia atropurpurea*), 0.5 mm thick, with a 3–4 mm thick hard seed shell (*Mezzetia parviflora*)*

Species (family)	Elastic modulus (GPa)	Fracture strength (MPa)	Fracture toughness (J/m^2)
Millettia atropurpurea (Leguminosae)	0.34	5	330
Mezzetia parviflora (Annonaceae)	7	67	2000

these properties change during development.[5] In summary – without sufficient care, rapid surveys of mechanical properties become dominated by error.

Seeds

The energy value of the seed is largely contained in the seed storage tissues, but to get at these, the outer covering must be removed. Colobines seem to eat large seeds with thin flexible coats. In a detailed study of seed-eating by *Presbytis rubicunda*, the study animals had a high intake of large seeds that were covered with thin, flexible testas (Davies, 1991), which differ from some lignified seed-coats. For example, Table 6.1 indicates mechanical differences between a thin unlignified testa (*Millettia atropurpurea* – Leguminosae) and a thick highly lignified seed coat (*Mezzetia parviflora* – Annonaceae). Seeds of *Millettia* spp. are eaten by *P. melalophos* (Bennett,

[5] Unlike in surveys of the chemical properties of foods, it is not generally necessary to test a large number of similar dietary items before arriving at a general conclusion, because the mechanical properties of foods can be correlated with and predicted by their anatomical structure (Vincent, 1990). For instance, the theory of cellular solids can be applied effectively to plant tissues when cell shape and tissue density, relative to the density of the cell wall, are known (Gibson & Ashby, 1988). Fractures of cellular tissues can run in two different ways: through the cell wall or between cells. Cell walls are stiff and tough, because of the arrangement of cellulose fibres within them. When a crack runs through the walls of cells, then the measured toughness depends linearly on the relative density (Gibson & Ashby, 1988). Higher-density plant tissues not only have more cellulose, which stiffens and toughens the wall, but are also usually lignified. Fractures run between such thick-walled cells where possible, and across them where not, because, in contrast to cell-wall fracture, fracture between these cells does not depend on density and costs little ($< 300\ J/m^2$): intercellular areas (the middle lamella) are lignin-rich and cellulose-free.

1983) and *P. rubicunda* (Davies, 1991). The properties of seeds of *M. atropurpurea* reported here, are similar to those of seeds of *Intsia palembanica* (Leguminosae), which are also important in the diets of the above colobines, but the latter are more difficult to test. Fruit tissue of *Mezzetia leptopoda* is eaten by *P. melalophos*, which avoids eating the seed (Bennett, 1983). The shell of *M. leptopoda* is ten times stiffer and stronger than the thin flexible covering of *M. atropurpurea*. This strength is attributable to the thick cell walls of the fibres in the shell. However, seed shells are much less tough than a comparison with wood predicts, probably due to the three-dimensional orientation of the component fibres to the load (Preston & Sayer, 1992).

The properties of the thin seed covering of *M. atropurpurea* are probably typical. The covering does not consist of fibres, but a thickened outer layer of palisade or stellate cells, which are, at most, lightly and unevenly lignified. The toughness of this arrangement is low because there are no mechanisms by which the propagation of a free-running crack could be obstructed. However, a significant contribution to mechanical resistance in such seeds comes from the seed storage tissues themselves. Though thin-walled, these tissues have fracture toughnesses of about 1000 J/m^2 (e.g. *M. atropurpurea* cotyledons and *Gnetum microcarpum* (Gnetaceae) endosperm, the latter eaten by *P. melalophos*; Bennett, 1983). This toughness is corroborated by evidence for the endosperm of *Zea mays* (Vincent, 1990). There is little that is remarkable anatomically about these tissues, and their high toughness, attributable to their plastic collapse at high strains (Gibson & Ashby, 1988), has not yet been adequately explained.

During the final stages of seed development, there is a significant increase in mechanical protection (see e.g. Corner, 1976), and this may explain the preference shown for unripe fruits by seed-eating colobines (see e.g. Davies, 1991), compared to the selection of ripe fruits by forest guenons (Gautier-Hion *et al.*, 1985).

Leaves

The simplest and most regular leaf known is that of *Calophyllum inophyllum* (Lucas *et al.*, 1991), in which the lamina has an even cross-section, and secondary and tertiary veins that are oriented at right-angles to each other. We can learn from this leaf, without any qualifications, what makes a leaf tough (Lucas *et al.*, 1991). The bulk of the lamina is spongy mesophyll, the cell walls of which are very thin. The mesophyll is capped by an upper and lower epidermis and their cuticles. These tissues as a group provide little resistance to fracture, with a toughness 200–300 J/m^2. This is about the average toughness found for such thin-walled plant tissue, though values as low as 60 J/m^2 are known (Vincent, 1990, 1991). Virtually all the true toughness of a leaf comes from thick-walled veins and similar tissue around them (generally termed the bundle sheath). These form only a small volume fraction of a leaf and yet have a toughness of about 6000 J/m^2, at least 20 times that of the other tissues in the leaf (Lucas *et al.*, 1991). The midrib has a similar toughness. In other words, the difference in toughness between veins and other tissues is much more crucial than the overall average toughness of the whole lamina. The network of veins in a dicotyledonous leaf produces a direction-dependent toughness which markedly reduces the ease with which it can be broken down into smaller pieces – particle-size reduction being a critical function of the mouth. In some leaves, these bundle sheaths extend to the epithelia and this effectively seals off compartments between the veins. This must have a marked effect on digestibility, unless these compartments can be opened by chewing (Vincent, 1991).

Some mature leaves do not have bundle sheaths (or they make them from thin-walled cells) and some, particularly light-demanding species with thin leaves that wilt as they dry out in the sun, do not have thick-walled veins. The average toughness of such leaves can be as low as 80 J/m^2 (Choong *et al.*, 1992).

Newly-expanded leaves do not have thick-walled veins and are generally just a compact mass of thin-walled cells, whose toughness is probably not direction-dependent and would lie between 60 and 200 J/m^2. Nothing really distinguishes this tissue from that of any thin-walled parenchyma, either from a mechanical or a digestional point of view. The development of 'toughness' in leaves has been followed by Kursar & Coley (1991) using a field method that cannot produce any comparative value and that is dependent on lamina thickness (Choong *et al.*, 1992). However, these results can be interpreted as showing that the toughening of the veins can be extremely rapid, mechanical changes being observed within a few days.

Dental biomechanics and colobine molars

From the previous two sections of this chapter, it can be seen that the lophs on colobine molars can indeed act as wedges in fracturing the storage tissues of large seeds. This can also be argued in a very general manner from the form of the upper molars.

All mammalian teeth are derived from the tribosphenic molar, which was, in occlusal view, a triangular tooth with the cusps merely being end-points of the long sharp ridges running buccolingually across the tooth (Crompton & Sita-Lumsden, 1970; see Figure 6.9(*a*)). Cross-sections at various points through the tooth of such a primitive, insectivorous mammal (Figure 6.9(*a*))

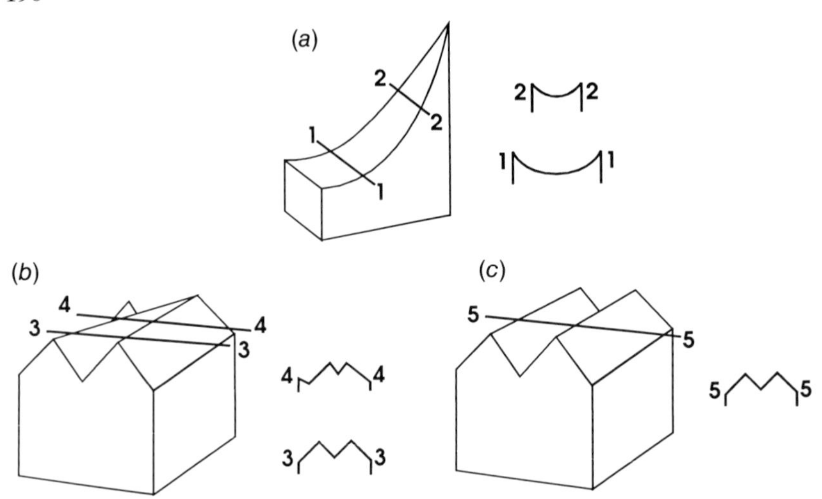

Figure 6.9. (*a*) Diagram of a tribosphenic molar with the outer edges of the long sides of its triangular form acting as blades. Cross-sectional profiles 1–1 and 2–2 show the lack of symmetry of these ridges. (*b*) Attempting to produce wedges from squared-up teeth retaining the oblique ridge (as in most anthropoids) is thwarted by the union of these wedges at one end of the tooth (as shown by the cross-sectional profiles). (*c*) Wedges can easily be formed by the loss of the oblique ridge to form two lophs which then have consistent cross-sections.

show profiles that are unmistakably blade-like rather than wedge-like, because the two sides of each ridge are very assymmetrical. They wore principally on their mesial and distal edges, which were aligned to the direction of jaw movement. Such a tooth is admirably designed for the mode-III fracture of thin sheets, such as the lightly sclerotized cuticle of insects.

Most primates have retained at least a remnant of this triangular structure. Anthropoid primates have developed a large fourth cusp, have brought this up to the level of the other three cusps and have 'squared' off the tooth. Nevertheless, substantial blades can be developed on such a tooth, as seen in some New World monkeys (Kay & Hylander, 1978). These would produce mode-III fracture in leaves, with the advantage that the work done in chewing could be transferred much more effectively into work of fracture than if fracture was in modes I or II.

However, it is impossible to redesign this tooth form as a wedge to produce mode-I fracture of a thicker structure while still retaining the triangular form (Figure 6.9(*b*)), because the wedges would coalesce at one end of the tooth and interfere with their ability to start independent cracks at low forces. It is only with the abandonment of triangularity, to form two separate, symmetrically sided ridges, that true wedges can be produced (see Figure (6.9*c*)). Colobine monkeys have evolved such a structure, with the lophs fairly symmetrical about the direction of movement, with an even cross-section, and being generally flatter than would be expected if they were blades.

To substantiate our argument, we investigated the behavior of *M. atropurpurea* cotyledon, an anatomically typical seed storage tissue (Corner, 1976), under load, including a semi-imitative dental loading. The seed is large enough to permit good specimens to be made for mechanical tests and is from a plant genus whose seeds are eaten by Asian colobines. To obtain fracture toughness, we used wedge tests (Vincent, 1990) on thick specimens, and scissor tests (Lucas *et al.*, 1991) on thin slices (< 1 mm thick). These were successful in running cracks through the tissue, and the toughness was measured as 900–1200 J/m^2, which is very high, and attributable to the amount of plastic deformation. Cobalt–chrome casts of the buccal segments (canine–third molar) of the worn maxillary and mandibular teeth of a *Trachypithecus* sp. were mounted on a universal testing machine and loaded onto cubes of fresh *M. atropurpurea* cotyledon. From a combination of compressive tests between flat plates and toughness tests, we had established that the Young's modulus of this material was approximately 10 MPa, the yield strength 2.5 MPa and the fracture toughness, 900–1200 J/m^2. The toughness was extraordinarily high for such a low-density tissue, and yet it is anatomically typical seed storage parenchyma (Corner, 1976), attaining its toughness by yielding at a 30% compressive strain followed by plastic collapse (Gibson & Ashby, 1988). It did not fail until about 50% compressive strain and the crack then formed did not spread. It is therefore clear that relatively flat teeth or those with pointed cusps will not fragment this material. Blades or wedges are required. Loading the cubes of cotyledon with the teeth did fragment the cubes, and damage to a cube prior to fragmentation shows why (Figure 6.10). Instead of converting external work directly into work of fracture, which is what blades do, this external work was first stored as elastic strain energy. Despite deep plastic indentations of the cusps, lophs and crests in the cube (Figure 6.10(*a*)), cracks started well ahead of the teeth, with fracture almost entirely in mode I. This is consistent with a wedge-like action. Although the deepest cracks were made by the buccal crests of the lower and the lingual crests of the upper molars (Figure 6.10(*b*)), the lophs also started cracks running at right-angles to these (Figure 6.10(*a*)). In combination, they produced a considerable reduction in particle size from a single loading.

This is a change of perspective from more traditional discussions of colobine dental functional morphology (see e.g. Kay, 1975, 1977*a*, 1978; Maier, 1977; Kay & Hylander, 1978) where the cross-lophs on colobine bilophodont molars are presented as nothing more than guides to occlusion. From our

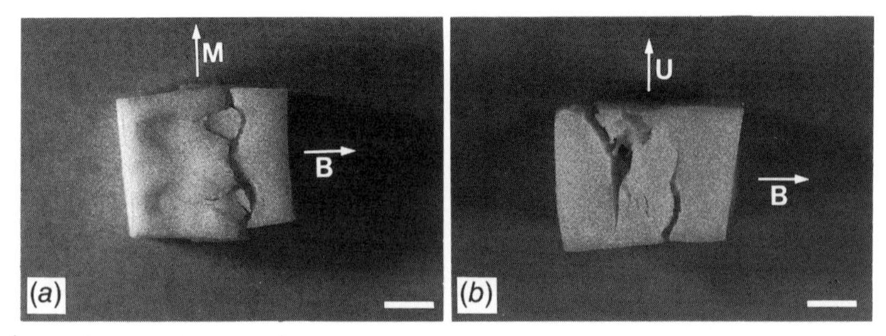

Figure 6.10. The result of loading a 6-mm edge-length cube of *M. atropurpurea* cotyledon with cobalt–chrome casts of the upper and lower molars of a *Trachypithecus* species (specimen no. ZRC 4.434 of the Zoological Reference Collection of the National University of Singapore) at 100 cm/min. Note that the cube was distorted by loading, so that its dimensions in the figure are no longer strictly the original. Letters give the orientation between the teeth: M = mesial; B = buccal; U = upper. (*a*) Occlusal view of the lower surface of the cube, showing extensive indentation by the cusps of the mandibular M2 with a 'ring' crack encircling the outline of the worn cusp tip. From the cusps, 'wedging' cracks have been produced by the buccal margin of this tooth and also by the lophs. (*b*) Mesial face of the cube showing the depth of the wedging cracks produced by the buccal crest of the lower and the lingual crest of the upper molars. Scale bar is 2 mm.

perspective, the cross-lophs can significantly aid in the fracturing of tough seeds, while (as suggested by the work of Kay and others) the more blade-like shearing crests along the buccal sides of the lower molars and the lingual sides of the upper molars can aid in the fracturing of both leaves and tough seeds. It may well be that the relative importance of crests and lophs changes with the reduction in food particle size occurring during mastication. In other words, crests, with their point contacts, might be most effective on larger food particles encountered earlier in the masticatory sequence (i.e. during puncture-crushing, to use the terminology of Hiiemae & Crompton, 1985) while lophs, with their line contacts, might be most effective on smaller food particles encountered later in the masticatory sequence (i.e. during chewing *sensu stricto*).

This combination of features makes mechanical sense in more ways than one. Since it is the fragmentation of foods that is crucial in chewing, following Ashby (1992) we have plotted the logarithms of two properties that, when taken together, give an indication of the fracture resistance of various food items and casings/coverings of food items (Figure 6.11). Dental tissues are also shown for comparison since they can also fracture. One of the properties is the elastic modulus (E) and the other is the intensity of the stress field in a solid close to the tip of a crack lying within it, or the critical stress intensity factor (K_{IC}). Very approximately, the fracture toughness is equal to K_{IC}^2/E. Lines of equal toughness are indicated on the graph by thin lines. There are

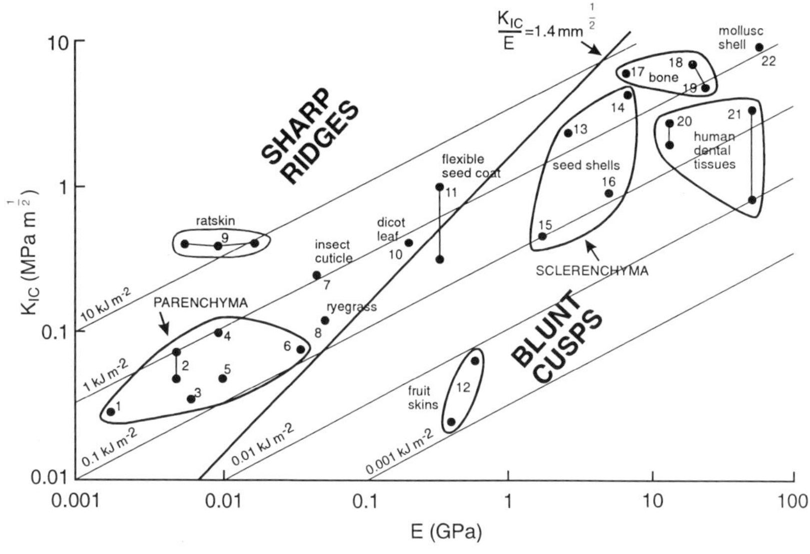

Figure 6.11. Food texture map (see text).

many potential complicating factors in the derivation and interpretation of such a figure, but one key point can be emphasized here: a certain value for the ratio of K_{IC}/E divides the sample of food items into those that crack easily ($K_{IC}/E < 1.4$ mm$^{1/2}$) and those that do not ($K_{IC}/E > 1.4$ mm$^{1/2}$). The former are generally known as stiff and brittle, whereas the latter are known as pliant and/or tough.

The selection pressures that act on primates with the dental types that we have described could now be suggested by referring to Figure 6.11. A primate with blunt-cusped teeth (e.g. a cercopithecine) might have its usual foods to the right of the thick line. In attempting to diversify its diet, it would find foods that lie close to the thick line difficult to fragment and could only process them at a significant cost of time and effort. A primate with sharp-ridged teeth (e.g. a colobine) that broadened its diet by approaching the thick line from the left, the domain of its usual diet, would be able to fragment foods to the right of the line (possibly at a faster rate than its usual dietary items) but would risk fracturing the sharp edges of the ridges. Such an animal would obviously need to be careful about the pattern of tooth contacts during chewing.

The proposed model also makes sense evolutionarily. Seed storage tissues, whether carbohydrate or lipid, are richer sources of energy than are young leaves (Davies *et al.*, 1988), and a greater surface area of tissue can be exposed by fracture per chewing cycle. However, seed storage tissues are also substantially tougher and therefore more difficult to chew than the brittle thin laminae of young leaves. Thus, an animal that could make efficient use of this resource might have a distinct advantage over an animal that could not use it. Furthermore, since protein is most easily obtained in forest environments through consumption of young leaf parts, the bilophodont molars of colobines could also give them access to a potential food resource which suffers less extreme seasonal changes in availability than fruit parts, and could also sustain colobines through periods of fruit scarcity.

Colobine and cercopithecine dental function compared

Both subfamilies of cercopithecid monkeys possess 'bilophodont' molars and yet colobines have very different diets as compared with cercopithecines. The important features of colobine molars that act in the fracture of leaves and seeds in the manner described above are de-emphasized in cercopithecines. Cercopithecine molars have blunter cusps, lower ridges or lophs and a much more constricted occlusal surface, which makes the ridges shorter.

Dental microwear comparisons shed light on some, but not all, of the functional implications of these differences. Initial analyses have shown that colobines have significantly more scratches and fewer pits on their molar facets as compared with cercopithecines (see Figure 6.12;) (Teaford, 1988; Teaford & Leakey, 1992). This might reflect a difference in the orientation of facets on colobine teeth, so that pits might be harder to form on colobine teeth for biomechanical reasons (Gordon, 1982, 1984). Another possibility, however, is that cercopithecines are processing more large, hard food items, more hard insects (Strait, 1991) or more hard abrasives in, or on, their food items (Ungar, 1992). In either case, some of the microwear patterns exhibited by colobines are just what one would expect, given the methods of food processing outlined above. Specifically, the incidence and orientation of scratches on phase-II facets would indicate that the colobines 'wedge' foods at the end of phase-I of chewing (in contrast to a heavier emphasis of puncture-crushing in cercopithecines).[6] In contrast, the incidence of scratches on phase-I facets probably cannot distinguish between the options of mode-III fracture versus mode-I wedging, as both might be expected to yield scratches. The fact that those scratches roughly parallel the orientation of normal

[6] Laboratory studies (see e.g. Hiiemae & Crompton, 1985; Hylander *et al.*, 1987) have shown that mammalian mastication can be broken down into two cycles: an initial cycle known as puncture-crushing, in which jaw movements are more vertical and tooth–food–tooth contact is the rule, and a subsequent cycle, known as chewing, during which jaw movements are more horizontal, and more precise tooth–tooth contacts are possible. In primates, as the teeth come into intercuspal range in chewing, two 'phases' are recognizable: phase I – as the tooth cusps and basins come together, and phase II – as the teeth move apart.

chewing movements indicates that these wear patterns probably occur during phase-I of chewing rather than when the teeth are far apart (i.e. during 'puncture-crushing').

As tooth wear progresses and enamel is worn away, areas of dentin are exposed, and differences in dentin exposure between colobines and cercopithecines are again what one might expect, given the model proposed here. Dentin exposure on cercopithecine molars quickly proceeds from small cusp tip exposures to broad dentin lakes surrounded by enamel rims (Welsch, 1967; Bramblett, 1969). In contrast, colobine molars have narrower dentin exposures focused more on the cross-lophs (Benefit, 1987) – probably due to an emphasis of occlusal contact along those lophs (Walker & Murray, 1975). Cercopithecine molars also lose much of their occlusal relief with wear. In contrast, even with large amounts of dentin exposed, colobine molars retain a fair amount of occlusal relief (Teaford 1983*a*; Benefit, 1987), as the crests on the lingual half of the lower crowns and the buccal half of the upper crowns remain readily identifiable.

In light of the above, this would seem to guarantee that tough seeds could continue to be processed by colobines even as wear progresses. Cercopithecines with heavy dental wear might be without such options and be forced to rely upon other types of food.

Theories of tooth size, which have yet to be tested, also support the hypothesis that colobine teeth are adapted for the consumption of seeds (Lucas, 1989). Some frugivorous primates, like gibbons, have much of the dentition disposed at the front of the mouth, because their dental arch is mesiodistally short and buccolingually wide. The focus of ingestion in these animals is often the removal of the peel, the thick protective covering of 'primate' fruits (Freeman, 1988). Once the fruit flesh and seed(s) are ingested, they are both quickly swallowed. The seeds are eventually passed out intact in the faeces. Such seed *swallowers* have relatively small dentitions, with diverging dental arcades. Symbolizing the size of anterior teeth as (A) and of postcanine teeth as (P), and using (+) to mean large and (–) to mean small, seed swallowers have A–P– dentitions. Such primates are limited in the range of fruits that they can handle by seed size. It is very difficult to swallow large seeds. To increase their dietary range, many primates process large seeds at the front of the mouth, holding them with the hands while taking off the flesh with the incisors, and then drop the seeds. These seed *cleaners* have similar dentitions to seed swallowers but larger incisors: they have A+P– dentitions.

Neither of these possibilities describe the dentitions of cercopithecines. These species have cheek pouches, which enable fruits to be harvested and stored prior to their return to the oral cavity, one by one, for removal of the

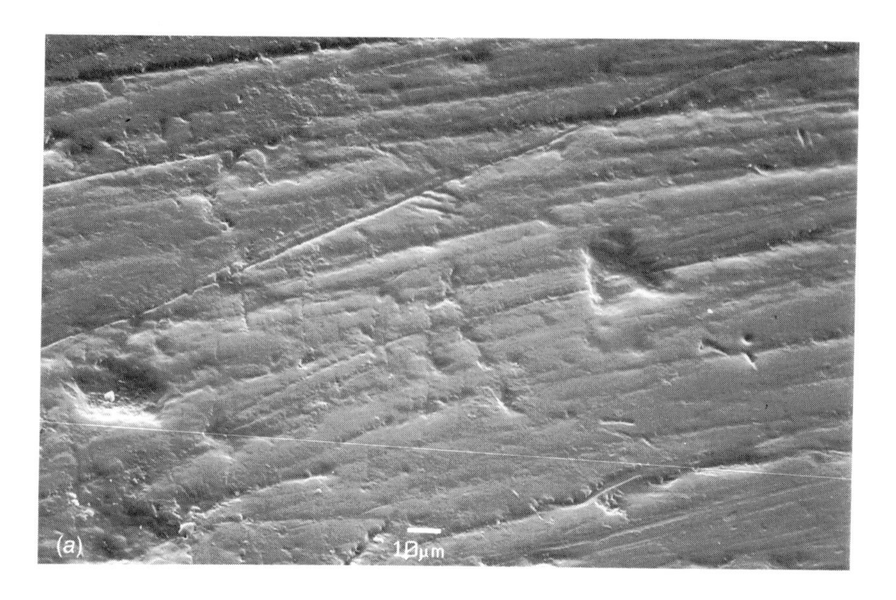
(a)
10μm

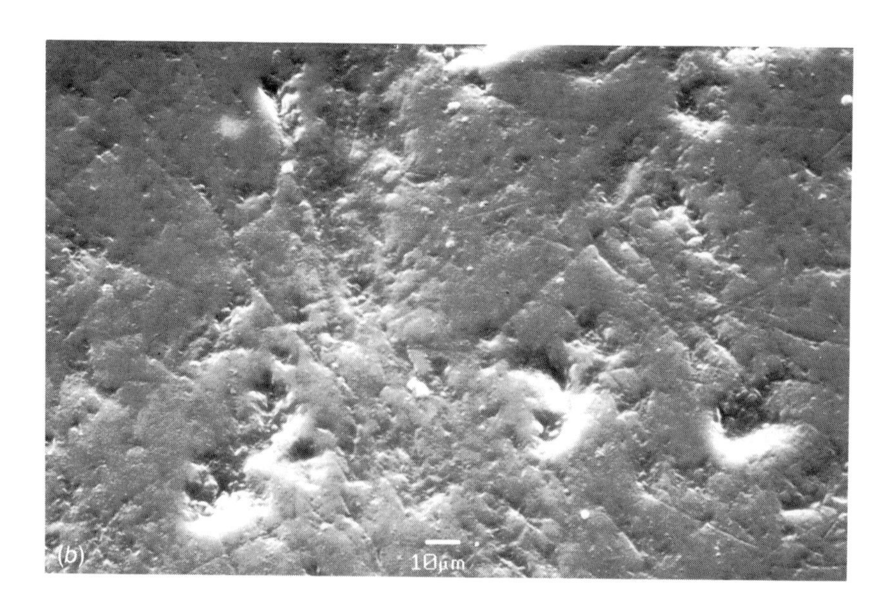
(b)
10μm

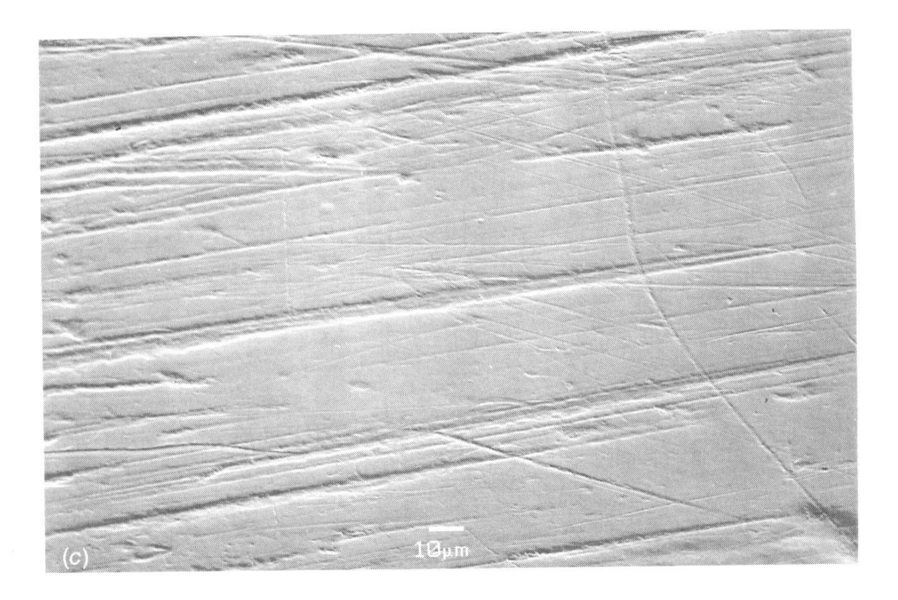

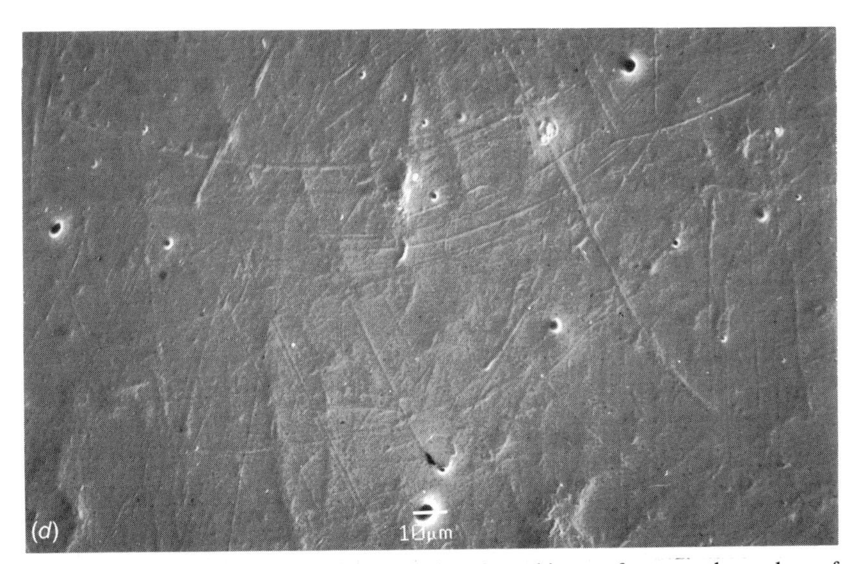

Figure 6.12. Scanning electron micrographs of crushing surfaces on the molars of modern cercopithecines and colobines. (*a*) *Macaca fascicularis*; (*b*) *Papio cynocephalus*; (*c*) *Nasalis larvatus*; (*d*) *Colobus guereza*. (*a* and *c* from Academy of Natural Sciences, Philadelphia; *b* and *d* from National Museums of Kenya, Osteology Collections).

fruit skin and fruit flesh. These animals do as much work in the front of the mouth, cleaning fruit skins, as they do inside the mouth, removing fruit flesh from the seeds. After cleaning off the fruit skin and flesh, they spit the seeds out (e.g. in long-tailed macaques analysed by Corlett & Lucas, 1990) – these seed *spitters* tend to have A+P+ dentitions.

Only one logical possibility is left: a primate with an A–P+ dentition. These are the seed *destroyers* (seed predators that use their teeth to break seeds), the colobines. Here the bulk of the work is done inside the mouth with the removal of the seed covering and the breaking up of the storage tissue. Colobines are often not dextrous enough to hold small objects in front of the mouth for incisal action. However, they probably do use the incisors to fracture larger, tough structures such as fruit pericarps by putting one end between the incisors, one end in a power grip in the hands, and pulling the hands and incisors away from each other (Davies, 1991).

Discussion

Our analysis of molar shape supports the hypothesis that colobines have not only evolved to consume leaves, they have also evolved to consume seeds. However, these seeds are not seeds with stiff hard coats made from fibres – such items would best be broken by a dentition characterized by blunt cusps rather than sharp ridges (Figure 6.11). Instead, the seeds in question are surrounded by relatively thin flexible coverings made from more isodiametric cells. Such coverings are not hard or tough. The contained storage tissues, however, *are* tough and very difficult to fragment. Wedges are necessary in order to fragment these tissues. In contrast, a young leaf is not tough and there is little structurally about it that suggests that it is very troublesome to chew. Its thin-walled tissue is similar to that of fruit flesh. The floppy sheet-like form of a leaf does not spread cracks, but the solution to this is to form blades, not wedges. The design criteria for blades and wedges differ in a clear and easily described manner. We interpret the colobine bilophodont molar as a pair of wedges bordered by blades along the buccal and lingual sides.

A logical, though semi-empirical, theory of tooth size has also been given that suggests that the relative sizes of the anterior and posterior teeth is a response to the destruction of seeds by chewing. However, it is possible on other grounds that such tooth sizes also suit leaf-eating (Lucas *et al.*, 1986*b*). Again, the key point for colobines remains that they can process both seeds and leaves.

Bilophodonty has evolved more than once. Primates such as *Propithecus* and *Alouatta* and phalangeroid marsupials possess an approximation to the true bilophodont form, and all are at least partly folivorous (Maier, 1977; Kay & Hylander, 1978). We suggest that a study of the shape of the lophs and their orientation relative to the direction of movement would indicate whether they are blades or wedges. This, in turn, might suggest whether these teeth are better suited for fracturing leaves or seeds.

Bilophodonty also evolved in early perissodactyls and is retained in living tapirs, which are largely folivorous (Collinson & Hooker, 1991). However, colobines are forestomach fermenters, not caeco-colic fermenters like perissodactyls. Caeco-colic fermentation appears to be adapted to a higher-fibre, and therefore tougher (Choong *et al.*, 1992) diet than forestomach fermentation (Alexander, 1991). It is curious that forestomach fermenting artiodactyls have selenodont, not bilophodont, molars. The fracture patterns on leaf specimens that are produced by such molars is complex (Wright, 1992).

Unfortunately, dental microwear comparisons of artiodactyls and colobines add little to this comparison. On the one hand, browsing artiodactyls generally show fewer microwear features and relatively more pits on their molars than do colobines (Teaford, 1988; Solounias & Moelleken, 1994). On the other hand, this is probably due to a combination of (1) biomechanical differences related to tooth shape, and (2) methodological differences between studies (i.e. micrographs were taken at distinctly different locations on the teeth).

The consumption of seeds with their bloated storage tissues seems to present a formidable mechanical challenge to the dentition. However, because the common ancestor of cercopithecoids was likely to have been a fruit-eater (Andrews, 1981; Collinson & Hooker, 1991), it was probably already ingesting seeds. The key question is 'Was it destroying them with its teeth?' Evidence from living primates, in particular *P. rubicunda* (Davies, 1991), suggests that seed predators show a real preference for large-seeded, fleshless fruits (contra Happel, 1988), and that treatment of seeds is strongly influenced by seed size (Corlett & Lucas, 1990).

Preliminary evidence from dental microwear analyses shows that the case for seed-eating in ancestral colobines is not as easy to prove as one would hope. *Victoriapithecus*, the Miocene predecessor of Plio-Pleistocene cercopithecoids, shows a mixture of microscopic pits and scratches on its teeth, as do a variety of modern primates with mixed diets (e.g. *Papio, Cebus, Cercopithecus*). This correlation is in accord with gross morphological analyses (Benefit, 1987), and it would certainly agree with the idea that the last common ancestor of cercopithecoids included significant amounts of fruit in its diet (Andrews, 1981; Collinson & Hooker, 1991). However, the Plio-Pleistocene cercopithecoids of Africa are another matter. They are the first monkeys that can be categorized into recognizable colobines and cercopithecines based on differences in tooth shape (Leakey, 1982; Benefit, 1987). However, their dental microwear is puzzling. The fossil geladas look fairly gelada-like – with subtle but significant differences between *T. brumpti* and *T. oswaldi* and also between *T. brumpti* and *T. gelada* (Teaford, 1993). The fossil colobines look fairly colobine-like, with perhaps more microwear on the teeth of the more terrestrial monkeys (Teaford & Leakey, 1992). However, the fossil cercopithecines, particularly *Parapapio*, do not look like their modern counterparts (Teaford & Leakey, 1992). Instead, they are more similar to the fossil colobines.

At the moment, there are three possible interpretations of these results. First, given the differences in dental morphology between the fossil colobines and cercopithecines, it may be that the biomechanical arguments presented in this chapter simply are not borne out by the evidence. On-going dental microwear work with living primates (see e.g. Teaford & Glander, 1991; Ungar, 1992) suggests that, while dental microwear analyses can detect subtle differences in diet and tooth use within and between species, interpretations are never easy. This may be due to the fact that most primate food items are not hard enough to scratch dental enamel (Lucas, 1991) (see Figure 6.13) and thus many dental microwear differences may be due to differing proportions of 'extraneous' grit in the diet (e.g. wind-blown dust, phytoliths, sand, etc.). With that in mind, two other options deserve serious consideration in interpreting the fossil results.

One would be that certain omnipresent extraneous abrasives had an overwhelming effect on dental microwear patterns exhibited by the Plio-Pleistocene fossil monkeys. For instance, all fossil forms may have been subjected to more wind-blown dust than are their modern counterparts. This might lead to relatively homogeneous microwear patterns across species, despite significant diet differences between species.

Another alternative would be that the Plio-Pleistocene *Papio*-like forms are indeed somehow different than their modern counterparts – for instance, more arboreal, or perhaps eating fewer plant roots. This might leave them without access to certain large-grained abrasives, such as sand, which might be causing a large proportion of the pitting on the teeth of modern baboons. From this perspective, early cercopithecoids may have been more like modern colobines, and modern cercopithecines may have changed in the interim. Clearly, further work on fossil and modern monkeys is needed before we can sort through these alternatives.

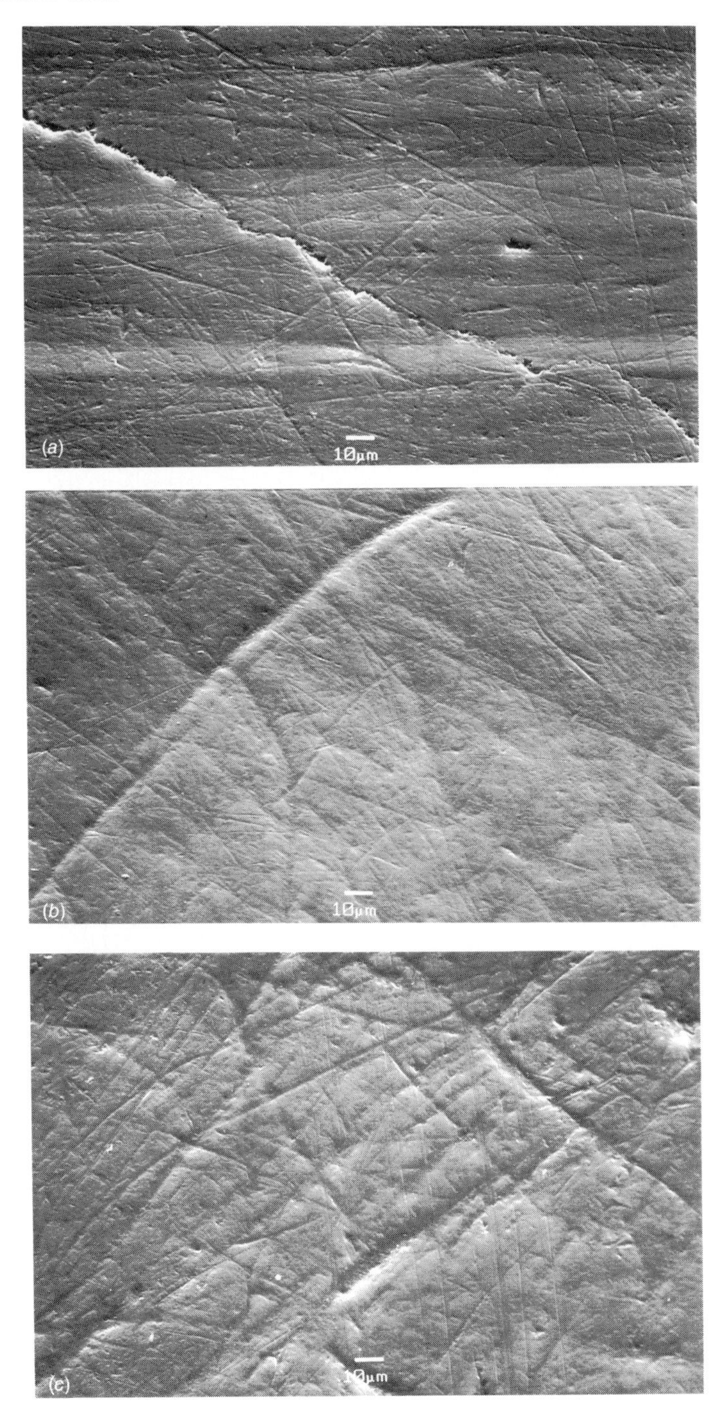
(a)
10µm
(b)
10µm
(c)
10µm

Figure 6.13. Scanning electron micrographs of crushing surfaces on the molars of Plio-Pleistocene colobines and cercopithecines. (*a*) *Rhinocolobus turkanaensis*; (*b*) *Parapapio*; (*c*) *Cercopithecoides williamsi*; (*d*) *Cercocebus*; (*e*) *Theropithecus oswaldi*. (From the National Museums of Kenya, Palaeontology Division).

Summary

Any use of the dentition routinely involves fracture, and the evolution of the dentition in any mammal is therefore usually a response to changes in the fracture properties of typical foods in the diet. Food structure is critical for understanding fracture because foods are heterogenous. However, fracture mechanics and fracture resistance are complicated topics, as is food breakdown in the mouth. Thus the results presented here are suggestive at best and merely meant to serve as preliminary indications of where further work might be done. Still, with all of these cautions in mind, one take-home message seems clear: colobine bilophodont molars make as much sense for processing seeds as they do for processing leaves. Were the earliest colobines seed-eaters? They probably were, although the seeds in question were probably not hard and brittle but instead pliant and tough.

Acknowledgements

Special thanks go to Glyn Davies and John Oates, not only for the invitation to contribute to this volume, but also for their patience and insightful comments during the review of the manuscript. We thank the Government of Kenya and the Governors of the National Museums of Kenya for allowing MFT to carry out research in Kenya at the National Museum. We also wish to thank David Burslem for help with the mechanical testing and Rose Weinstein for taking many of the SEM micrographs used in this paper. The following people kindly provided access to specimens in their care: Meave Leakey, Emma Mbua, Alfreda Ibui, Mary Muungu and Alice Maundu (National Museum of Kenya), Richard Thorington and Linda Gordon (Smithsonian Institution), Mrs C. M. Yang and her staff (Zoological Reference Collection of the National University of Singapore), and Fred Ulmer and Ted Deshler (Academy of Natural Sciences of Philadelphia). This work was supported in part by NSF grants BNS-8904327 & DBC-9118876.

7

Functional anatomy of the gastrointestinal tract

DAVID J. CHIVERS

Introduction

Monkeys of the sub-family Colobinae are unique among primates for the complexity of their stomach, partly as a response to the chemical problems of digesting leaves, as well as to neutralizing the effects of digestion inhibitors and toxins. These vegetative parts of plants are the commonest but least digestible of available foods, with much fibre and many secondary compounds (see Chapter 9).

Fermentative digestion in an enlarged stomach compartment is a strategy that has been adopted independently by several groups of mammals (Moir, 1968). Besides in colobine monkeys and about 170 species of ruminant artiodactyls (Ruminantia and Tylopoda), forestomach digestion is also well developed in some suids (peccary and hippopotamus), sloths and macropod marsupials. Only the ruminants 'chew the cud' (ruminate), so this process is evidently not an essential feature. A further perspective on the colobine condition is given by Langer's (1986) ecophysiological division of mammals with microbial fermenting systems into:

(1) Arboreal, caeco–colic-fermenting frugivores and browsers (concentrate feeders), e.g. ceboid primates, marsupials, tree hyrax;
(2) Arboreal, forestomach-fermenting frugivores/folivores, e.g. sloths, colobine monkeys, tree kangaroos;
(3) Terrestrial, caeco–colic-fermenting 'intermediate feeders' (frugivore/folivores), e.g. pigs, horses, proboscids, non-ruminant artiodactyls; and
(4) Terrestrial, forestomach-fermenting 'bulk/roughage' feeders (folivores), e.g. peccaries, ruminant artiodactyls, macropod marsupials.

There are three important misconceptions prevalent in the literature that need correcting. First, no mammal has more than one stomach; what varies is the complexity of the stomach, so that 'compound' and 'simple' are more correct than 'polygastric' and 'monogastric', respectively. Second, it is not just the caecum that provides a fermenting chamber ('caecal fermentation'); the first part of the colon (primitive right colon, embryologically) is equally, often more, important, since it is usually, especially in the primates, much more voluminous than the caecum. Third, these parts of the gut are not 'hindgut', which refers only to the descending (left) colon and rectum, which are very similar in all mammals, and have a different blood supply (because of the different embryological origin) from the caecum and right colon, which develop from the 'midgut' loop. Hence, the need to refer to 'caeco–colic', 'midgut' or 'intestinal' fermentation, in contrast to 'forestomach' fermentation.

Comparative anatomy of the gut

The basic descriptions of colobine monkey digestive anatomy and physiology are those of Hill (1958), Kuhn (1964), Bauchop & Martucci (1968), Bauchop (1978), Parra (1978) and Langer (1986) with special comparisons with ruminants and other mammals with enlarged stomachs, and to mammals with expanded caecum and colon (as the alternative, or additional, site for the microbial fermentation of cellulose).

The pioneering review of the anatomy of the primate gastrointestinal tract (Figure 7.1) was by Hill (1958). He described the stomach as a single smooth-walled sac in cercopithecine monkeys, in contrast to the more globular sac of apes, and the larger, more complex stomach of colobine monkeys, with much distension and sacculation proximally, and a U-shaped tube distally (sacculated along the proximal part of the greater curvature). These sacculations are produced by the reduction of the longitudinal muscle into bands, a design feature of the wall of the large intestine in specialized folivores (seen also in Old World primates). Kuhn (1964) described the stomach of African colobines as more elongated, with the fundic and pyloric parts of the distal tube bent back more noticeably on the sac than in Asian colobines, where the sac is more spherical.

In the Old World monkeys, the duodenum – the first part of the small intestine – is C-shaped, in contrast to the elongated U-shape of other mammals, and retroperitoneal (not suspended in a mesentery). The long coils of jejunum and the short ileum are basically the same in all species, and the colon – the main part of the large intestine – is much elongated and sacculated compared to that of New World monkeys, especially those eating more leaves. The caecum has a globular base, a short and capacious body and a

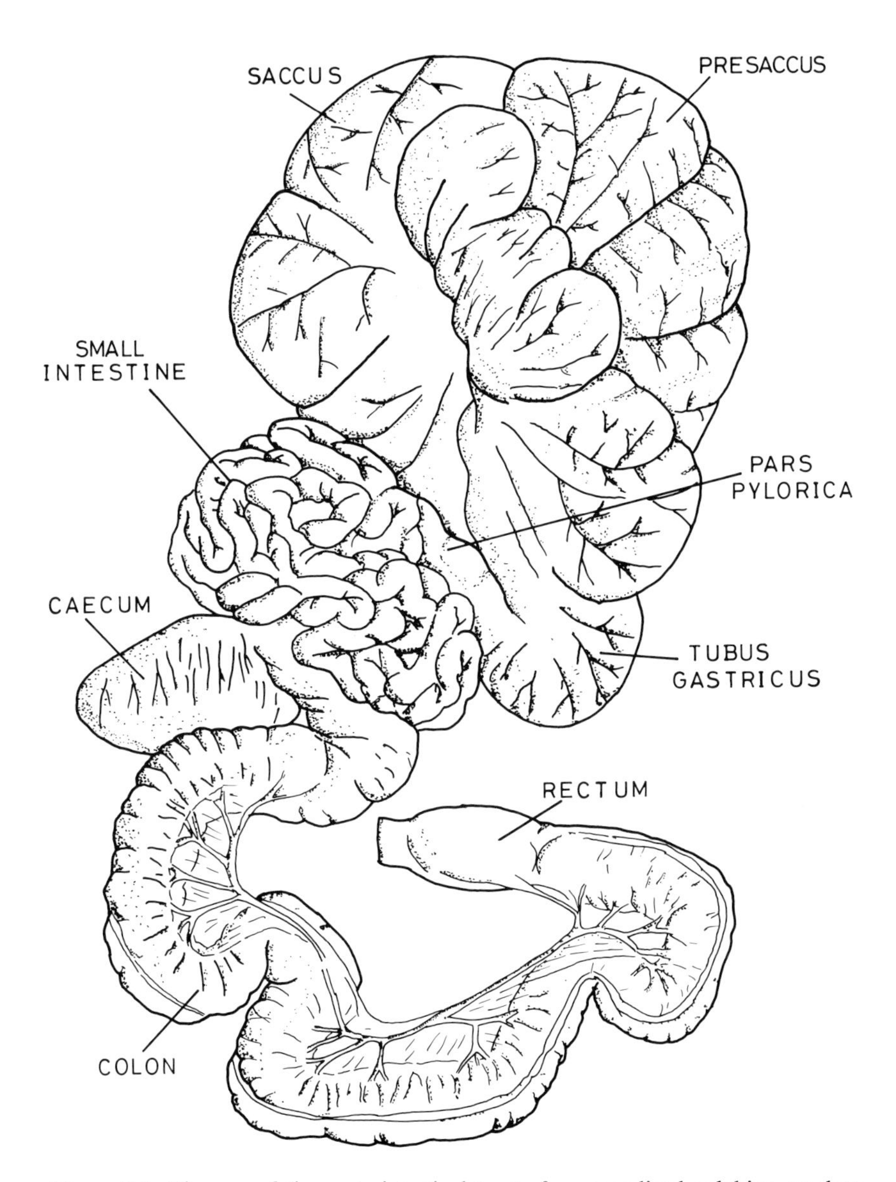

Figure 7.1. Diagram of the gastrointestinal tract of a generalized colobine monkey (after Chivers, 1992).

blunt but conical apex, with a terminal vermiform appendix (compact mass of lymphoid tissue only in hominoids (Hill, 1958)).

Gut complexity and increased body size are correlated with increased folivory in primates, as quantified by Chivers & Hladik (1980). Whereas the large intestine is much expanded and the small intestine reduced in folivorous prosimians – converging with specialist folivores such as lagomorphs and equids – it is the stomach that is much expanded in colobine monkeys as the site for microbial fermentation – converging with artiodactyl ruminants, as well as some edentates, hippopotamus, camelids, proboscids and macropod marsupials. The large intestine is also expanded, as in the more folivorous cercopithecine monkeys (e.g. *Macaca sylvanus*) and hominoids (e.g. *Hylobates syndactylus, Gorilla gorilla*), as well as in the larger ceboids (e.g. *Alouatta* spp.).

Andrews & Aiello (1984) argue that the ancestral Old World monkeys veered towards folivory in more open habitats to avoid competition with the diverse and frugivorous arboreal apes of the Miocene. Whereas the cercopithecid monkeys, having acquired some tolerance to secondary compounds in plant parts other than ripe fruit, were then able to out-compete the apes, mainly by eating unripe fruit to which the apes lacked such tolerance, the colobine monkeys developed leaf-eating and leaf-digesting abilities much further, including tolerance to secondary compounds. As indicated by present-day diets, they may also have focused on seed-eating, which would make tolerance or detoxification of secondary compounds even more important.

As in ruminants, the colobine stomach is divided into four chambers, but here the similarity ends. There is no direct equivalent of the reticulum and omasum, or of the rumen, in terms of elaborate muscular pillars dividing and shaping dorsal and ventral sacs, or of dense large papillae increasing the surface area as much as ten-fold for the absorption of volatile fatty acids (the main product of microbial fermentation). The stomach of colobine monkeys has a sacculated and expanded fermenting chamber – the presaccus and saccus – a long tubus gastricus and a short pars pylorica (Kuhn, 1964). In the absence of a stratified epithelium and papillae, the sacculated part of the stomach, and the first two-thirds of the gastric tube, are lined with cardial glands, found at the entrance to the ruminant abomasum, which is otherwise lined with fundic and, distally, pyloric glands. Bauchop (1978) emphasizes how the structure of the colobine stomach allows the separation of ingesta between proximal and distal parts, and thus between alkaline and acidic environments. Nevertheless, the well-defined gastric canal from the oesophagus along the lesser curvature of the stomach might allow rapid passage of liquid and small food particles to the distal glandular part of the stomach (Kuhn, 1964).

Parra (1978) demonstrated cautiously that there were no major differences in the capacity of the gastrointestinal tract or fermentation contents between foregut and caeco–colic fermenters. He also stressed the functional importance of passage time in interpreting gastrointestinal anatomy, which is affected by factors such as gut capacity and architecture, intake level, diet composition and rate of digestion.

It is against this background that the gross anatomy of the gastrointestinal tract of colobine monkeys is now examined. Apart from the original, quantitative analyses of Chivers & Hladik (1980), the relationship of colobine monkeys to other primates, and to mammals in general, can be discerned from R. D. Martin *et al.* (1985) and MacLarnon *et al.* (1986). Furthermore, in their comparison of gut morphology and diet in New World monkeys, J. M. C. Ayres and D. J. Chivers (unpublished observations) draw comparisons with Old World monkeys, in which the distinctions between colobine and cercopithecine monkeys are highlighted.

Colobine material

The gastrointestinal tracts of 26 colobine monkeys have been measured (22 described and analysed by Chivers & Hladik, 1980), 17 from the wild and nine after death in captivity (Table 7.1). These represent nine species – one colobus, six langur/leaf-monkeys and two odd-nosed monkeys.

In each case, body weight and length were recorded, and each part of the gastrointestinal tract – stomach, small intestine, caecum and colon – were weighed after removal of the gut contents. The surface area of each compartment was calculated from measurements of length and breadth; volumes were calculated on the assumption that the gastric (fundic) and pyloric parts of the stomach, and the small and large intestines, are cylindrical, while the presaccus and saccus are regarded as spherical. *Volume* is relevant to fermenting capability, *surface area* to absorptive ability and *weight* to muscular activity required for digestion. The functional surface area for absorption must involve more than that of the small intestine, therefore, on the basis of best-fit models (Chivers & Hladik, 1980), half the surface areas of stomach and large intestine were added.

An indication of the relative importance of each part of the gut is achieved by considering area, weight or volume as a percentage of the gut total, especially in monkeys of similar size. More accurate comparisons are achieved,

Table 7.1. *Sources of data on the anatomy of the colobine gastrointestinal tract. Imm, immature (♂ and ♀)*

	Wild			Captive			All			
	♂	♀	Imm	♂	♀	Imm	♂	♀	Imm	Total
Colobus guereza										
guereza monkey		2[a]		1			1	2		3
Semnopithecus entellus										
Hanuman langur	2[a]						2			2
Trachypithecus vetulus										
purple-faced langur	1[a]	1[a]					1	1		2
T. cristatus										
silvered langur		1			1			2		2
T. obscurus										
dusky langur	2		2		2		2	2	2	6
Presbytis melalophos										
banded leaf-monkey	2	3	1				2	3	1	6
P. rubicunda										
red leaf-monkey				1	1		1	1		2
Pygathrix nemaeus										
douc monkey					2			2		2
Nasalis larvatus										
proboscis monkey				1			1			1
Total	7	7	3	3	6	0	10	13	3	26

[a]Data from C. M. Hladik.

however, by relating the size (area, weight or volume) of each compartment to body size – *Nasalis* and *Semnopithecus* are larger than the rest, *Pygathrix* is smaller.

Colobine anatomy

Comparison of wild male and female *Presbytis melalophos* showed no significant differences in the surface area, weight and volume of each gut compartment; body size was very similar (Table 7.2). The stomach was slightly smaller in the male and the large intestine slightly larger in the female. Despite similar body size and gut proportions, each compartment was absolutely larger in females in all respects. While one of the three females was far advanced in pregnancy, its gut compartments were smaller than those of the non-pregnant females.

Table 7.2. *Comparison of mean gut dimensions in wild male and female* Presbytis melalophos *and wild male (and immature female) and captive female* Trachypithecus obscurus

	Body weight (kg)	Body length (cm)	Intestinal length (cm)	Surface areas (cm^2)					Weight (g)				
				St	SI	C	Col	Total	St	SI	C	Col	Total
P. melalophos													
2 wild ♂	6.7	50	543	1049	1388	51	614	3102	102	66	5	47	220
(%)				34	45	2	20		46	30	2	22	
3 wild ♀	6.9	50	585	1435	1837	55	671	3998	138	79	7	61	285
(%)				36	46	1	17		46	28	2	22	
5 wild adults	6.8	50	564	1281	1657	54	647	3639	125	74	7	56	262
(%)				35	46	2	18		48	28	3	21	
T. obscurus													
2 wild ♂ 2 wild immature ♀	5.8	48	487	1151	1504	66	648	3369	138	63	6	59	266
(%)				34	45	2	19		52	24	2	22	
2 wild ♂	7.6	52	540	1357	1961	76	786	4180	167	70	7	71	315
(%)				32	47	2	19		53	22	2	23	
2 captive ♀	5.7	58	444	1403	1194	77	1132	3806	188	44	7	64	303
(%)				37	31	2	30		62	15	2	21	
4 adults (2 wild, 2 captive)	6.6	55	492	1380	1578	77	959	3994	178	57	7	68	310
(%)				35	40	2	24		57	18	2	22	

Note: St, stomach; SI, small intestine; C, caecum; Col, colon.

The captive females of *Trachypithecus obscurus* were longer but lighter than the wild males, probably a result of ill-health prior to death. Gut dimensions were surprisingly similar, however, although the wild males had smaller stomach and colon and larger small intestine (Table 7.2). These differences were supported when two wild immature females were included with the two wild males, so the differences would seem to be a consequence of captivity, and perhaps ill-health, rather than sexual dimorphism.

Despite the dietary differences (see below) the gut proportions (by area, volume and weight) were almost identical between the four wild (two immature) *T. obscurus* and the five wild *P. melalophos*.

Comparison of wild and captive female *T. cristatus* revealed few differences in compartment area and volume – only a lighter small intestine and heavier colon in the wild one. The captive male *P. rubicunda* had a proportionately larger stomach and smaller intestine than the captive female. The heavier captive female *C. guereza* had a proportionately larger stomach and colon than the two wild ones.

After calculation of the area, weight and volume of each gut compartment, and the production of a mean score for each species (Table 7.3), the following comparative observations can be made. These are augmented by comparisons (Table 7.4) of: gut (intestinal) length (GL) in relation to body length (BL); gut weight (GW) in relation to body weight (BW); functional absorptive area (FAA) and potential fermenting volume (PFV) as proportions of gut total area and volume, respectively, and in relation to body size (BL and BW, respectively); and coefficients of gut differentiation (CGD) for area, weight and volume – the ratio of measures for potential absorbing compartments (stomach, caecum and colon) to the main absorbing compartment (small intestine).

Guereza (Colobus guereza)

The stomach was large in area and volume, but of low weight; the small intestine scored low in all parameters, but the colon was large, scoring highest of the nine species for area. The small intestine was the shortest, relative to intestinal length, the caecum and colon the longest. The FAA was very low, PFV medium, but the lowest in relation to BW; the ratio of GL/BL was somewhat low, and GW/BW was medium.

Table 7.3. *Mean gastrointestinal tract dimensions of colobine monkeys*

	n	Body weight (kg)	Body length (cm)	Intestinal length (cm)	Surface areas (cm^2)					Weight (g)					Volume (cm^3)		
					St	SI	C	Col	Total	St	SI	C	Col	Total	St	C + Col	Total
Colobus guereza	3	8.5	62	483	1275	1053	36	932	3296	190	99	6	69	364	2818	514	3615
(%)					39	36	3	28		52	27	2	19		78	14	
Semnopithecus entellus	2	(10)	64	445	1512	1420	123	824	3879						3706	766	4955
(%)					39	37	3	21							75	15	
Trachypithecus vetulus	2		54	334	1177	718	139	548	2582						2586	393	3149
(%)					46	28	5	21							82	12	
Trachypithecus cristatus	2	6.1	52	540	1435	1629	60	787	3911	224	63	7	79	373	3520	503	4518
(%)					37	42	2	20		60	17	2	21		78	11	
Trachypithecus obscurus	6	5.8	51	472	1235	1401	73	809	3518	155	57	6	61	279	2665	527	3650
(%)					35	40	2	23		56	21	2	22		73	14	
Presbytis melalophos	6	6.5	49	547	1183	1560	49	601	3393	122	74	6	54	256	2417	299	3168
(%)					35	46	1	18		48	29	2	21		76	9	
Presbytis rubicunda	2	5.2	52	548	922	1553	50	607	3132	123	50	10	46	229	2259	337	3029
(%)					29	50	2	19		54	22	4	20		75	11	
Pygathrix nemaeus	2	4.1	57	526	1337	1557	58	716	3668	169	55	5	48	277	3201	374	4063
(%)					36	42	2	20		61	20	2	17		79	9	
Nasalis larvatus	1	15.9	64	886	1978	3120	100	1234	6432	357	153	6	82	598	6523	721	8371
(%)					31	49	2	19		60	26	1	14		78	9	

Note: St, stomach; SI, small intestine; C, caecum; Col, colon; parentheses indicate estimate.

Table 7.4. *Features of gut dimensions in colobine monkeys*

	Gut length		Gut weight		Absorptive area[a]			Fermenting volume[b]			Coefficient of gut differentiation[c]		
	GL (cm)	GL/BL	GW (g)	% of BW	AA (cm^2)	% of total	AA/BL	FV (cm^3)	% of total	FV/BW	Area	Weight	Volume
Colobus guereza	483	7.8	363	5.8	2175	68	41	3332	92	392	2.13	2.68	11.77
Semnopithecus entellus	445	7.0			2650	68	41	4473	90	447	1.73		9.28
Trachypithecus vetulus	334	6.2			1650	64	31	2977	95		2.60		17.22
Trachypithecus cristatus	540	10.4	372	6.1	2770	71	53	4023	89	660	1.40	4.92	8.13
Trachypithecus obscurus	472	9.3	278	4.8	2460	70	48	3192	87	550	1.51	3.89	6.97
Presbytis melalophos	547	11.2	256	3.9	2477	73	51	2715	86	418	1.18	2.46	6.00
Presbytis rubicunda	548	10.5	229	4.4	2343	75	45	2598	86	499	1.02	3.58	6.00
Pygathrix nemaeus	526	9.2	276	6.7	2613	71	46	3575	88	872	1.36	4.04	7.33
Nasalis larvatus	886	13.8	598	3.8	4776	74	75	7244	87	456	1.06	2.91	6.43

Notes: GL, gut length; BL, body length; GW, gut weight; BW, body weight, AA, absorptive area; FV, fermenting volume.

[a]Area of small intestine + $\frac{1}{2}$ (stomach + caecum + colon).

[b]Volume of stomach + caecum + colon.

[c](Stomach + caecum + colon)/small intestine.

Hanuman langur (Semnopithecus entellus)

The stomach was moderately extensive in area, but the volume was low. The small intestine scored low for area and volume, but the large intestine scored the highest for volume (as percentage of gut total). The intestine was of medium length, but very short in relation to BL; the FAA was low, the PFV medium, but both were low in relation to body size.

Purple-faced langur (Trachypithecus vetulus)

The stomach scored highest in terms of area and volume, the small intestine the lowest of the nine species (in proportion to the gut total). The caecum and colon had a large area, but only an average volume; although the colon was long, the small intestine was short, and the ratio of GL/BL was the lowest of the nine species. The FAA was also the lowest, but PFV was quite high.

Silvered langur (T. cristatus)

The stomach was of average area, large volume, but the heaviest weight of the nine species (relative to gut totals). The small intestine was of average area and volume, but average weight. The small intestine was quite long, the caecum and colon short, but the ratio of GL/BL was quite high. The FAA was average, but high relative to BL; the PFV was high, as was the ratio of GW/BW.

Dusky langur (T. obscurus)

The stomach was of average area and weight, but of lowest volume. The small intestine was average, but the large intestine was the heaviest of the nine species, with high scores for area and volume, relative to gut totals. The intestine was of medium value, as were all the other parameters.

Banded leaf-monkey (Presbytis melalophos)

The stomach scored lowest of nine species for weight, relative to the gut total; the area was low, the volume average. The small intestine dominated, with the highest weight and volume, and a large area, whereas the large intestine scored lowest for area, was of small volume, but quite heavy. The long small intestine and short caecum and colon were reflected by a high ratio of GL/BL. The FAA was average, high in relation to BL; the PFV was quite high, but very low in relation to BW; the ratio of GW/BW was low.

Red leaf-monkey (P. rubicunda)

The stomach was relatively small by all measures, as was the caecum and colon (apart from weight), but the small intestine had the largest relative area and volume of the nine species, although only of average weight. The small intestine was the longest part of the gut, of the nine species, the caecum and colon the shortest; the ratio of GL/BL was average. The FAA was average; the PFV was somewhat low, but the highest relative to BW (it is the smallest *Presbytis* species).

Douc monkey (Pygathrix nemaeus)

Although of only average area, the gut was relatively the heaviest and most voluminous of this smallest of the nine species. The small intestine was of average dimensions, although rather light. The caecum and colon scored low for area, weight and volume. The FAA was average, the PFV somewhat low, but the highest relative to BW, as was the ratio of GW/BW; GL/BL was average.

Proboscis monkey (Nasalis larvatus)

The stomach, in this largest of the nine species, had the smallest area relative to gut total, but it was quite heavy and of only average volume. The small intestine was relatively the largest of the nine species in terms of area; it was quite heavy and voluminous. The caecum and colon had a small relative area and low volume; they were the lightest of the nine species. The ratio of GL/BL was the highest of the nine species, although the caecum and colon were rather short. The FAA was very high, the highest of the nine species relative to BL, but the PFV was low, and the ratio of GW/BW was very low.

Gut morphology

It is important to remember that *Nasalis*, *Pygathrix* and *Presbytis rubicunda* (the last very similar to the closely related *P. melalophos*) were represented only by captive specimens; thus, their distinctive features, given the malleability of the gastrointestinal tract, may partly reflect captivity (and terminal ill-health) rather than adaptations to wild diets. Chivers & Hladik (1980), however, found surprisingly little difference between wild and captive colobines (compared with more frugivorous primates).

The gut and body size

After it was shown (R. D. Martin *et al.*, 1985) that the surface areas of stomach, small intestine, and caecum plus colon (but not colon alone) scale to body weight according to Kleiber's Law, whereby larger animals have lower metabolic needs per unit body weight than small ones, lines with fixed slope of 0.75 (representing the three-quarters power of body weight) were fitted to the data. Quotients were calculated for each gut compartment for each species, according to the deviation (positive or negative) from the expected value for that body weight. Log indices were found to be preferable to absolute indices, because they treat equally both positive and negative deviations from the fixed slope (absolute indices are strongly biased in favour of gut compartments that are larger than expected).

Colobine monkeys had significantly larger stomachs than expected (Figure 7.2), forming a distinct grade along with ruminant artiodactyls, sloths and kangaroos; quotients for the small intestine were clustered just below the regression line, and those for the colon even more so, but that is typical of primates.

The gut and diet

Examination of the dietary data (Table 7.5) reveals *Trachypithecus, Nasalis* and *Colobus* to be more folivorous (according to the populations sampled), and *Presbytis* and *Semnopithecus* to be more frugivorous. There are no obvious morphological correlates of this dichotomy, apart from an increased potential fermenting volume in the more folivorous species, which are, apart from *Semnopithecus*, the largest. In these species, stomach weight and colon volume tend to be larger than in the more frugivorous species, perhaps indicative of a secondary fermenting role for the distal part of the gut. Otherwise, there is considerable overlap in the morphological parameters analysed. From these data, *Pygathrix*, despite its small body weight (not length), would appear to be in the more folivorous category. It is generally the smaller species, especially *P. rubicunda*, that eat more seeds.

There is a distinctive increase in stomach size in relation to the amount of foliage in the diet – from *Presbytis*, through *Trachypithecus* and *Nasalis*, to *Colobus* (Figure 7.3(*a*)) The relationship between small intestine area and frugivory highlights the expansion of the gut in *Presbytis*, with *Colobus* aberrant, because of small area and low intake (Figure 7.3(*b*)).

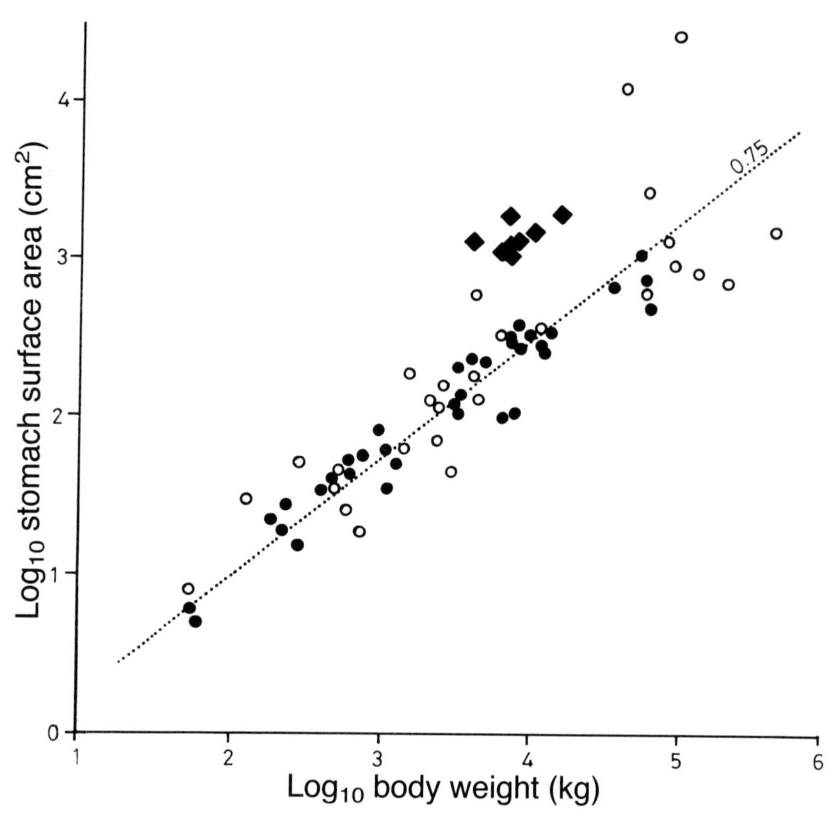

Figure 7.2. Relationship between stomach size and body size, showing colobines (diamond points) consistently with relatively large stomachs (after R. D. Martin *et al.*, 1985).

Colobine diets, in comparison with those of other primates (Figure 7.4), are characterized by very little animal matter and high intake of foliage, varying to moderate intake of foliage and moderate intake of fruit, mainly unripe, or seeds. The convergence with ceboids is interesting here. In both the seed-eating adaptations of pithecines and the leaf-eating adaptations of ceboids such as *Alouatta* spp., the enlargement of the caecum and colon must be functionally significant.

The gut compartment quotients were subjected to multi-dimensional clustering techniques, based on the derived matrix of Euclidean distances between them (R. D. Martin *et al.*, 1985). While dendrograms (branching relationships) distinguished clearly between primate and non-primate fore-stomach fermenters and caeco–colon fermenters, mammals with less spe-

Table 7.5. *Percentage diet of colobine monkeys*

	Leaves	Flowers	Fruit	Fruit + seeds	Seeds	Other	Animals	Source
Colobus polykomos	58	3	3	32		3		Dasilva (1989)
Colobus guereza	81	2		14		2	–	Oates (1977*a*)
Semnopithecus entellus	48	7		45		0	–	Hladik (1977*b*)
Trachypithecus vetulus	60	12		28		0	–	Hladik (1977*b*)
Trachypithecus auratus	54	14	14		13	5	–	Kool (1992)
Trachypithecus obscurus	58	7	32		3	0	–	Curtin (1980)
	49	13	26		10	2	–	Hardy (1990)
Presbytis melalophos	35	6	48		8	3	+	Curtin (1980)
	34	17	9	11	26	3	–	Bennett (1983)
Presbytis rubicunda	37	12	19		30	2	+	Davies *et al.* (1988)
Nasalis larvatus	47	3	35		15	0	–	Bennett (1986*a*)
	52	3	20		20	5	+	Yeager (1989)

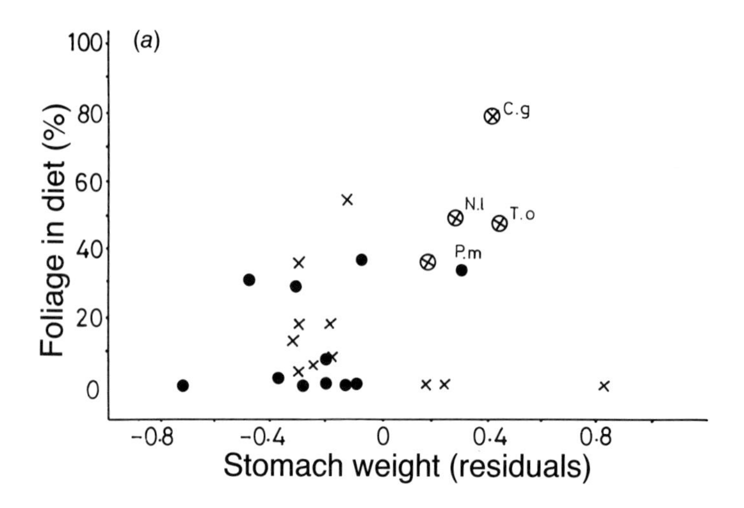

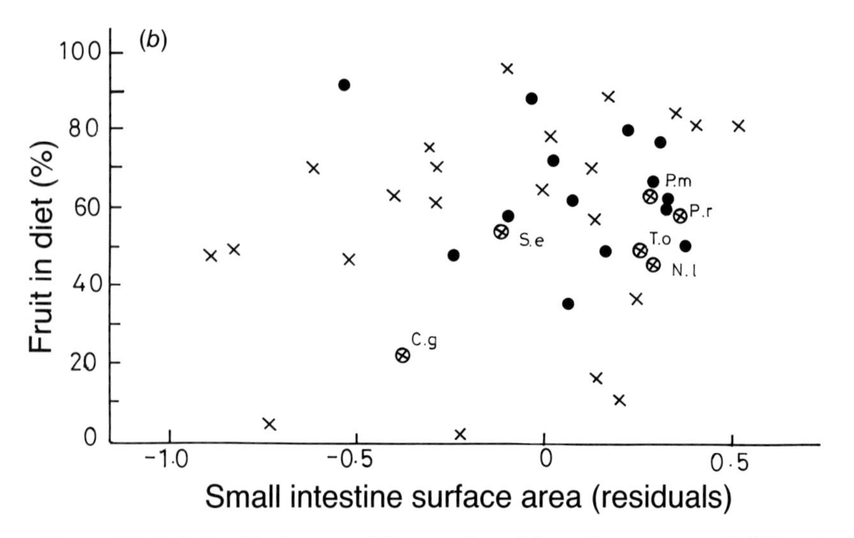

Figure 7.3. Relationship between (*a*) stomach weight and percentage of foliage in annual diet and (*b*) small intestine surface area and percentage of fruit parts in annual diet of primates. New World, solid circle; Old World, cross; colobines, cross in circle; C.g., *Colobus guereza*; N.l, *Nasalis larvatus*; P.m, *Presbytis melalophos*; P.r, *P. roxellana*; S.e, *Semnopithecus entellus*; T.o, *Trachypithecus obscurus*. From Ayres and Chivers, in preparation).

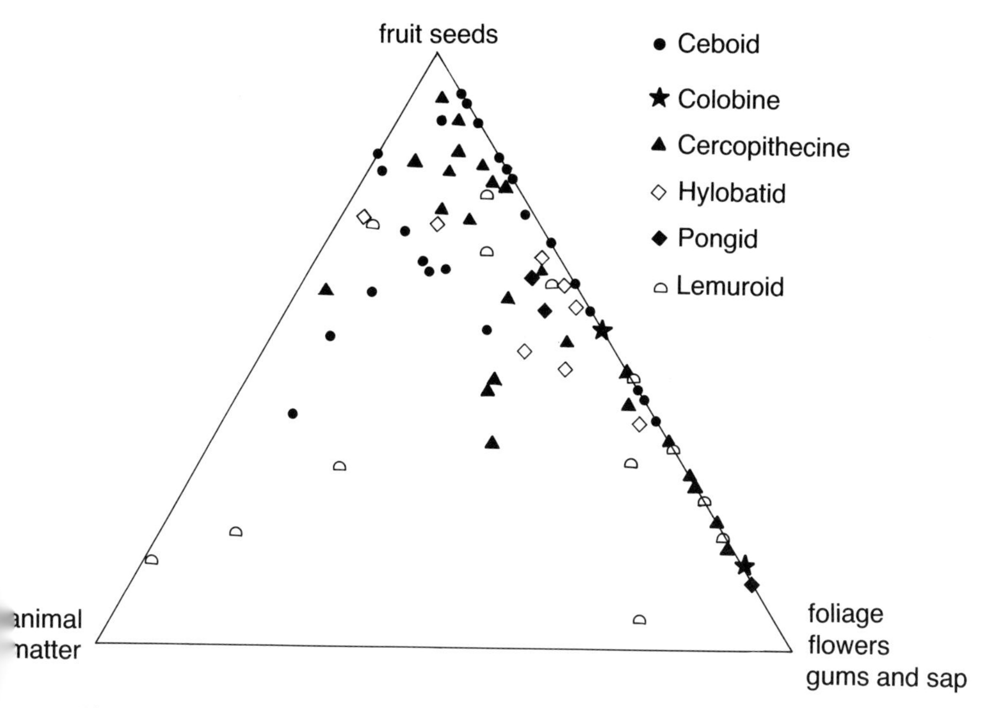

Figure 7.4. Proportion of three types of food item in the annual diet of primates (from Chivers, 1992).

cialized diets, including faunivores (consumers of animal matter), were difficult to classify. Furthermore, altering the data-set in this way introduced an unacceptable level of lability; the addition or subtraction of species' values changed groupings among the non-folivores for no good biological reason.

Multi-dimensional scaling, however, proved much more robust (R. D. Martin *et al.* 1985). Plots spread out in three directions from the central core of 'frugivores' (Figure 7.5) – foregut fermenters (below), midgut fermenters (upper left) and faunivores (upper right). Since log indices are undefined for caecumless species, they can only be included if values for caecum and colon are combined (which makes more sense in functional terms).

In their study of gut morphology and diet in Neotropical primates, J. M. C. Ayres and D. J. Chivers (unpublished observations), included Old World primates for comparison. This highlighted the dramatic increase in stomach area with increasing stomach weight from plots of residuals (the compartment quotients described above); simple-stomached Old World monkeys may have heavier stomachs, but there is little increase in area (Figure 7.6(*a*)). Compared

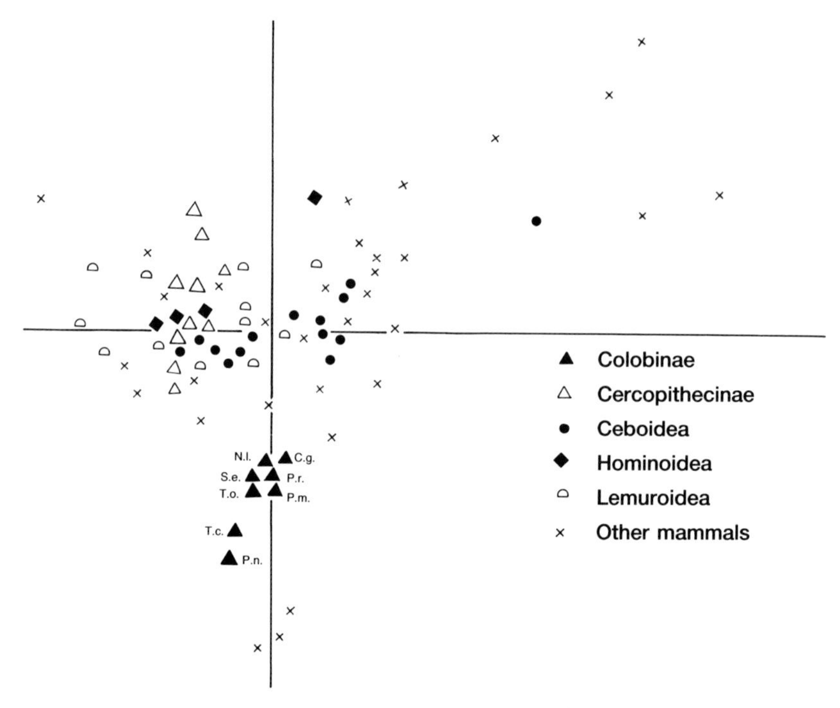

Figure 7.5. Multi-dimensional clustering of gut compartment quotient measures, showing distinctive overall structure of colobine guts. C.g, *Colobus guereza*; N.l, *Nasalis larvatus*; P.m, *Presbytis melalophos*; P.n, *Pygathrix nemaeus*; P.r, *P. roxellana*; S.e, *Semnopithecus entellus*; T.c, *Trachypithecus cristatus*; T.o, *T. obscurus*.

with New World monkeys, Old World monkeys show a very variable relationship between small-intestine area and weight, probably reflecting their more variable diets. While most colobines cluster with large area and weight, *Colobus guereza* has a small area but large weight, and *Presbytis rubicunda* has a large area but small weight.

New World monkeys have particularly extensive and heavy caeca, but *C. guereza* and *P. rubicunda* again fall well away from the colobine cluster (Figure 7.6(*b*)). There is a distinctive increase in colon weight and area among the primates studied, with colobines generally showing even greater area per unit weight, especially in *Nasalis* and *P. rubicunda*.

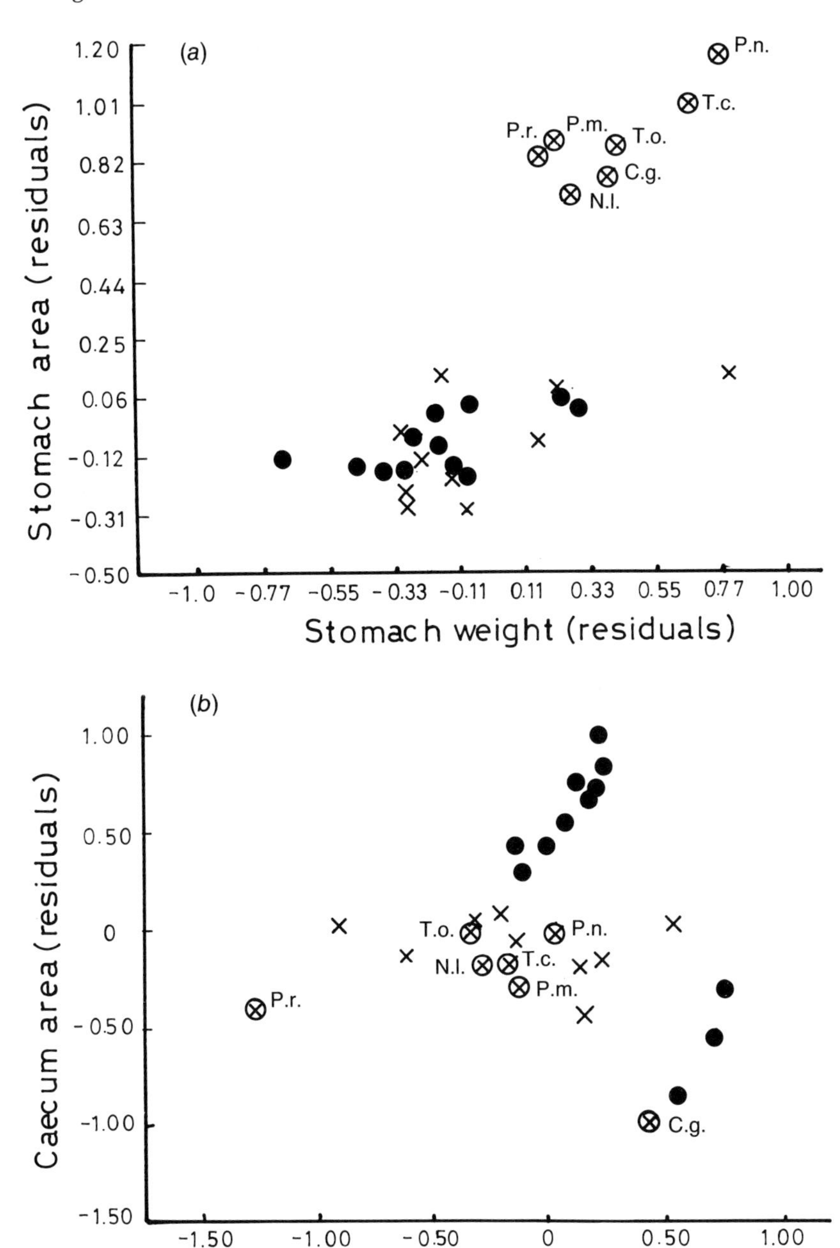

Figure 7.6. Relationship between (*a*) stomach weight and area and (*b*) caecum weight and area for primates. New World monkeys, solid circle; cercopithecines, cross; colobines, cross in circle; C.g, *Colobus guereza*; N.l, *Nasalis larvatus*; P.m, *Presbytis melalophos*; P.n, *Pygathrix nemaeus*; P.r, *P. roxellana*; S.e, *Semnopithecus entellus*; T.c, *Trachypithecus cristatus*; T.o, *T. obscurus*.

Discussion

The colobines represent a unique group of primates, through the dramatic elaboration of their stomach, historically believed to be an adaptation to folivory. While leaves may constitute 60% or more of the diet of some colobids, e.g. *Colobus*, *Trachypithecus* spp., they contribute less than 50% for *Semnopithecus entellus* and *Nasalis larvatus* and less than 40% for the *Presbytis* spp. studied (Chapters 4 and 5). Seed intake approaches 30% in those *Presbytis* spp. and 20% in *Nasalis larvatus*, as well as being 53% in *Colobus satanas* (McKey *et al.*, 1981). One should therefore think anew as to the significance of the colobine stomach, and, indeed, of compound stomachs in general (including those of ruminants), especially in view of Bodmer's (1989) study of frugivory in Amazonian ungulates.

In arguing that compound stomachs are adaptations to folivory, the need for a link into leaves is conveniently forgotten. How can frugivorous mammals suddenly start consuming leaves in significant amounts? Will they require a prior enlargement of a part of their gut to provide the necessary fermenting chamber? What food encourages provision of this enlargement? Folivores have always been divided into 'browsers' ('concentrate feeders'), 'intermediate feeders' and 'grazers' ('bulk feeders') (Jarman, 1974; Hofmann, 1973), with unselective bulk feeders having large body size. The smaller browsers have always been regarded as more selective in their feeding, given their greater need for more nourishing food per unit body weight, without clear ideas as to what this more nourishing food was; the early suggestions were that it was young, rather than mature, leaves; however, two studies of forest ungulates indicate that fruit parts are preferred foods.

In his detailed study of six duikers in the forests of West Africa, Dubost (1984) showed clearly that fruit parts, and especially seeds, contributed the bulk of these ruminants' diets: 69–82% of food by weight. Bodmer (1989, 1991), who observed seed-eating in peccaries and deer of the Amazon, argued that in some cases fermentation is necessary to digest the seed coat, so that the rich nutrients of the cotyledons can be released. This at last makes real sense of the large, sacculated stomach in the smaller foregut-fermenting 'folivores' and Bodmer (1989) advises that the category of 'frugivore' has to be added to the three categories that have classically described ungulates (see e.g. Jarman, 1974), preceding 'browser' (typically consuming a mixture of fruit and leaves) in terms of body size and niche.

Thus, as fruit is the general link between animal matter and foliage, so are seeds the specific link between fruit and leaves. Develop the adaptations for processing seeds (and unripe fruit), and young leaves, at least, are then accessible. Increase in body size allows for a more capacious gut, and a prolonged retention of foodstuffs, allowing digestion of mature leaves. While the seed-eating habit is seen as the alternative when leaves are digestively inaccessible, perhaps it represents the preceding adaptation. Ruminant 'frugivores' are a distinct part of the continuum through to 'grazers'; like 'browsers', they have less-developed rumens, a more rapid passage of ingesta, more post-ruminal digestion, and thus large abomasa and small intestines, but a smaller body size.

Certainly, the anatomical results presented here demonstrate that the stomach is largest in all, or at least some, parameters in the more folivorous species, and smaller, with an enlarged small intestine, in the more frugivorous (seed-eating) species. *Colobus* has the largest large intestine; the more folivorous species have greater fermenting capacity and less absorptive ability than the seed-eating species. *Pygathrix* has the most voluminous and heaviest gut of the colobines studied; despite its small size, it seems to be one of the more folivorous species. In contrast, the largest species studied, *Nasalis*, has the smallest stomach and largest small intestine – low fermenting capacity and the highest absorbing ability – which accords with its more frugivorous diet (Chapter 5).

As Ripley (1984) pointed out, caeco–colic fermentation allows selective fermentation of those foods that require it, increasing the passage rate of more digestible foods. Forestomach fermentation does not allow for such changes in digestion and passage rates (Janis, 1976; Langer, 1987), with less flexibility in the system, and seeds providing the only real alternative to leaves. This explains the continuing 'popularity' of caeco–colic fermentation among mammals (see also Bauchop, 1978; Parra, 1978). Whereas this first evolved to cope with large quantities of poor-quality foliage, it can now cope with a great array of foods, including those of good quality.

Returning to the colobines, it is difficult to equate this dichotomy of 'leaf-eaters' (*Colobus* – but not *C. satanas* or *C. polykomos* – *Trachypithecus* and *?Pygathrix*) and 'seed-eaters' (*Semnopithecus, Presbytis* and *Nasalis*) with Caton's (1990) observations that the saccus of *Presbytis, Trachypithecus, Semnopithecus* and *Colobus* is augmented by a presaccus only in *Nasalis, Pygathrix, P.* (*Rhinopithecus*) and *Procolobus* (Kuhn, 1964; Figure 7.7). *Rhinopithecus* is presumed to rely on leaf parts, at least in some seasons; *Procolobus* eats 14–48% fruit parts annually and few mature leaf parts; *Nasalis* eats few mature leaves, specializing in fruit and young leaf parts. So the expansion of the forestomach could be related to a seed-eating adaptation,

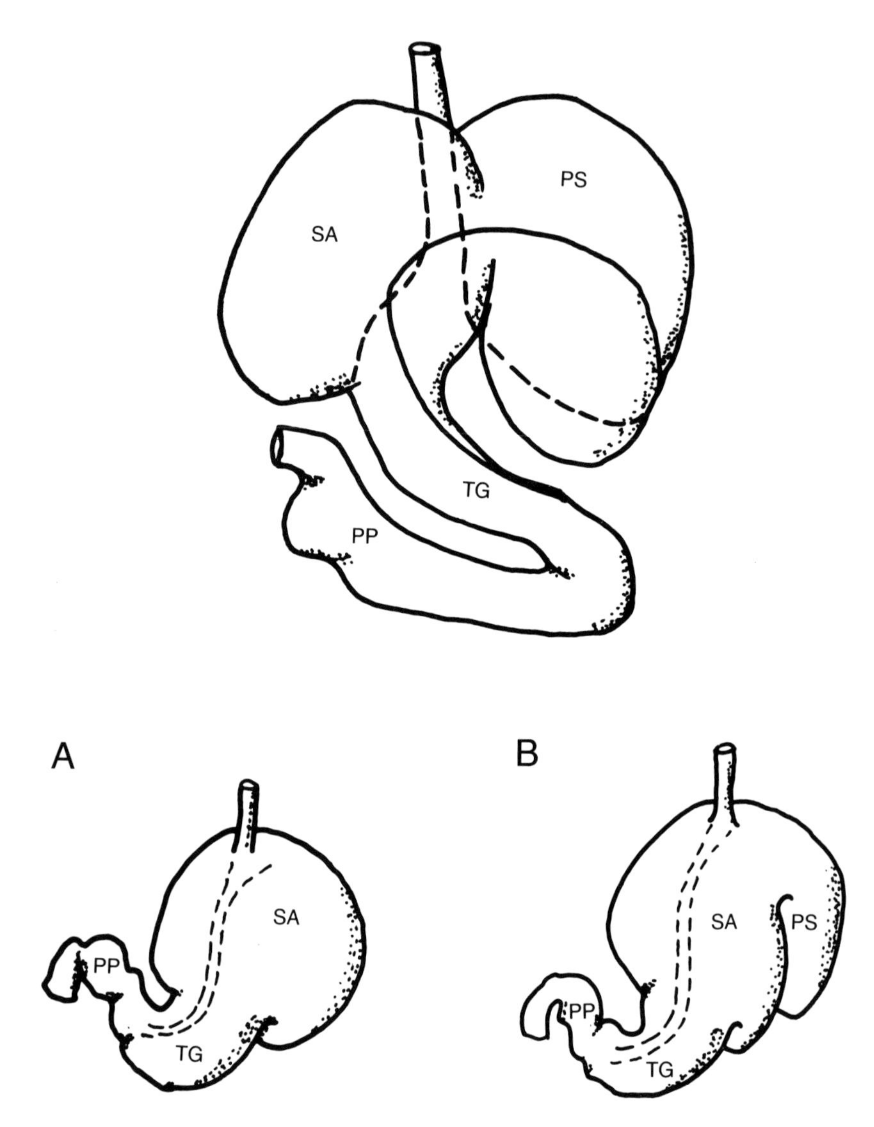

Figure 7.7. A stylized colobine stomach, with 'A' type lacking presaccus and 'B' type with sizeable presaccus (after Caton, 1990). SA, saccus; PS, presaccus; TG, tubus gastricus; PP, pars pylorica.

but given the 'seed-eaters' with only a saccus, it would be more likely that the addition of the presaccus represents a 'leaf-eating' adaptation, which is not related to subsequent shifts to, or back to, 'seed-eating'.

The considerable amount of data available on colobine diets, and their flexibility according to seasonality and food availability, are gradually being augmented by comparable data on gut morphology. Certain patterns are emerging, but more investigations are needed to clarify the full scope of such variations, and the significance of differences in structure and function.

Acknowledgements

I thank Glyn Davies and John Oates for inviting me to participate in writing this volume, and for their help in finalizing my manuscript, and Marcel Hladik, Bob Martin and Anne MacLarnon for years of collaboration on this topic. Marcel Hladik, in particular, has been an inspiration since our first meeting in 1967 for his multi-disciplinary approach, with his wife Annette, to plant science, behavioural ecology, gut morphology (gross and microscopic) and digestive physiology.

8

Digestive physiology

ROBIN N. B. KAY and A. GLYN DAVIES

Introduction

Colobine monkeys digest their food in much the same way as do ruminant animals. They have an enlarged and sacculated forestomach, described in Chapter 7, which houses a multitude of microbes that ferment the ingested food to produce volatile fatty acids, which are absorbed readily. The forestomach contents, food and microbes, then pass on for further digestion by the acid and enzymes of the last gastric compartment and the small intestine. Finally, the residues are fermented once again in the lower intestine. What advantages and opportunities does this mode of digestion offer to colobine monkeys, and what are its limitations?

Digestion and metabolism are now well understood in domesticated ruminants (see Milligan *et al.*, 1986) and some information is available on wild ruminants of diverse feeding habits (Hofmann, 1973, 1989; Kay *et al.*, 1980; Van Soest, 1982), but our knowledge of colobine physiology is limited. The digestion of leaves by arboreal folivores was previously reviewed by Bauchop (1978), who considered not only colobines but also sloths and tree kangaroos.

In this chapter, the various facets of forestomach digestion will be summarized with reference to ruminant animals, and what little is known of colobine digestion will be considered in this context. The digestive tract and digestive processes of colobines are shown diagrammatically in Figure 8.1.

Digestive function

Morphophysiological adaptation to diet

Ruminant animals occupy a diversity of feeding niches. Some large species (cattle, buffalo) have a strongly muscular and well-compartmented forestom-

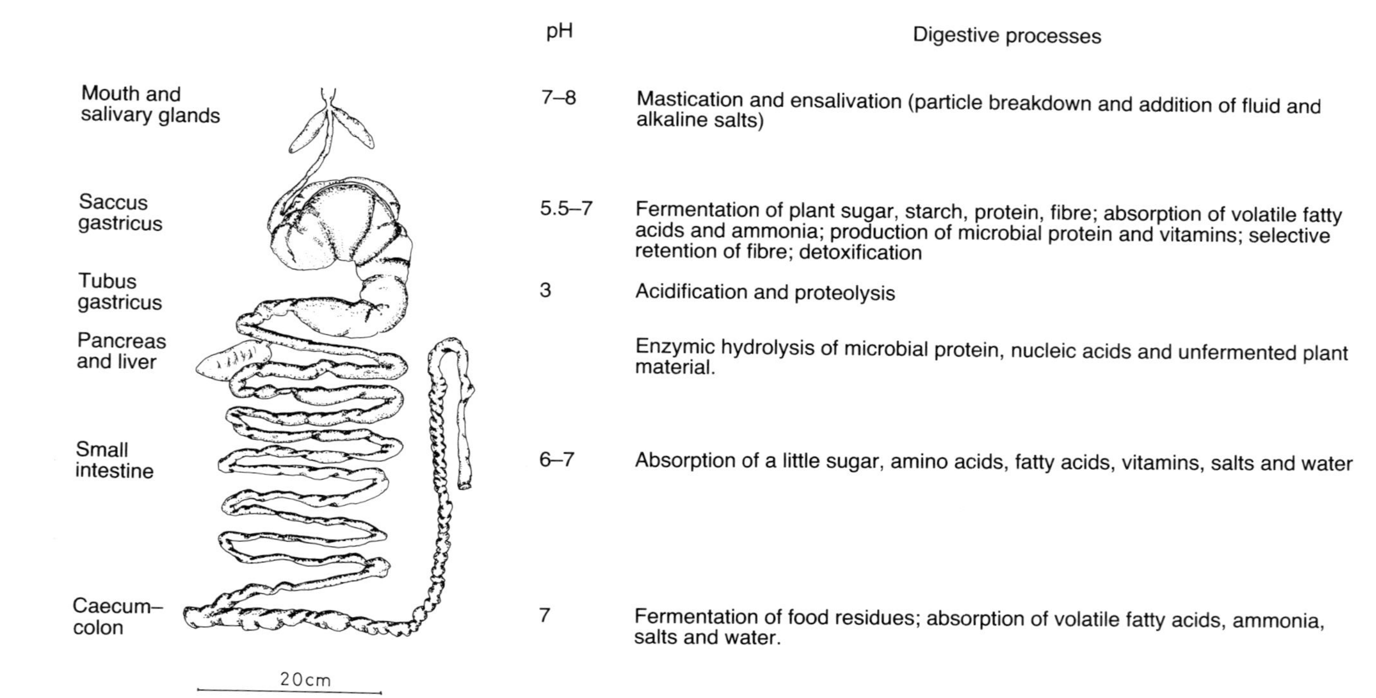

Figure 8.1. Diagram of the alimentary tract of *Colobus guereza* from Mount Kenya (Ohwaki *et al.*, 1974), showing the pH and digestive processes in different parts.

ach and other digestive adaptations that allow them to graze on mature grasslands, which provide bulk fibrous forage of low digestible energy content. Other species of many sizes, from mousedeer and dikdik to moose and giraffe, are adapted to browsing on digestible dicotyledonous forage. They have a simpler forestomach and a rather better developed distal fermentation chamber in the caecum–colon, perhaps resembling the early ruminants that existed before extensive grasslands developed (Hofmann, 1973, 1989). It is to this group that we should turn for comparison with colobine monkeys. All ruminants are quite versatile, able to adapt morphophysiologically and metabolically to substantial change of diet whether caused by weaning, by seasonal change in vegetation, or by capture and domestication by man; some species have become especially adapted to dietary change (Hofmann, 1989).

Within each feeding niche, ruminants will, when they have the opportunity, tend to select for better quality food (lower lignification, greater digestibility, higher protein content) as far as mouth size and prehensile organs – lips, teeth or tongue – will allow. This tendency is accentuated when metabolic requirements are high, as during juvenile growth and lactation. A pair of hands, even with reduced or absent thumbs, clearly allows colobines to select and gather forage in a manner not possible for hoofed animals.

Certain deficiencies may lead herbivores to develop specific appetites. Dehydration leads to the selection of succulent food, sodium deficiency to choice of sodium-rich forage or drinking water, and phosphorus deficiency to the bone-chewing behaviour seen in South African cattle. Water, sodium and other minerals may sometimes be scarce in the arboreal habitat of colobines and this may contribute to forage selection and the peculiar earth-eating behaviour that has been observed (see pp. 247–248).

Ensalivation

Ruminants have large salivary glands which provide the copious volume of liquid and mucus needed to fluidize and lubricate food before it can be swallowed and to maintain it in fluid suspension during digestion in the forestomach. Secretion of body fluid rather than consumption of drinking water to fluidize ingested food is especially advantageous to animals that eat large amounts of often dry food in habitats sparsely provided with drinking water, whatever the mode of digestion. In some Indian localities, Hanuman langurs (*Semnopithecus entellus*) are able to maintain themselves successfully in habitats where they take no drinking water for months on end during the dry season (Jay, 1965).

In ruminants the prominent salivary gland is the parotid, which has an unusual innervation and structure and continuously secretes an isotonic saliva, rich in bicarbonate and phosphate (Kay, 1966). The parotids are about three times larger, relative to body weight, in ruminant species adapted to high-quality forage browse (1.5 to 2.2 g/kg body weight) than in those eating bulk roughages (0.5 to 0.7 g/kg) (Hofmann, 1973, 1989; Kay, 1987). Hanuman langurs have quite large salivary glands (Ayer, 1948); and both parotids together weighed 1.3 and 2.2 g/kg in two *Colobus guereza*[1] specimens (Kay *et al.*, 1976). However, nothing is known of the volume of saliva secreted nor its composition.

Digestion in the forestomach

The microbes of the rumen and their fermentative activity are richly documented (Hobson, 1988). They include bacteria, protozoa and fungi, which are adapted to the anaerobic conditions of the rumen. A number are cellulolytic micro-organisms, which produce the enzymes that can hydrolyse the β-linkages of the cellulose and hemicellulose of the plant cell wall. This is the uniquely valuable contribution of microbial activity to the digestive process in herbivorous mammals, providing access to the resource of plant fibre, and to the cell contents sealed within cellulose walls.

The main products of microbial fermentation are: the volatile fatty acids (VFA) – acetic, propionic and butyric; microbial cell material (especially microbial protein); the gases carbon dioxide, methane and hydrogen (the last being more or less removed by reduction of carbon dioxide to methane); and ammonia. The VFA are selectively absorbed through the rumen epithelium in their associated (acidic) form, the absorptive surface area being enlarged by papillation, which is most luxuriant in species that select the richest diets (Hofmann, 1989). Further absorption occurs as forestomach fluid flows on through the omasum; the concentration of VFA in the abomasum is only about 5% of that in the rumen.

The ruminal microbes produce many vitamins of the B group which are of use to the host animal. Ruminants have high serum levels of vitamin B_{12} and colobines have a much higher serum concentration of vitamin B_{12} than do other primates (Oxnard, 1966). The microflora also hydrolyse di- and triacylglycerides in the diet, fermenting the galactose and glycerol moieties to VFA and saturating the fatty acid moiety to cause the notably hard and saturated nature of ruminant adipose tissue. A further microbial activity of importance to the host is the partial breakdown of certain toxic constituents

[1] Previously reported as *C. polykomos.*

of the diet – some alkaloids, amines and organic acids, for example (see below). Other plant components, such as essential oils and tannins, not necessarily toxic to the host, may have an indirectly harmful effect through their bacteriostatic action in the rumen (Oh *et al.*, 1968). Robbins *et al.* (1987) have shown that this defence by the plant against herbivory has been circumvented by browsing deer, which secrete a tannin-inactivating protein in their saliva (and see Chapter 9, p. 257).

Microbial activity in the saccus gastricus of colobine monkeys quite closely resembles that of a ruminant animal eating a concentrate diet (Table 8.1). Drawert *et al.* (1962) found the concentrations of VFA in the saccus and tubus of *Procolobus verus* to be greater than those found in cattle but similar to values in small selectively browsing antelopes (Hoppe, 1977*b*). Bauchop & Martucci (1968) took samples by stomach tube from the saccus of captive langurs (*Trachypithecus cristatus* and *Semnopithecus entellus*) that were given a diet of fresh alfalfa with yams, green beans and a cereal preparation. They identified a number of bacteria including *Methanobacterium ruminantium* and two cellulolytic *Bacteroides* spp. resembling those in the rumen. Protozoa, however, were not recorded. The stomach gases included methane, carbon dioxide and a little hydrogen. After feeding, the VFA concentration increased to around 180 mmol/l.

Ohwaki *et al.* (1974) examined the stomach contents of two *Colobus guereza* shot in forest flanking Mount Kenya. The animals were evidently selecting seeds and fruit, for the saccus contained a finely comminuted and almost white paste, and one of them appeared to have eaten just before it was killed. A range of bacterial species was found in the contents of the saccus, similar to that of ruminants on such a diet, but no cellulolytic bacteria or protozoa were detected. The material had a pH of 5.5 or less, and contained acetate, propionate and butyrate, together with a little lactate. The forestomach gas contained hydrogen but no methane. Much lower concentrations of VFA were found in two captive *Colobus guereza* consuming small amounts of a leafy diet (Kay *et al.*, 1976).

Rapid rates of fermentation by saccus contents *in vitro* have also been recorded in three studies of colobines (Table 8.1). The ability to limit VFA concentration, osmolarity and pH within the range permitting continuing microbial activity, even when fermentation rates are very high, indicates that colobines have developed mechanisms for absorbing VFA and buffering their acidity that are as effective as those in ruminants. Effective VFA absorption is also indicated by the low VFA levels found in the tubus, the next stomach compartment, especially since colobines lack any organ analogous to the omasum of ruminants interpolated between saccus and tubus. While the

Table 8.1. *Fermentative characteristics in the saccus gastricus, tubus gastricus and caecum–colon of various colobines*

	Saccus gastricus						Tubus gastricus		Caecum–colon		
	pH	Volatile fatty acids (mmol/l)	Acetic acid (mol %)	Propionic acid (mol %)	Butyric acid (mol %)	Fermentation rate *in vitro* (μmol gas/g wet wt.h)	pH	Volatile fatty acids (mmol/l)	pH	Volatile fatty acids (mmol/l)	Reference
Procolobus verus		230						24			Drawert *et al.* (1962)
P. badius						12					Kuhn (1964)
Trachypithecus cristatus	7					38		2–3			Kuhn (1964)
T. cristatus	5.0–6.1	95–133[a]	47–56	24–26	10–18	63–79[b]					Bauchop & Martucci (1968)
Semnopithecus entellus	5.4–6.5	89–233[a]	46–50	22–23	14–23						Bauchop & Martucci (1968)
Colobus guereza[c]	5	107[d]	57	18	17	28					Ohwaki *et al.* (1974)
		434	78	11	7	179					
C. guereza[c]	6.5	65					2.4	4	7.2	80	Kay *et al.* (1976)
	7.0	53					2.2	3	7.2	47	

[a]μmol/g wet matter.
[b]Calculated from volatile fatty acids per gram dry matter, assuming 15% dry matter in contents.
[c]Values for two animals.
[d]2 and 10 μmol lactate/g also found.

rumen epithelium is keratinized and heavily papillated, that of the saccus gastricus is unpapillated and glandular (Kuhn, 1964). Nevertheless, colobines seem quite as well able to absorb VFA from their forestomach as ruminants, just as the llama is able to absorb VFA rapidly through its predominantly glandular forestomach (Engelhardt & Sallmann, 1972).

Forestomach pH and its control

The VFA produced in the forestomach would acidify the contents to an extent incompatible with a healthy fermentation if they were not buffered so as to maintain pH within the range normally found (pH 5.5–7.0). Only a little alkali is present in herbage diets but in ruminants the saliva contains high concentrations of both bicarbonate ($\approx$110 mmol/l) and phosphate ($\approx$30 mmol/l). In sheep and cattle, enough saliva is secreted to neutralize about one-third of the VFA produced (Kay, 1966), the remaining VFA being selectively absorbed in the acid state to be neutralized by blood buffers until metabolized.

When diets rich in starch or sugar are eaten, rumen pH is in the lower half of the normal range (5.5–6.2), cellulolytic activity is slight, protozoa are reduced or absent, and molar proportions of propionate and/or butyrate are elevated at the expense of acetate. Similar observations were made for colobines receiving a concentrate-type diet (Bauchop & Martucci, 1968) or selecting a seed-rich diet (Ohwaki *et al.*, 1974). When ruminants eat diets rich in fibre, rumen pH is between 6.2 and 7.0, cellulolysis is active and protozoa are present, and it appears that a similar situation pertained in the leaf-eating colobines of Kay *et al.* (1976), although no microbiological study was made.

The extent to which dietary alkali, salivary alkali, possible alkaline secretions of the saccus mucosa, and selective absorption of VFA in their acid form contribute to the preservation of favourable pH conditions in the colobine saccus cannot yet be assessed.

Forestomach capacity and retention time

The capacity of the forestomach and the speed with which food residues are eliminated from it will limit the amount of forage that can be consumed, while the time food is retained will be an important factor influencing the extent of degradation of slowly digested components such as plant fibre.

The capacity and morphological characteristics of the digestive tract are clearly related to diet and this relationship has been carefully documented both in ruminants (Hofmann 1973, 1989) and in primates (Chivers & Hladik, 1980; see also Chapter 7). For African herbivores fermenting their food in either a foregut or a hindgut chamber, Van Soest (1982) expressed the relationship of weight of fermentation contents (Y) to body weight (X), in kilograms, as:

$$\log_{10}Y = 1.032 \log_{10}X - 0.936 \qquad (r = 0.99).$$

For ruminants, Parra (1978) gives the relationship of fermentation contents (Y) to body weight (W) as:

$$Y = 0.1050\ W^{1.0453}.$$

Thus, fermentation contents vary little as a fraction of body weight in animals of a wide range of sizes, and the relatively greater metabolic requirement of small herbivores is met by selection of a rich and rapidly digested diet, with consequently the short retention time that permits a large food intake (Van Soest, 1982; Kay, 1985). The contents of the reticulorumen usually account for some 8–16% of body weight, the value being some 3% less in selective feeders than in bulk grazers (Van Soest, 1982). However, some domestic breeds such as Bangali cattle traditionally maintained on rice straw (Mould *et al.*, 1982) and the Heidschnucken sheep from the heaths of northern Germany (Weyreter *et al.*, 1987) have a much more capacious rumen, which allows them both to eat a large amount of their fibrous diet and to subject it to prolonged digestion.

The weights of stomach contents in colobines recorded in five reports are summarized in Table 8.2. They range from 4% of body weight in an underfed *Colobus guereza*, to 17% in a *Trachypithecus cristatus*. The values cover the range typical for ruminant species.

No record of retention time in the stomach of colobines can be found. However, from the data given by Kay *et al.* (1976), values can be derived from the lignin content of various digestive compartments. If one assumes lignin to be indigestible, and divides the values by an approximate lignin intake of 3.2 g/day, the retention times of lignin in the saccus gastricus, caecum–colon, and whole digestive tract (excluding small intestine) can be calculated to have been roughly 14 h, 8 h and 38 h, respectively. These are rather short periods compared with those in domestic ruminants, more resembling those of a small selective antelope (Hoppe, 1977*a*). A still shorter retention time would have been expected had the colobines been eating a more substantial amount of food. A rapid flow of fluid through the forestomach,

Table 8.2. *Weight of contents of gut compartments of various adult colobines*

		Saccus gastricus contents			Caecum–colon contents			
	Body weight (kg)	Wet weight (g)	Wet weight (% body wt.)	Dry matter (%)	Wet weight (g)	Dry matter (%)	Condition and diet	Reference
Procolobus badius	7.3	950	13.0				Wild	Kuhn (1964)
P. badius	7.25–10.0	642	7.4				Wild	Kuhn (1964)
P. verus	5.7	395	6.9				Wild	Kuhn (1964)
P. verus	3.5	453	12.9				Wild	Kuhn (1964)
Trachypithecus cristatus	5.4	938	17.4				Captive – alfalfa leaves with yams, green beans and cereals	Bauchop & Martucci (1968)
Colobus guereza	7.0	436	6.2				Wild	Jones (1970)
Colobus guereza[a,b]	9.5	690	7.3	19			Wild – mainly seeds	Ohwaki et al. (1974)
	12.3	960	7.8	32				
Colobus guereza[a]	6.2	426	6.9	5	100	10	Captive – succulent plants at submaintenance intake	Kay *et al.* (1976)
	3.4	130	3.8	9	34	21		

Notes:
[a]Values for two animals.
[b]Calculated from weight of stomach and contents less stomach weight as 3.26% of liveweight (Kay *et al.*, 1976).

together with unfavourably acid conditions, would discourage or eliminate slowly-dividing and acid-sensitive microbes such as cellulolytic bacteria and protozoa, as in ruminants (Hobson, 1988).

A point to remember when comparing grazing and browsing ruminants is that grass contains long stringy fibres that are slowly digested and form a floating mat within the rumen; they are moved away from the reticulum at just the moment the reticulo-omasal orifice opens. However, the dicotyledonous leaves and seeds eaten by selective feeders, including colobines, break down to small rounded fragments so that the stomach contents are more homogeneous, reducing the opportunity for selective retention of fibre in the stomach.

Digestion in the tubus gastricus and small intestine

So little is known of this phase of digestion in colobines that only two points deserve mention, both relating to the digestion of micro-organisms. Dobson *et al.* (1984) point to an intriguing similarity between colobines and ruminants. Both groups show a high activity of the bacteriolytic enzyme lysozyme in the fundic mucosa of the tubus gastricus and abomasum, respectively, the tissue responsible for secretion of gastric acid. On the other hand, herbivores lacking a forestomach show little such activity. Moreover, the colobine and ruminant fundic lysozymes have a sharp peak of activity at pH 5, whereas lysozyme from other tissues and other species has a broad plateau of activity between pH 5 and pH 7. However, Dobson *et al.* (1984) mistakenly interpret data from Smith (1965) as indicating that the pH in the anterior part of the abomasum is about 6 (Smith's value actually refers to rumen contents). In fact, the pH in abomasal contents, as in colobine tubus contents (Kuhn, 1964; Kay *et al.*, 1976), is normally between 2 and 3. In this range, fundic lysozyme shows very little activity and so the function of this enzyme awaits further explanation.

The second point relates to the digestion of the nucleic acids which form an important part of the microbial nitrogen fraction but are present in only low concentrations in forage diets. Beintema *et al.* (1973) found that ribonu clease activity in the pancreas of ruminants was very high (> 50 μg/g and sometimes > 500 μg/g) as it was in some non-ruminants that fermented their food in a forestomach. The value for a colobine was rather lower (between 5 and 50 μg/g) but still much higher than in primates without a forestomach.

Digestion in the large intestine

The food residue passing to the large intestine of all mammals is anaerobically fermented by micro-organisms with the production of VFA, ammonia, microbial matter and vitamins, much as in the forestomach. While the VFA and ammonia are readily absorbed, most of the microbial matter is excreted in the faeces and so is lost, unless the animal practises coprophagy (Hörnicke & Björnhag, 1980). Coprophagy does not occur in colobines. The quantitative importance to digestion of microbial fermentation in the lower intestine varies from a trivial scavenging process in carnivores to the major digestive strategy in many herbivores that lack a forestomach, such as horses, rabbits and leaf-eating primates such as howler monkeys, lemurs and gorillas (Chivers & Hladik, 1980).

Among the ruminants, the caecum–colon is quite small in those that eat bulky roughage, only about 1/30 the size of the reticulo-rumen, but it is rather larger, 1/10 of the reticulo-rumen, in concentrate selectors (Hofmann, 1989). At first sight this seems surprising, for concentrate selectors consume little indigestible matter to enlarge the bulk of the contents of the lower intestine. However, work by Hoppe (1977*a*) on the dikdik (*Madoqua kirkii*) and suni (*Neotragus moschatus*) suggests that the fibre ingested by small selective ruminants is poorly digested in the rumen, presumably due to rapid passage and unfavourably acid conditions, so that much remains to be fermented in the caecum–colon.

The relative advantages and drawbacks of siting microbial digestion of food either before or after the true stomach and small intestine, in the forestomach versus the large intestine, have been fully discussed by Parra (1978) and by Stevens *et al.* (1980). Most herbivores practise both options to some degree. The two processes should be regarded as complementary rather than alternatives, with their relative contributions being adjusted to allow optimal utilization of various types of diet. Such adjustment may occur not only between species but also in response to change of diet by the individual. Sheep changed from chopped hay to pelleted hay, for example, increase the fraction of the diet digested in the large intestine two- or three-fold (Beever *et al.*, 1972).

The relative weight of the colobine lower intestine and its contents resembles that of small concentrate-selecting ruminants, suggesting that this is a quite significant fermentation chamber. The two leaf-fed *Colobus* studied by Kay *et al.* (1976) had about half as much dry matter in the caecum–colon as in the saccus, and VFA concentrations were much the same at each site; moreover, a substantial amount of cellulose disappeared after digesta left the saccus. Colobine diets, which consist mainly of seeds and fruit, can apparently depress cellulolysis in the forestomach (Ohwaki *et al*, 1974). A reasonably large caecum–colon may allow a versatile compromise between fermentation in the forestomach and large intestine, and permit opportunistic alternation between leaf and seed diets.

Digestibility of the diet

Ruminant animals can maintain themselves on diets with dry-matter digestibility ranging from 90% for the highly selective lesser mousedeer (*Tragulus javanicus*) (Nordin, 1978) to around 40% for bovines tolerant of bulk roughage. Normally, about two-thirds of the digestible organic matter and energy of the diet is absorbed from the forestomach, about 15% of the digested energy being eliminated as methane or fermentation heat.

Watkins *et al.* (1985) formulated a high-fibre biscuit (34.4% neutral detergent fibre, NDF) for colobines, to resemble the natural diet. Offered to two guerezas (*C. guereza*) together with fruit and vegetables it provided a diet containing 25% NDF. The apparent digestibility of the diet, assessed by total faecal collection for five days, was: dry matter, 87%; crude protein, 78%; gross energy, 85%; NDF, 81%; acid detergent fibre, 69%; and lignin, 21%. Assessed by use of lignin as a reference substance these values become 84%, 72%, 82%, 76%, 61% and 0%, respectively. The authors cite unpublished results of Ullrey *et al.* that indicate similarly high values (81–90%) for dry matter digestibility in other captive colobines given diets containing 15–22% of NDF.

In the two specimens examined after death by Kay *et al.* (1976), progressive loss of dietary components down the gut was assessed from their falling concentration relative to lignin. Some 10–20% of dietary dry matter, nitrogen and cellulose, appeared to be lost from the saccus gastricus (relative to lignin) in one animal and 30–50% in the other; overall digestibilities of dry matter and cellulose were 52% and 61%, respectively, in the second animal. In this second animal, at least, digestion in the saccus appeared to have been about as important as in the rumen. The animals had been restricted to a single species of plant, the succulent *Commelinum benghalense*, which proved to be rather indigestible despite the low intake achieved.

Nitrogen metabolism

In ruminants, most of the protein and other nitrogenous material in the diet is degraded to ammonia in the rumen, although a small but potentially important fraction of dietary protein flows to the intestines in undegraded form. The ammonia is used by the microbes for synthesis of their own protein. In turn, this is digested in the small intestine and usually forms the greatest source of amino-nitrogen for the host animal. Blood urea secreted in saliva or diffusing across the rumen wall adds to the rumen ammonia pool. If dietary nitrogen is insufficient, relative to fermentable energy, to maintain microbial activity, especially of cellulolytic organisms, the recycled urea nitrogen will allow additional microbial protein to be synthesised and encourage fibre digestion. If microbial activity is not nitrogen limited, however, the urea will merely be reabsorbed as ammonia and converted back to urea by the liver at some metabolic cost.

What scraps of information there are concerning nitrogen metabolism in colobines indicate a similar pattern. When saccus contents from langurs given a protein-rich alfalfa diet were incubated *in vitro*, high ammonia concentrations indicative of protein degradation were maintained (Bauchop & Martucci, 1968). The fairly high concentration of pancreatic ribonuclease in colobines, already mentioned, will help to digest the substantial nucleic acid component from microbial protein. The relatively large quantity of fermentable substrate reaching the lower intestine of colobines will allow moderate amounts of ammonia to be converted to microbial protein in this compartment and eliminated in the faeces, thus reducing the water-costly need to excrete this fraction as urinary urea.

Food selection and digestive physiology

Having established the characteristics of colobine digestive physiology, and seen how closely they compare with small, concentrate-feeding ruminants, we can consider how the digestive system may affect the way in which monkeys capture nutrients from their environment while avoiding the negative effects of plant defence chemicals which may be toxic or reduce digestibility (Freeland & Janzen, 1974; and see Chapter 9). To help pin-point distinctive characteristics of colobine diets, we have compared them with diets of simple-stomached primates that lived sympatrically, and had access to the same food resources. To the extent that the animals' nutritional requirements are similar, any consistent differences in the diets are likely to indicate food preferences resulting from adaptations or limitations of the food-processing systems.

Insectivory

One conspicuous dietary difference is the paucity of insects in colobine diets. At Kuala Lompat, for example, long-tailed macaques (*Macaca fascicularis*) spent much time searching for insects, which accounted for 20% of the annual diet and the more robust pig-tailed macaques (*M. nemestrina*) also consumed many insects (Caldecott, 1986). Sympatric lar gibbons (*Hylobates lar*) forage through mature leaves and will feed extensively when they encounter a mass of insects, although their annual intake (6% of the diet) is lower than that of the macaques (Raemaekers, 1978). In contrast, neither the dusky langur (*Trachypithecus obscurus*) nor banded leaf-monkey (*Presbytis melalophos*) in the same forest had more than 1% animal matter in their diets (Curtin, 1980; Bennett, 1983). The red leaf-monkeys (*P. rubicunda*) at Sepilok had a similarly low intake, and the animal matter in their diets comprised only termites (Davies, 1984).

Hanuman langurs at Mount Abu and Ranthambore consumed grasshoppers, termite alates and fly larvae (Moore, 1985*b*), and the same species at Kanha fed on lepidopteran larvae and cecropid nymphs for 2.8% of feeding time (Newton, 1985). In the Kibale forest, red colobus (*Procolobus badius*) spent up to 4% of their time foraging for, but only occasionally consuming, insects in rotten branch stumps (Struhsaker, 1978). However, these figures are still very low compared with the small-bodied (2–4 kg) African cercopithecines which include 10–30% insects in their annual diets (see e.g. Gautier-Hion, 1980).

Animal protein can readily be fermented in the forestomach as exemplified by fish meal and chitinous krill being satisfactory protein supplements for farm ruminants. The low level of insectivory in colobines, therefore, does not indicate that insects are indigestible, but that the supply of microbial protein through fermentation of vegetable substrates is sufficient to satisfy maintenance and growth requirements.

It is probably more profitable for forestomach fermenters to acquire protein by digesting nitrogen-rich foliage, and any insects within it, rather than to spend energy seeking elusive insects. The arthropod exoskeleton is of uncertain digestibility, and could lower the nutritional value of insects. Furthermore, very high intakes of animal protein would doubtless lead to excessive ammonia production with potentially damaging effects on the blood and liver (Wolter, 1980, 1982).

Seed eating

A very important feature of colobine diets is seed eating (Chapters 4 and 5). At several sites, colobine frugivory has been distinguished by the high proportion of large, dry seeds taken from dull-coloured fruits. In Cameroon and Gabon, with *Colobus satanas* (McKey, 1978*b*; Harrison, 1986), in Borneo, with *Presbytis rubicunda* (Davies, 1991) and in Sierra Leone, with *C. poly-*

komos (Dasilva, 1992), seeds contributed 58, 47, 49 and 30%, respectively, to the annual diets.

Seed eating is less common at other sites, but remains a characteristic feature of colobine frugivory. At Kuala Lompat, banded leaf-monkeys and lar gibbons had similar intakes of fruit; the former specialized in seed eating, especially from the common leguminous trees (Bennett, 1983), but the gibbons ate very few unripe fruits, showing a strong preference for ripe figs and neglecting hard, dry seeds (Gittins & Raemaekers, 1980). In the semi-evergreen forests of Polonnaruwa, toque macaques (*M. sinica*) eat succulent fruits while the purple-faced langurs (*T. vetulus*) and Hanuman langurs (*S. entellus*) tend to eat dry or coriaceous fruits (Hladik, 1977*b*), although the highly adaptable Hanuman langur can also consume moderate amounts of succulent fruits.

In Africa, the red and black-and-white colobus at Kibale are highly folivorous, but eat seeds and unripe or leathery fruits (Struhsaker, 1975; Oates, 1977*a*) in preference to the juicy, ripe fruits eaten by mangabeys (*Cercocebus albigena*), red-tailed monkeys (*Cercopithecus ascanius*) and chimpanzees (*Pan troglodytes*) (Waser, 1975; Struhsaker & Leland, 1979; Ghiglieri, 1984). A similar distinction occurs between the seed-eating colobines (three species) on Tiwai and the fleshy-fruit-eating *Cercopithecus diana* (Whitesides, 1991; A. G. Davies, J. F. Oates and G. L. Dasilva, unpublished observations).

Thus, seed eating is a common characteristic of colobine frugivory, but to what extent is it a consequence of forestomach fermentation?

If a large quantity of concentrate diet is consumed, the rate of VFA production in the forestomach may become too rapid for the VFA to be absorbed or buffered, resulting in an undesirable fall in forestomach pH. Such 'acidosis' has caused fatalities in captive colobines (Goltenboth, 1976), but wild populations seem less prone to this condition. For example, red leaf-monkeys in peak periods may have a monthly dietary intake of 80–94% seeds, which may be as much as 70% digestible, but the monkeys show no sign of digestive disorders (Davies, 1991).

The pH of the forestomach may be lowered more directly by consuming fruit flesh which is acidic. Ripe fruits tend to contain organic acids, while succulent flesh would allow rapid swallowing with little addition of alkaline saliva. If the acids can be fermented or absorbed quickly, no problem will arise. However, if intake is high and the concentration of acid from fruits builds up, pH may be reduced with deleterious results.

These two problems may occasionally limit fruit choice, but they do not offer strong enough explanations for the consistent lack of succulent ripe fruit in colobine diets. The answer probably lies in the nutritional value of succu-

Table 8.3. *Comparison of seeds and fruit flesh as foods at two sites*

	% Fibre (ADF)	% Protein (N×6.25)	Energy (kJ/g)
Tiwai			
Seeds ($n=9$)	35	20	22.6
Fruit flesh ($n=4$)	43	4	2.9
Sepilok			
Seeds (n = 20)	15.8	9.6	18.8
Whole fruits ($n=8$)	39.8	8.2	17.2
Arils ($n=7$)	30.2	6.2	

Note: ADF, acid detergent fibre
Sources: Tiwai: Dasilva, 1992. Sepilok: Davies, 1991; Davies *et al.*, 1988.

lent fruit flesh. In a comparison of seeds and fruit flesh at two sites, Tiwai and Sepilok (Table 8.3), it is clear that fruit flesh is often of much lower quality than seeds and could only sustain a small primate if large amounts passed quickly through the gut. Food passage rates in primates are poorly understood, but the estimate made for captive guereza (see above) of 38 h is an order of magnitude greater than the 3 h measured for spider monkeys eating fleshy fruits (Milton, 1981); spider monkeys are similar in size to guerezas, but have a simple-stomached digestive system like cercopithecines and gibbons. The colobine digestive system is unlikely to allow very rapid food passage rates and many pulpy fruits may therefore be unsuitable food.

It is important not to overlook the fact that some types of starch, usually the main storage carbohydrate in seeds, may be digested much more fully in a forestomach than in the small intestine of a simple-stomached primate. The flinty starch of maize seed and the starch granules of raw potatoes, for example, are not readily digested by pancreatic amylase. The colobine digestive system, therefore, is well adapted to the digestion of seeds, which may be preferred to more fibrous whole fruits or fruit flesh.

These examples offer an insight into the range of interactions between a forestomach digestive system and fruit composition which may affect food selection. However, fruits are notoriously variable in the composition of different parts, so care must be taken not to over-generalize.

Folivory

Despite the frequent references to the capacity of colobine monkeys to digest foliage, few have a high intake of mature leaves (Figure 8.2). Exceptions like

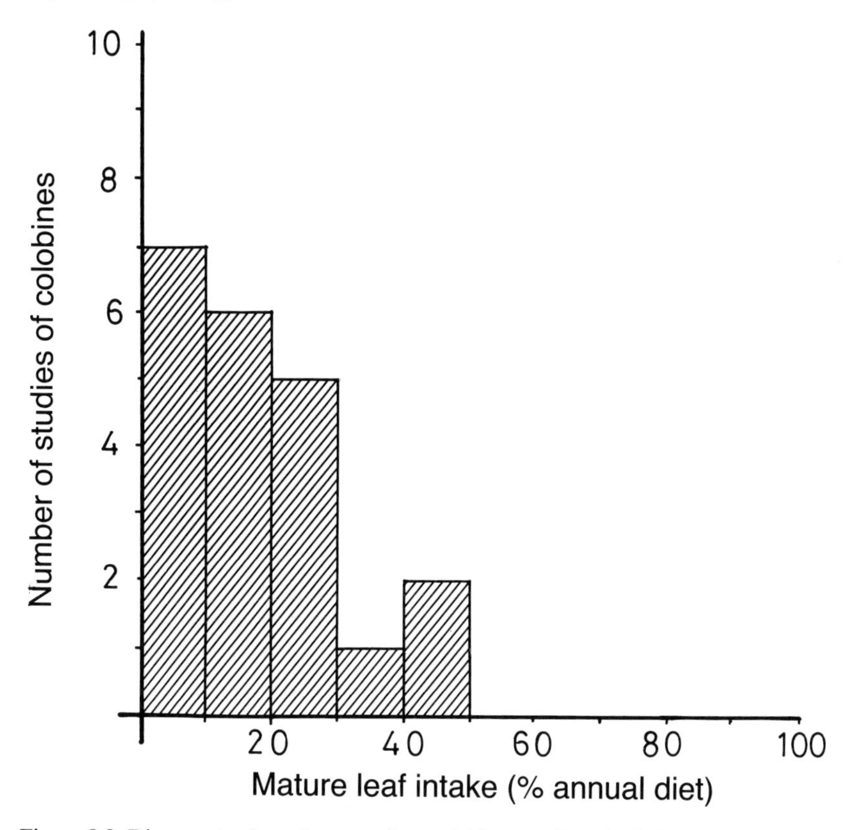

Figure 8.2. Diagram to show that very few colobine monkeys include more than 30% mature leaves in their annual diet. Sources: Clutton-Brock, 1975*a*; Struhsaker, 1975; Gatinot, 1977; Hladik, 1977*b*; Oates, 1977*a*, Curtin & Chivers, 1978; Oates *et al.*, 1980, 1990; Marsh, 1981*b*; McKey *et al.*, 1981; Ruhiyat, 1983; Bennett, 1986*a*; Davies *et al.*, 1988; Stanford, 1991; Starin, 1991; Kool, 1992.

the red or black-and-white colobus of Kibale (Struhsaker & Oates, 1975), with their enormous intake of foliage, are rare and even these two species show a preference for young leaf parts or high-quality mature leaf parts (Waterman & Choo, 1981). The availability of alternative foods and the quality of foliage will influence the extent of folivory, so forestomach fermentation does not necessarily lead to a folivorous habit.

The critical feature of forestomach fermentation that facilitates digestion of foliage is the capacity of cellulolytic bacteria to break down cellulose and hemicellulose molecules. Such breakdown serves a vital function in breaking open plant cell walls and liberating their readily digested contents, but the extent to which cellulose can be used as a nutrient, in addition to cell contents, will depend on:

(1) The cellulose content of the food.
(2) The pH of the forestomach.
(3) The length of time that items are retained in the forestomach, and subsequently in the caecum–colon.

Cellulolysis, and its products, have been identified in captive langurs (Bauchop & Martucci, 1968) and captive guerezas (Kay *et al.*, 1976). However, wild-shot guerezas (*C. guereza*) (Ohwaki *et al.*, 1974) had relatively low forestomach pH and showed no evidence of cellulolysis. In this case the specimens had been eating fruits and seeds, which were rapidly fermented, with a consequent reduction in both pH and cellulolytic activity. This will occur in colobines whenever foods contain little cellulose and forestomach pH is low.

The retention time of food will be affected by the quality and quantity of items ingested and the size of fermentation chambers. These in turn will relate to body size. The energy and protein requirements of mammals scale to approximately the 0.75th power of body weight (Kleiber, 1961) so small animals must consume high-energy proteinaceous foods. Colobines are no exception to this rule; small species such as *Presbytis melalophos* (5.9 kg), *P. rubicunda* (6.2 kg) and *Procolobus verus* (4.7 kg) all specialize in high-quality young leaves and readily digested seeds. Despite having enlarged forestomachs, therefore, they eat very few mature leaves, which would require prolonged retention for digestion.

Colobines are expected to be like other small, active concentrate-feeding mammals, which tend to have quicker food passage rates than large species (Parra, 1978; Kay, 1985). This reduces the opportunity for prolonged retention of fibrous foodstuffs for fermentation. Increases in body size might allow longer retention time but could become incompatible with an arboreal lifestyle. Most colobines, therefore, eat moderate- to high-quality leaves and cannot maintain themselves on low-quality mature leaves for long periods.

Plant toxins

In addition to facilitating the digestion of specific plant parts, it has been suggested that forestomach fermentation may have evolved to allow detoxification of plant secondary compounds (McKey, 1978*b*). One group of secondary compounds are the alkaloids, some of which are toxic to mammals and are known to deter herbivores. Despite this deterrent capacity, however, colobine monkeys have been recorded eating foods containing alkaloids.

Procolobus badius at Kibale and on Tiwai, and *Colobus satanas* at Douala-Edéa eat seeds and leaf parts of *Erythrophleum ivorense* which are known to have high levels of alkaloids (McKey, 1978*b*). At Polonnaruwa, *Trachypithecus vetulus* eat toxic *Strychnos* fruits which contain alkaloids that are avoided by *Macaca sinica* (Hladik, 1978). The reason that colobines can eat these items without suffering ill-effects is probably that the alkaloids can be broken down through fermentation before they are absorbed by the monkey.

Forestomach fermentation can therefore confer considerable advantages to colobines over cercopithecines in respect of these plant toxins, but it also confers a disadvantage in that colobines are susceptible to bacteriostatic compounds which may inhibit microbial activity in the forestomach. One example would be the oleoterpenes in dipterocarp trees of South-East Asia, which have a bacteriostatic resin (Waterman *et al.*, 1988). This resin probably has a negative influence on *Presbytis rubicunda*, which eats very few of the super-abundant dipterocarp plant parts available in the forest.

Although there is a myriad of different plant toxins (Chapter 9), no systematic study has been made of their influence on colobine digestion. Detrimental effects of bacteriostatic compounds can be avoided be rejecting plants that contain them or by keeping their intake within tolerable limits, so that no toxin accumulates to noxious levels (Hladik, 1977*b*; Oates, 1977*a*), but the protective effect of forestomach microbes makes it difficult to predict which compounds are toxic and which can be tolerated.

Geophagy

Soil consumption by colobines, often involving the soil of termite mounds (Chapters 4 and 5), is a conspicuous but rare activity which is still poorly understood. Early explanations suggested that the clay ingested served an antacid function by buffering the pH of the forestomach (Poirier, 1970*b*). Subsequently, physicochemical functions were suggested, relating to taste and absorption of toxins (Hladik & Guegen, 1974). However, there is insufficient evidence from any study to reject the possibility that colobines eat soil as a mineral supplement, especially where species have monotonous diets, and studies of non-colobines have indicated that soil feeding is related to mineral requirement (Nagy & Milton, 1979).

Eating of soil from termite mounds by *Presbytis rubicunda* is most easily understood as mineral acquisition (Davies & Baillie, 1988), in a similar way to *C. guereza* selecting pond plants with high sodium concentrations (Oates, 1978). Soil eating, therefore, could serve both physiological and nutritional functions (Davies & Baillie, 1988).

Conclusion

To conclude, the digestive systems of colobine monkeys do not constrain them to folivorous habits. Like the ruminant duikers (Dubost, 1984) and the mousedeer (Nordin, 1978), which dwell on the forest floor, they can digest seeds efficiently and often make use of these high-energy food sources. But whereas the ungulates forage in the dark herb and shrub layers of the understorey, where poor illumination limits leaf production, colobines utilize the well-lit canopy, where there is prolific leaf growth. They therefore have the opportunity to exploit both seeds and foliage, allowing them to attain high biomass at some sites (see Chapter 10).

The large majority of mature leaves are of poor quality, however, and remain unexploited; strategies different from those used by colobines are needed to exploit them. One way is to increase body size, as have the folivorous gorillas, but this is an impediment to efficient arboreal foraging. An alternative strategy has been adopted by sloths, which are abundant in the forests of the Neotropics. They ferment their leafy diet in a large and sacculated forestomach, but have a reduced metabolic rate, which, through reducing nutritional requirements, allows the ingestion of fibrous mature leaves that are retained for a long time and thoroughly digested. Colobine monkeys are far more active than sloths, but some evidence of energy economy has been found in species with high foliage intakes (e.g. *C. guereza*, *C. polykomos*, *T. vetulus* and *T. pileatus*); long rests, early-morning sun-bathing and positional changes observed in these species serve to reduce heat loss and therefore conserve energy (Oates, 1977*a*; Hladik, 1977*b*; Stanford, 1991; Dasilva, 1992).

A third strategy for accommodating low-quality foliage is that of *Indri indri* (Pollock, 1977) and howler monkeys (*Alouatta* spp.) which lack enlarged forestomachs, but which have considerable caecum–colon capacity (Hladik, 1978; Milton & McBee, 1983). Hindgut fermentation in these species may allow more complete digestion of low-quality foliage than forestomach fermentation in colobines, if the analogy with horses and cattle (Janis, 1976) can be extrapolated. However, more research will be needed before firm conclusions can be drawn.

The flexibility of the colobine digestive system may therefore be its most important feature, since other strategies can be used to digest foliage.

Throughout this chapter we have separated the different aspects of digestion, and how these may relate to different types of diet, to give an overall impression of the capacities of the colobine digestive system. In reality, of course, no colobine monkey eats just leaves or just seeds in a given day, so digestive plasticity is even greater than we have indicated; fruit parts, leaf parts and the occasional insect may all be ingested in a single feeding period and digested effectively. The precise details of the digestive processes are still largely unknown, but the distinctive features of colobine diets correspond with those of other small 'forestomach fermenters', reinforcing conclusions about some of the influences of forestomach fermentation on food selection that have been proposed in this chapter.

Acknowledgments

Helpful comments on several drafts of this chapter were received from David Chivers, Barbara Decker, Jim Moore, Martyn Murray and John Oates. Their contribution is gratefully acknowledged.

9

Colobine food selection and plant chemistry

PETER G. WATERMAN and KAREN M. KOOL

Introduction

Because colobine monkeys rely almost entirely on plant material to meet their nutrient and energy requirements, they face special challenges in selecting an optimal diet. Plants and plant parts are not simply discrete packets of nutrients, they also contain a range of metabolites that are, variously, refractory to digestive processes, capable of lowering the efficiency with which nutrients can be obtained, or actually harmful to the animal, because they interfere with normal physiological processes.

The significance of plant metabolites in lowering food value to herbivores through interfering with digestive processes has long been recognized, particularly the importance of fibre in the feedstuff of domestic ruminants (Van Soest, 1963, 1982). In contrast, the potential role of plant metabolites as active deterrents of selection through effects on palatability or overt toxicity has been examined seriously only since the early 1970s (for general reviews, see Freeland & Janzen, 1974; Rosenthal & Janzen, 1979; Harborne, 1988; Freeland, 1991; Rosenthal & Berenbaum, 1991). This chapter outlines the present state of our knowledge of the modes of action and distribution of plant-derived metabolites with the potential to act as feeding deterrents and discusses the information we have on the ways in which the feeding ecology of colobines may be influenced by plant chemistry.

In this exercise it must be remembered that the pathways of plant secondary metabolism that we observe today have been evolving since the Devonian (Swain, 1978) and that all major extant plant groups have probably existed for over 100 million years. With the evolution of the most recent group, the angiosperms, there was an expansion and diversification of secondary metabolism. Kubitzki and Gottlieb (1984) suggest that this expansion of metabolic effort has resulted from increased interaction with herbivores and pathogens. They argue that within the angiosperms there has also been a change in metabolic emphasis, away from those metabolites that arise from the shikimic acid biosynthetic pathway (lignins, lignans, condensed and hydrolysable tannins, flavonoids, isoquinoline alkaloids) and towards an increasing role for metabolites originating from the acetate/mevalonate route (iridoids, indole and steroidal alkaloids, saponins, sesquiterpene lactones, polyacetylenes).

Colobines therefore inhabit environments that are often chemically inhospitable, and in response, they have developed a range of behavioural, structural, and physiological mechanisms. They are generally highly selective in their choice of food items (Struhsaker, 1975; McKey, 1978*b*; Oates *et al.*, 1980; McKey *et al.*, 1981; Davies *et al.*, 1988; Oates, 1988*b*; Kool, 1992) and exhibit counter-adaptations to potentially harmful plant chemicals. Such adaptations include not only internal detoxification processes such as mixed function oxidases (Brattsten, 1979) but also the microbial interface gained from the development of a preliminary anaerobic fermentation chamber in the stomach (Freeland & Janzen, 1974).

Chemical feeding deterrents

Although various authors had previously recognized and discussed actual or anticipated anti-feedant activity of secondary metabolites, the first coherent hypotheses drawing together plant chemical defence strategies can be attributed to Feeny (1976) and Rhoades & Cates (1976). These authors proposed that two types of defence chemical are produced in plants, referred to here as quantitative and qualitative defences. The quantitative (digestion-inhibitor) defence strategy involved substances capable of interfering with digestive processes, either passively, due to their refractory nature, or through active disruption; the effectiveness of such substances, it was presumed, should correlate positively with the quantity present. Qualitative (toxin) defence was considered to be more directly deleterious, ingested compounds exerting a specific action harmful to the normal functioning of the animal. Such a defence, which requires entry of the chemical into the body, will be liable to counter-adaptation (detoxification), but when effective will be successful at much lower dose levels than a quantitative defence.

Feeny (1976) related quantitative and qualitative defences to plant life history through 'apparency' theory. This predicted that long-lived plants and plant parts, likely to be encountered frequently by herbivores, would be best defended through a large investment in quantitative defences. On the other hand, short-lived plants and plant parts, which through their 'unapparency' could escape herbivory in space and time, would find the production of small amounts of the less certain but potentially completely deterrent qualitative defences more cost-effective.

The hypotheses of Feeny (1976) and Rhoades & Cates (1976) have been the subject of considerable experimental investigation in the last 15 years. As a result, many of their propositions have now been largely discredited. In particular, the 'apparency' concept has found little support and assumptions concerning the 'universal' mechanism of action of some quantitative defence chemicals have been shown to be far too simplistic (see below). New hypotheses have been advanced to explain the distribution of plant secondary metabolites (Coley *et al.*, 1985; Haslam, 1985; Waterman & Mole, 1989) in terms of nutrient/energy balance and its effect on secondary pathways. However, the basic division between quantitative and qualitative defence chemistry remains conceptually useful in separating types of anti-feedant activity exhibited by plant chemicals.

Digestion inhibitors

Plant structural components

Cellulose (Figure 9.1(*a*)), the basic polysaccharide building block of the cell wall (Neville & Levy, 1985), is largely indigestible to herbivores other than those with gut adaptations that allow anaerobic bacterial fermentation. There is substantial evidence (Chapter 8) that colobines have considerable capacity for cellulose digestion and are more capable than most other primates of obtaining energy and nutrients from a diet rich in cellulose.

As with ruminants, the foregut microflora of colobines does not confer any ability to digest lignin. This ubiquitous polymer of cinnamyl alcohol and related phenylpropenes (Figure 9.1(*b*)) is associated with cell-wall cellulose (Harkin, 1973; Northcote, 1985) and, by 'protecting' the polysaccharide from cellulolytic enzymes, inhibits or at least impairs the rate of cellulose fermentation to an extent where it is no longer cost-effective. Levels of lignification are variable between plant species and between different plant parts. Marked variation is also seen with respect to age, particularly in leaves. Lignin would appear to be an entirely passive quantitative defence and possesses no biological activity per se.

Numerous chemical methods have been proposed for the analysis *in vitro* of cell wall polysaccharide and lignin (Bailey, 1973). The most widely used are the detergent fibre methods developed by Van Soest (1963, 1977, 1982); neutral detergent fibre (NDF), acid detergent fibre (ADF) and acid detergent lignin (ADL) methods allow estimation of progressively more refractory cell-

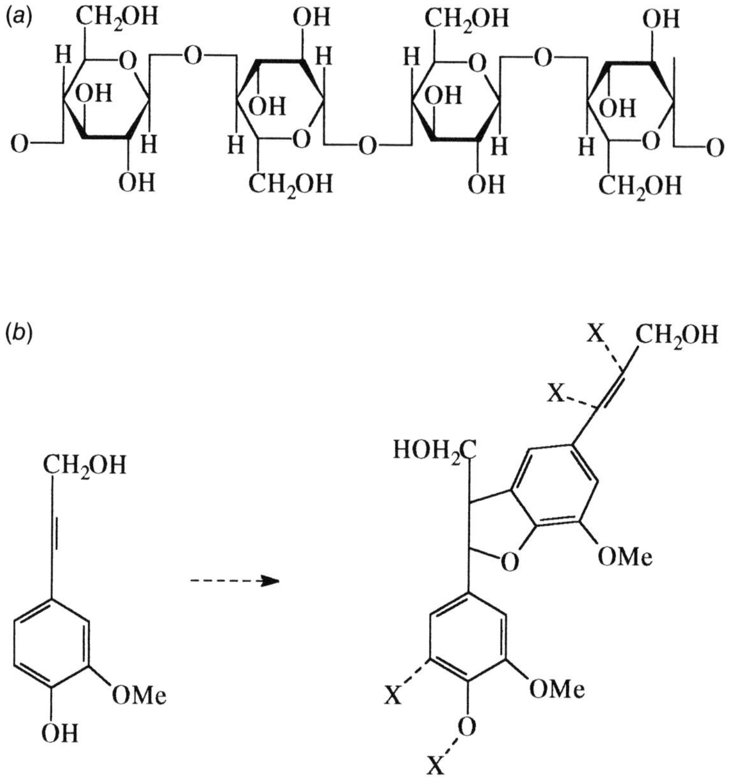

Figure 9.1. Partial structures of (*a*) cellulose and (*b*) simple phenylpropene units that polymerize to give lignin (X = point of attachment of further phenylpropenes in polymerization to lignin).

wall fractions (Figure 9.2). All have been used to some extent in studies on food plants of colobines and other primates (Oates *et al.*, 1980; Milton, 1984; Waterman, 1984; Calvert, 1985; Davies *et al.*, 1988; Oates, 1988*b*; Rogers *et al.*, 1990; Kool, 1992). It has been widely assumed that breakdown and rate of digestion of cellulose in foliage is a function of both lignin concentration and measures of refractory fibre content (e.g. ADF). Numerous studies in which digestion has been measured enzymatically, either *in vivo* or *in vitro*, have confirmed that digestibility is inversely correlated with fibre content (Van Soest, 1977; Aerts *et al.*, 1978; Waterman *et. al.*, 1980; Choo *et. al.*, 1981; Faithful, 1985; Kool, 1992). Woody plants have, on average, a lower cell-wall content than do the leaves of herbs and grasses (Cork & Foley, 1991) but the proportion of lignin is generally higher in the former so

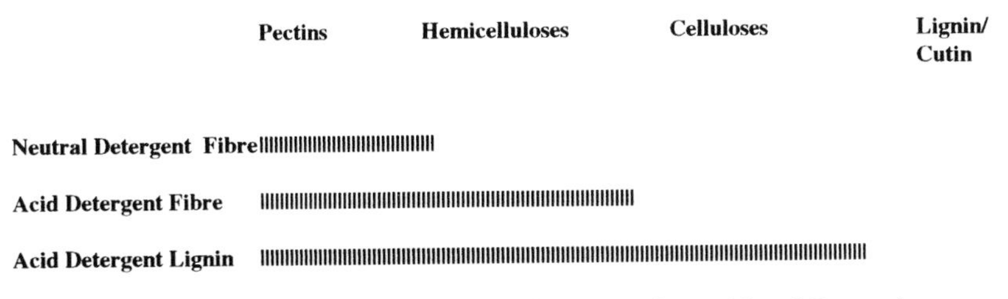

Figure 9.2. The elements of plant cell wall that are estimated by different detergent digestibility methods (after Van Soest, 1963, 1977, 1982).

that their digestibility is not, on average, greater than that of herbs (Robbins *et al.*, 1987). Moreover, an assessment of handling time of plant materials by ruminants (Spalinger *et al.*, 1986) suggested that for leaves from a number of different species there was an inverse relationship with lignin concentration but a positive relationship with mean cell-wall thickness. It is proposed by these authors that high levels of lignification cause cellular rigidity and consequent brittleness, leading to more rapid comminution than would occur in material with lower lignin levels. This may be an important factor in comminution during rumination rather than during fermentation, in which case it will have little importance for a colobine.

Polyphenols

The polyphenolic plant products known as tannins have received more attention in relation to defence chemistry than have any other group of secondary metabolites (Swain, 1979; Zucker, 1983; Haslam, 1989; Hagerman & Butler, 1991). They are of widespread occurrence, particularly in trees, sometimes being found in high concentration. There are two major types in higher plants, the flavan-derived condensed tannin polymer (Figure 9.3(*a*)) and the less widespread hydrolysable tannins (Figure 9.3(*b*)), which are esters between simple sugars and gallic acid and derivatives.

Both types share the ability to interact with proteins to form insoluble complexes. This was once considered a non-specific and irreversible reaction, through which tannins acted as defence compounds by their taste properties (astringency) and their ability to interrupt protein digestion and the activity of digestive enzymes in general (Feeny, 1976; Rhoades & Cates, 1976; Swain, 1979; Haslam, 1989). Detailed investigations have now revealed this view of the reaction between tannin and proteins to be far too simplistic. Tannins do not interact equally with all proteins (see e.g. Hagerman & Butler,

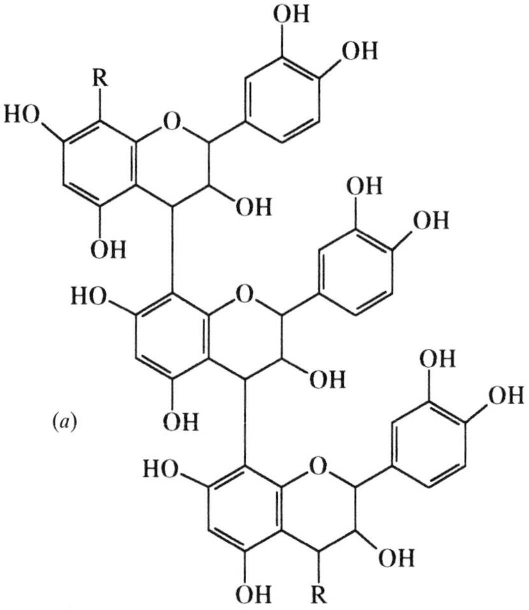

Figure 9.3. Structures of polyphenols: (*a*) a condensed tannin polymer (R = point of attachment of further monomers); (*b*) a simple hydrolysable tannin.

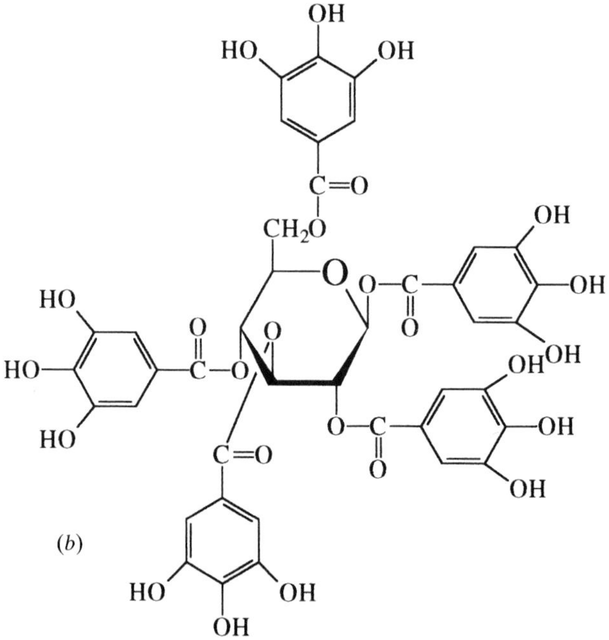

1981; Mole & Waterman, 1987*a*). Surfactant substances have long been known to be able to interfere with the formation and precipitation of tannin–protein complexes, and it has recently been shown that endogenous surfactant substances found in the guts of insects (M. M. Martin *et al.*, 1985; Martin *et al.*, 1987) and mammals (Mole & Waterman, 1985) are capable of similar action. Furthermore, disruption of tannin–protein complexes by these surfactants leads to the regeneration of a denatured protein that is liable to more rapid proteolysis than the original (Mole & Waterman, 1985), so that under some circumstances the presence of tannins can actually enhance the rate at which protein is hydrolysed. Proline-rich salivary proteins have a particularly high binding affinity for tannins. Evidence has recently emerged (Mehansho *et al.*, 1985; Mehansho *et al.*, 1987; Austin *et al.*, 1989; Mole *et al.*, 1990) that these proteins can play a crucial role in preventing ingested tannins from acting in a deleterious manner in the gut. This suggests a possible general method for tannin 'deactivation' through the sacrifice of salivary proteins, although this in itself must place a nitrogen burden on the herbivore.

The uniform quantitative role envisaged for tannins as defence compounds, attractive though it was, does now appear to have been discredited (Bernays *et al.*, 1989). A survey of reports on food selection by vertebrate herbivores in which the role of tannins had been considered (Mole & Waterman, 1987*b*) revealed a wide variation in the levels to which they were tolerated. For each species surveyed, there appeared to be a marked reluctance to select food items with tannin concentrations above that tolerance level. In some cases excess tannin appeared to have the effects on nitrogen digestion anticipated from protein complexation, but in others overt toxicity appeared to occur. Thus, despite the demise of the general digestion-inhibitory role of tannins, their potential to have an impact on food selection remains. In general this impact is negative but situations can, and almost certainly do, occur where selection for some tannin in the diet will be beneficial (Cork & Foley, 1991).

A further complication in assessing the role of tannins is that the methods available for their analysis remain far from satisfactory. Colorimetric procedures inform on content to varying degrees of specificity (Mole & Waterman, 1987*c*) but such methods are generally rather non-specific and tell little about biological activity (Martin & Martin, 1982, 1983; Mole & Waterman, 1987*d*). Biochemical methods involving protein precipitation are very dependent on reactants and on conditions (Hagerman & Butler, 1981, 1991; Schultz *et al.*, 1981; Mole & Waterman, 1985, 1987*d*). A review of methods of tannin analysis has recently been published (Waterman & Mole, 1994).

Miscellaneous substances

Other substances that can be considered as quantitative defences include cutin and suberin and some inorganic cellular inclusions. Cutin and suberin are complex polymers made up primarily of straight-chain acids, alcohols and aldehydes. Both are hydrophobic and non-reactive, suberin being associated mainly with cork and endodermal cells, and cutin forming a coating on the epidermal cells of leaves (Kolattukudy, 1980). Cutin is insoluble in the detergent analyses and is measured together with lignin. Among cellular inclusions silica represents a strongly refractory element and, where present in quantity, it can cause extensive wear on teeth (McNaughton *et al.*, 1985). So far, none of these substances has been implicated as a factor affecting colobine food choice, although the Ulmaceae (*Celtis* spp., *Chaetacme aristata*), the foliage of which are important food sources for colobus at Kibale, Uganda, are rich in silica (Waterman *et al.*, 1980).

Toxins

Of the vast array of secondary metabolites produced by plants, many can be regarded as having potential value as defences against herbivores. The general *modus operandi* of these 'qualitative' defence compounds is to enter the body via the gut and then to act at a specific site in a way that disrupts normal physiological processes. For example, pyrrolizidine alkaloids are generally hepatotoxic, the steroidal cardenolides act on the heart and non-protein amino acids act as anti-metabolites by replacing normal amino acids in proteins (Figure 9.4). For reviews of many of the major classes of qualitative defences see Keeler *et al.* (1978), Rosenthal & Janzen (1979), Harborne (1988) and Rosenthal & Berenbaum (1991).

The specificity of action of qualitative defences means that only small amounts (often much less than 1% dry weight) are sufficient for deterrency as long as the herbivore is susceptible. This specificity does, however, open the way for counter-adaptation by particular or general detoxification systems in the herbivore. General detoxification systems in mammals are predominantly based on oxidation and/or conjugation reactions that modify the foreign compound (Millburn, 1978). In contrast, such systems may be highly specific, as demonstrated by the ability of the bruchid beetle *Caryedes brasiliensis* to detoxify the non-protein amino acid canavanine (Rosenthal & Janzen, 1985) and use it as a major source of nitrogen. Nothing that definite has been established for colobines, but it has been observed (McKey *et al.*, 1981) that the black colobus (*Colobus satanas*) consumes, at certain times of the year,

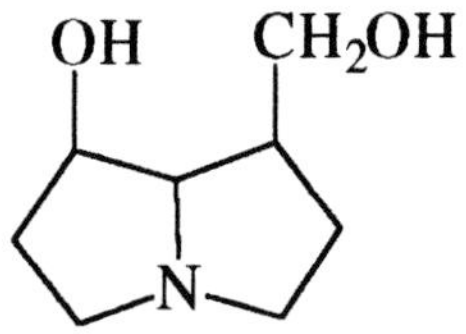

(*a*)

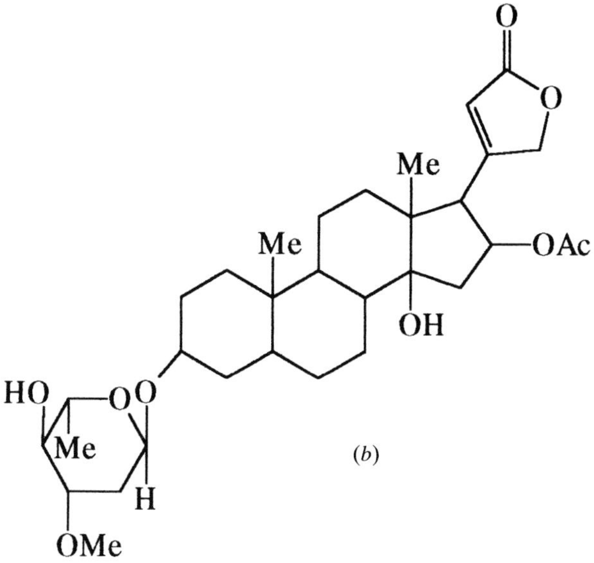

(*b*)

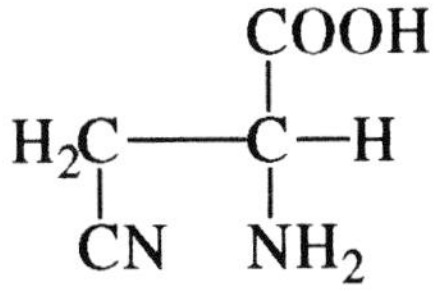

(*c*)

Figure 9.4. Some examples of potential qualitative anti-feedants: (*a*) the pyrrolizidine alkaloid heliotridine; (*b*) the cardenolide oleandrin; (*c*) the non-protein amino acid β-cyanoalanine.

appreciable quantities of the alkaloid-rich leaves of *Rauvolfia vomitoria* (Apocynaceae) and certainly requires to detoxify large amounts of these alkaloids, very small amounts of which would be fatal to a non-adapted herbivore.

Ingested compounds can, of course, be changed by the action of the gut microflora. The potential significance of this is obviously greatest in ruminants and ruminant-like herbivores, such as colobine monkeys, where the initial interface will be between the diverse bacterial flora and the qualitative defence chemical. Only limited evidence exists indicating that qualitative defence compounds can be altered in the rumen (Allison, 1978) but it seems probable that, as Freeland & Janzen (1974) suggest, the phenomenon is widespread. For example, the pyrrolizidine alkaloid-containing plant *Echium plantagineum* (Boraginaceae) is lethal to rats at levels between one-third and one-twelfth that required to show marginal toxic effects in sheep (Culvenor *et al.*, 1984). The black colobus–*Rauvolfia* interaction may well be explained in this manner.

Of course, although the initial interaction between ingested phytochemical frontation with a toxic substance, it does put the well-being of the microflora at risk. To ruminants and colobines the most dangerous qualitative defences may be those that are strongly anti-microbial rather than those that are overtly toxic to the mammal itself.

Although there is no direct evidence to support its occurrence in colobines, the possibility that the toxicity of an individual compound can be potentiated by synergistic effects between it and other ingested compounds must not be discounted (Berenbaum & Neal, 1987). The opposite effect, i.e. nullifying the toxicity of one compound by eating another, is well illustrated by the observations that simultaneous feeding on saponins and tannins by mice reduces the deleterious effects caused by eating either alone (Freeland *et al.*, 1985*a*). In a similar set of experiments, Freeland *et al.* (1985*b*) found that although a high tannin diet depleted sodium levels, the negative effects that resulted could be reversed by mineral supplementation in the diet. Tannins could actually play a beneficial role in neutralizing several groups of qualitative defences, notably alkaloids. The practice of geophagy using clay-rich soils could represent another form of detoxification through absorption of potential poisons in the gut. T. Johns (1986) presents evidence that geophagy using clays capable of absorbing steroidal alkaloids may have been important in the early phases of domestication of the potato. Geophagy has been reported for many colobines but as yet there is no compelling evidence to associate it with detoxification of qualitative defences or with gastric disorders (Oates, 1978; Davies & Baillie, 1988).

One further interesting possibility to be considered is that apparent qualitative defences may, if taken in appropriate form and concentration, be beneficial in acting to correct some physiological disorder or excessive parasite load (i.e. taken as 'medicines'). Rodriguez *et al.* (1985) have pointed to one possible example of this in that chimpanzees have been observed to swallow, without chewing, leaves of *Aspilia* spp. (Compositae). These leaves are a rich source of the antibiotic thiarubine A, which, in addition to its anti-bacterial properties, is an effective antifungal and worming agent.

Methods for the accurate analysis of qualitative defences on a routine basis are, unfortunately, still highly unsatisfactory for most classes of compound. For a few, such as cyanide production, the concentration of a known toxin can be readily estimated. For most, general spot tests that indicate the presence of a particular class of compounds can be used (Harborne, 1984). However, these tests do not provide information about structure or biological activity, such that a positive alkaloid spot test could indicate the presence of the relatively benign caffeine or the highly toxic strychnine. Only prolonged and detailed analysis with extraction and identification of active compounds (for examples see Waterman, 1986) give reliable identification.

From the previous discussion it is apparent that our understanding of the impact of qualitative plant defences on colobines remains extremely fragmentary. It is difficult to see how this situation will improve without significant experimentation on the monkeys, and in the absence of such studies our best models will be domestic ruminants (Keeler *et al.*, 1978).

Distribution of metabolites within and between plant parts

As shown in Chapters 4 and 5, colobines are, with a few notable exceptions, typically arboreal herbivores of tropical moist forests with a diet containing a large foliage component. Such herbivores are generally faced with a wide range of potential foodstuffs to choose between. Community-wide comparisons of plant chemistry in tropical moist forest ecosystems have been attempted and quite extensive data are available for quantitative defences and nutrients (for reviews see Waterman, 1984; Waterman & McKey, 1989). Information on the distribution of qualitative defences is less well documented for the reasons outlined above, but some evidence suggests a negative correlation between content of alkaloids and tannins (Lebreton, 1982; Janzen & Waterman, 1984 – discussed later). From currently available data we can at best make some general statements about the biochemical spectrum of compounds present in plants and the factors influencing their distribution. Those data are here examined with respect to three variables: (a) plant part,

(b) plant habit, and (c) environmental factors; and in terms of various biochemical factors that may be important to food selection: (a) nitrogen content, (b) food digestibility, (c) fibre and lignin content, and (d) potential quantitative phenolic defence compounds.

Distribution in different plant parts

Leaves and flowers

There have been a number of comparisons of the levels of nitrogen/protein and quantitative defence in young and mature foliage in rain-forest plants (Waterman & McKey, 1989; Oates *et al.*, 1990; Kool, 1992). These all confirm that the normal pattern is towards lowering of nitrogen content, often by as much as 50%, and to an increased emphasis on structural components (fibre and lignin) during maturation. In some species structural defences develop rapidly and investment is complete before the leaf has expanded to full size, while at the other extreme leaves may become fully expanded before the build-up of fibre and lignin commences. We know very little about this aspect of the development of young leaves other than that it does seem to be a variable phenomenon (Waterman & McKey, 1989) and this variability may be an important factor in relation to that leaf's palatability to a herbivore.

The significance of polyphenolics in the defence of leaves, particularly young leaves, remains unresolved. Some studies indicate that investment is higher in young leaves (Waterman & McKey, 1989 – and references therein) but others indicate little variation with leaf age (Bennett, 1983) or higher levels in mature leaves (Feeny, 1976; Rhoades & Cates, 1976). Higher concentrations of phenolics in young leaves than in mature leaves of *Shorea* spp. (Dipterocarpaceae) were considered to be the cause of greater protein-precipitating ability of young-leaf extracts (Becker & Martin, 1982), but it remains a valid generalization that young leaves are usually more digestible than their mature counterparts (Waterman *et al.*, 1980; Choo *et al.*, 1981). It has been suggested that apparent high levels of condensed tannins in young leaves may reflect relative ease of extraction (Gartlan *et al.*, 1980). Comparisons of the relative levels of phenolics, fibre and nitrogen in young and mature leaves are given in Table 9.1.

Despite the ambiguous role of polyphenols it seems safe to assume that young leaves (and leaf buds) as a class will generally offer a herbivore a better nutrient package at lower processing cost than do the mature leaves of the same species. However, young foliage is, in its availability, a patchy resource in space and time. Synchronous flushing with consequent periods of

Table 9.1. *Comparison of investment in total phenolics, condensed tannins, acid detergent fibre, cellulase digestibility and protein (=6.25×N) in young (Y) and mature (M) leaves of the same species. Collections were made from six sites where colobine monkeys have been studied.*

	Stage of higher investment	Douala-Edéa	Jozani	Kakachi	Kibale	Kuala-Lompat	Sepilok	%
Total phenolics	Y	5	5	1	6	1	8	29.2
	n/c	4	4	4	13	8	12	50.6
	M	2	5	0	3	4	4	20.2
Condensed tannins	Y	6	5	3	3	0	5	24.7
	n/c	3	3	0	14	7	8	39.3
	M	2	6	2	5	6	11	36.0
Acid detergent fibre	Y	0	1	0	1	0	1	3.4
	n/c	6	12	4	12	10	11	61.8
	M	5	1	1	9	3	12	34.8
Cellulase digestibility	Y	6	5	2	6	6	15	44.9
	n/c	5	9	3	15	6	9	52.8
	M	0	0	0	1	1	0	2.2
Protein	Y	7	9	3	14	5	16	60.7
	n/c	4	5	1	8	8	8	38.2
	M	0	0	1	0	0	0	1.1
n		11	14	5	22	13	24	

Note: n/c, change of less than 20%.
Source: From Gartlan *et al.*, 1980; Oates *et al.*, 1980; Waterman *et al.*, 1980; Choo *et al.*, 1981; Waterman *et al.*, 1988; F. Omari, personal communication.

Table 9.2. *Comparison of investment in acid detergent fibre (ADF), protein (PROT) and the ratio of PROT/ADF for leaf lamina (L) and petiole (P) from 21 species. A change in investment was considered to be a minimum of 5% difference for ADF and 3% for PROT*

	L>P	L≈P	L<P
ADF	8	9	4
PROT	18	2	1
PROT/ADF	17	2	2

absence (Janzen, 1974), or the regular production of small populations of young leaves, are strategies that limit the ability of a large herbivore like a colobine to focus on this resource.

The leaf petiole can be selected as a food resource, the lamina being discarded. This trait seems to be of particular importance for the red colobus (*Procolobus badius*) at Kibale. A comparison of protein levels in the petiole and lamina of *Markhamia platycalyx* (Bignoniaceae) taken on a monthly basis (Baranga, 1983) showed that throughout the year the concentration in the leaf lamina was approximately twice that of the petiole. Comparisons of a number of species in our own laboratory have confirmed that protein levels are generally much lower in the petiole than in the lamina. Relative production of fibre and polyphenols are much more variable (Table 9.2). In contrast, at least in *Markhamia platycalyx*, the levels of some ions (potassium, calcium and sodium) appear to be higher in the petiole than in the leaf lamina (Baranga, 1983).

Both Milton (1984) and Waterman (1984) indicate that the chemistry of flowers is more variable than that of leaves, but that it is more comparable to that of foliage than to any other plant part. Where flowers are nectar-rich they will be major sources of simple, readily digestible carbohydrates, a form of nutrient that is not normally to be found in the diet of colobines.

Seeds

Among seeds of sufficient size to be attractive to a colobine there is the potential for far greater biochemical variability than is found in leaves. Leaves tend to be good-to-adequate sources of protein but are poor sources of non-structural carbohydrate or lipid. Nitrogen is found in high concentrations in some seeds, notably those of Leguminosae, but seeds are more often a rich source of non-structural carbohydrates such as starch and/or lipids. These nutrients, supplied by the plant as resources for the developing seedling, do not necessarily occur in a form which can be utilized by a colobine. For example, many legumes store seed nitrogen in the form of highly toxic non-protein amino acids, while lipids can be stored as similarly toxic cyanolipids.

Quantitative defences tend to be much lower in most seeds, although they may be emphasized in the protective cover (shell, pod, etc.). Polyphenols, where present, are often confined to the testa, where they can form a highly concentrated layer; such layers may well have evolved to inhibit the penetration of fungal hyphae. Polyphenols have been found to be present throughout the whole seed in relatively few cases, such as large seeds with a dispersal mode (e.g. water-borne) that does not require a compact nutrient-rich package (Waterman & McKey, 1989). In contrast qualitative defences are as widespread in seeds as in leaves, and can often attain a relatively high concentration. A notable example is the widespread distribution of amines, non-protein amino acids, flavonoids and isoflavonoids in the seeds of Papilionoideae (Leguminosae) (Gomes *et al.*, 1981; Evans *et al.*, 1985).

Fruits

Included here are all parts of the fruiting body external to the seed. These may be purely protective, in which case they usually rely on a highly fibrous or lignified structural defence. Polyphenols are often present and may be particularly significant in wind-dispersed seeds (Gartlan *et al.*, 1980). Some types of qualitative defences occur, notably latex and resinous materials, e.g. resins in *Hymenaea* spp. (Leguminosae) (Langenheim *et al.*, 1978). To overcome these defences and obtain the seed, the prime requirements of a herbivore would appear to be the strength and/or manipulative skill necessary to break open the container.

Of far more interest in terms of relating feeding ecology to plant biochemistry are fruits designed to play an active role in seed distribution (e.g. fleshy fruits, arillate seeds). In such cases there is the need for the dual role of deterring seed-predators whilst attracting distributors at the correct time. This is a subject that has been discussed in some detail (Howe & Smallwood, 1982; Janzen, 1983) and is only touched upon briefly here. From the viewpoint of a colobine the following points are worthy of particular consideration:

(1) Fruit chemistry often changes greatly during maturation. This is particularly notable in relation to levels of polyphenols, which often reduce

drastically with ripening, a factor that may be important with respect to primate selection (Milton, 1981; Wrangham & Waterman, 1983).

(2) Nutrient rewards offered by fruits are generally either carbohydrate- or lipid-based. These can be concentrated, as in arils, or in dilute packages, as in large fruits with watery flesh (Davies *et al.*, 1984).

(3) Toxins may restrict the appeal of fruits to herbivores not especially adapted to them. Toxins can also be localized within a part of the fruit (e.g. rind) and so offer a barrier that can be overcome only by some particular manipulative skill.

Stems, pith, bark and wood

Green stems have rarely been analysed but what information is available (Goodall, 1977; Hladik, 1977*a*; Waterman *et al.*, 1983; Calvert, 1985; Rogers *et al.*, 1990) suggests they are generally less protein-rich than are leaves. They do not appear to be significant for colobines but can be important in the diets of other primates, notably the apes.

Plant habit

Both plant apparency and resource allocation hypotheses predict that the foliage of climax tree species is likely to be a richer source of quantitative defences than that of colonizing trees (Feeny, 1976; Coley *et al.*, 1985). That this is indeed the case is supported by the extensive study of Barro Colorado vegetation carried out by Coley (1983). However, it must be stressed that this is only a generalization. Many colonizing species are rich in phenolic and structural quantitative defences whereas many climax species are low in phenolics and fibrous material.

There may be a number of reasons for this. The distribution of most types of secondary metabolite follows a phylogenetic pattern (Harborne & Turner, 1983) and condensed tannins are largely restricted to trees (Bate-Smith, 1962). A comparison of evergreen and deciduous tree species of climax forest in Costa Rica (Janzen & Waterman, 1984) revealed that long-lived leaves contained significantly more polyphenolics and were more fibrous than short-lived leaves. Certain combinations of defence chemistry appear to be contra-indicated; for example, studies in both West Africa (Gartlan *et al.*, 1980; Lebreton, 1982) and Costa Rica (Janzen & Waterman, 1984) revealed a negative relationship in the distribution of alkaloids and tannins. This is presumed to be because these two classes of compound will interact in the gut and form insoluble complexes, thus neutralizing the defence potential of the alkaloid, for which its entry into the body is a necessity.

Nitrogen and moisture occur at higher concentration in the leaves of colonizers than in those of climax species (Coley, 1983). Janzen & Waterman (1984) found moisture levels to be higher in short-lived than in long-lived foliage, whereas among climax species those producing qualitative defences such as alkaloids tended to be richer in nitrogen than those in which quantitative defences predominated.

What emerges from these data is a dichotomy in defensive biochemistry in the leaves of tree species, between the employment of quantitative and qualitative defences. The qualitative strategy appears potentially to offer the more nutritionally valuable food package to a herbivore. It would be anticipated that other fast-growing plants, notably lianes, should tend to be low in quantitative defences. Some limited comparative data from rain-forests in Cameroon (McKey *et al.*, 1981) and Sabah (Davies, 1984) support this contention; foliage from climbers is usually lower in both polyphenols and fibre. Epiphytes, on the other hand, tend to be slow growing and it is anticipated that these should be well defended. Unfortunately, no general survey of epiphytes has ever been undertaken to test this hypothesis.

Ground-cover under closed-canopy rain-forest is sparse but herbs become common in light gaps and around the forest edge. A survey of quantitative defences in the herb layer from the Afro-montane habitat of the mountain gorilla (*Gorilla gorilla beringei*) revealed a far lower investment in quantitative defences and a virtual absence of polyphenols (Waterman *et al.*, 1983). This contrasts markedly with the more typical rain-forest environment faced by the western lowland gorilla (*Gorilla gorilla gorilla*) (Rogers *et al.*, 1990). There is no reason to expect that this is not generally true, and that, whereas simple phenolics can occur in high concentrations in the foliage of herbs, polyphenols are to be found relatively rarely. Fibre levels are usually lower in herbs than in leaves of tree species but nitrogen levels tend to be higher (Waterman *et al.*, 1983; Waterman & McKey, 1989).

These generalizations suggest that the climax forest as a habitat is more demanding for a folivore in terms of food selection for an adequate diet, because of a heavier emphasis on quantitative defences. In contrast, secondary forest, and in particular the forest edge and light gaps, is likely to offer a greater choice of nutritionally acceptable foliage. This does not take account of the potential impact of qualitative defences which, theoretically, should pose greater problems in those areas where quantitative defences are less well developed, and may be of particular significance to a colobine where they are strongly anti-microbial in their actions. Finally, it should be stressed that this section has been concerned only with foliar chemistry; similar generalizations cannot be made in relation to fruit and seed chemistry, where adaptations toward different methods of dispersal must to a large extent govern allocation of defence chemicals.

Environmental factors

It has long been known that external factors can influence plant biochemistry. The potential ecological consequences of this were examined by Janzen (1974) who proposed that highly acidic soils produced conditions of nutrient stress, under which plants would invest heavily in chemical defence, as losses to herbivory were more 'expensive' to replace under these conditions. That such biochemical differences do occur has been shown in comparisons between rain-forest sites in Africa (Gartlan *et al.*, 1980) and in Malaysia (Waterman *et al.*, 1988). From these and other investigations of mature foliage obtained from tropical forest sites (Table 9.3) it is now apparent that folivores inhabiting different forests are faced with a food resource whose overall biochemical profile differs appreciably in terms of both quantitative defence and nutrient chemistry. For example, a colobine in the Douala-Edéa Forest Reserve in Cameroon will have to survive in an environment in which common species have invested more in quantitative defences but contain less nitrogen and qualitative defences than will a counterpart in the Kibale Forest of Uganda (Gartlan *et al.*, 1980; McKey *et al.*, 1981). A similar dichotomy exists between the Malaysian colobine sites at Sepilok and Kuala Lompat (Waterman *et al.*, 1988).

Such gross biochemical differences between sites may reflect both species-specific differences between the species that dominate at individual sites and phenotypic variation within the same species. In a recent study comparing the biochemistry of the mature leaves of 11 tree species common to both Lopé (Gabon) and Douala-Edéa (Cameroon), M. Harrison (personal communication) found significant differences in both polyphenolic and fibre content between the two sites.

The question of why plants pursue different biochemical strategies at different sites is undoubtedly complex and certainly not well understood. Janzen's (1974) hypothesis that it stems from nutrient stress through soil acidity and consequent nutrient deficiency is a plausible contributory factor. Furthermore, it can be argued that it is self-sustaining, in that the leachate from tannin-rich foliage will further increase soil acidity and suppress nitrification (Rice & Pancholy, 1974), so further lowering soil quality. More recent suggestions (Coley *et al.*, 1985; Haslam, 1985) look towards an imbalance in metabolic pathways to explain the emphasis placed by different species on secondary metabolite production through different biosynthetic routes. They

Table 9.3. *Comparison of total phenolics (TP), condensed tannins (CT), acid detergent fibre (ADF), cellulase digestibility (CDIG) and protein (PROT) levels (expressed as % dry weight) and the ratio PROT/ADF for mature foliage from seven rain-forest sites where colobines occur. Data are for the abundance weighted profile (AWP) (for method of calculation, see Waterman* et al. *(1988)*

	AWP	*n*	TP	CT	ADF	CDIG	PROT	PROT/ADF
Douala-Edéa	56.3	38	11.4	4.2	55.2	26.9	11.0	0.20
			(7.3)	(5.4)	(47.0)	(33.6)	(12.3)	(0.26)
Kakachi	88.1	14	7.2	9.0	44.5	27.0	10.4	0.24
			(6.3)	(5.4)	(39.4)	(33.0)	(11.7)	(0.31)
Kibale	87.4	23	3.6	1.3	34.6	50.1	16.5	0.51
			(4.1)	(4.7)	(37.1)	(47.1)	(16.4)	(0.49)
Kuala Lompat	49.2	33	5.1	6.7	51.1	34.4	12.6	0.24
			(4.3)	(4.8)	(46.1)	(39.5)	(12.5)	(0.28)
Pangandaran	62.1	9	2.4	1.0	36.7	46.5	11.0	0.41
			(3.4)	(0.8)	(36.7)	(46.5)	(13.8)	(0.37)
Sepilok (flat site)	79.8	19	5.5	9.0	63.4	24.3	10.6	0.17
			(6.0)	(10.3)	(61.2)	(26.8)	(9.9)	(0.16)
Sepilok (ridge site)	68.0	17	7.3	8.0	62.7	27.0	9.9	0.16
			(6.3)	(10.3)	(60.8)	(26.3)	(8.9)	(0.15)
Tiwai	70.9	20	n/a	15.3	43.9	n/a	14.2	0.35
				(14.7)	(41.6)		(13.6)	(0.42)

Notes: Numbers in parentheses are the means for the same species. Basal areas in AWP computations is based on basal area except for Kibale, which is based on the number of stems.
n/a, not available.
Source: Douala-Edéa: J. S. Gartlan, personal communication. Kakachi: Oates *et al.*, 1980. Kibale: Struhsaker 1975. Kuala Lompat and Sepilok: Waterman *et al.*, 1988. Pangandaran: Kool, 1989, 1992. Tiwai: Oates *et al.*, 1990.

suggest that this imbalance will usually arise through the rate of carbon fixation being in excess of the rate of nutrient assimilation rather than nutrient deficiency and that formation of polyphenolics represents a common route for shunting of this excess carbon photosynthate under these conditions. This 'carbon overload' can be induced by various forms of stress including not only soil acidity but also variation in light intensity.

The potential importance of light-stress in influencing investment in phenolic compounds is illustrated by the investigation of *Barteria fistulosa* (Passifloraceae) in Cameroon by Waterman *et al.* (1984). Individual trees growing in a sunny environment in light gaps were found to have foliage significantly richer in phenolics than did individuals growing under the forest canopy. In a more refined study of four rain-forest species from Tiwai, Sierra Leone (Mole *et al.*, 1988; Mole & Waterman, 1988), differences in phenolic levels were found from leaves within individual trees that related closely to the light environments of those individual leaves. These differences were observed over relatively small portions of the light continuum. For example, *Diospyros thomasii* (Ebenaceae) is an understorey species of closed-canopy forest, for which maximum light levels were estimated at only 25% of total incident light at the forest canopy, levels of polyphenolics still varied markedly and in direct relationship to incident light intensity. Such variations may account, at least in part, for individual tree selectivity by *Trachypithecus* in Java (Kool, 1989) and may also explain why some primates can be very selective in feeding within the crown of an individual tree (Glander, 1982).

One other external factor that can have a pronounced effect on the food quality of foliage is the occurrence of nitrogen-fixing root symbionts. A comparison of the amino acid content of mature foliage of Leguminosae (mostly nitrogen-fixing) with those of other Dicotyledonae has shown that the former had significantly greater amounts of amino acids (Waterman, 1994). This, potentially, makes forests rich in legumes particularly important for colobines, but it should be remembered that nitrogen can also be translated into toxins such as non-protein amino acids.

Phytochemical parameters of food selection by colobines

Interpreting biochemical data in relation to colobine food choice

The digestive physiology of colobines is discussed in Chapter 8. Their ruminant-like forestomach with its capacity for fermentation of structural polysaccharides must be a major determinant of food choice. Effective functioning of the forestomach requires that an optimum (not necessarily minimum) level of fibre be present to regulate emptying rate. Tannins may have a net positive effect, if present in the right concentrations, through their ability to inhibit bloat (Jones & Mangan, 1977) or to enhance the digestibility of proteins (see above) but they remain likely anti-feedants at higher levels of intake (Mole & Waterman, 1987*a*). As already noted, the potential toxicity of qualitative defences may be altered by the rumen microbes or by other substances ingested into the rumen.

The approach that has been adopted for colobines to date has been, of necessity, based on nutrient and quantitative defence intake derived from the comparison of what is selected with what is available. The impact of qualitative defences, for which virtually no hard data are available, is discussed in a separate sub-section below.

Biochemical parameters of folivory

Extensive sampling of foliage in relation to colobine food selection has so far been undertaken at nine tropical forest sites: Kibale Forest, Uganda (*Procolobus badius tephrosceles* – Struhsaker, 1975; Gartlan *et al.*, 1980; McKey *et al.*, 1981); Douala-Edéa Forest Reserve, Cameroon (*Colobus satanas* – Gartlan *et al.*, 1980; McKey *et al.*, 1981); Jozani Forest, Zanzibar (*Procolobus badius kirkii* – F. Omari, unpublished observations); Tiwai Island, Sierra Leone (*Colobus polykomos polykomos, Procolobus badius badius, Procolobus verus* – Oates, 1988b; Oates *et al.*, 1990; A. G. Davies and G. L. Dasilva, personal communication); Kakachi, Agastya Malai range, Tamil Nadu, India (*Trachypithecus johnii* – Oates *et al.*, 1980); Kuala Lompat, Krau Game Reserve, W. Malaysia (*Presbytis melalophos* – Chivers, 1980; Bennett, 1983; Davies *et al.*, 1988); Sepilok Virgin Jungle Reserve, Sabah (*Presbytis rubicunda* – Davies, 1984; Davies *et al.*, 1988; Davies, 1991); Lopé Reserve, Gabon (*Colobus satanas* – M. Harrison, personal communication) and Pangandaran Nature Reserve, Java (*Trachypithecus auratus sondaicus* – Kool, 1989, 1992). The methods used for determining the chemical profile of vegetation in these studies have been consistent and results are therefore directly comparable.

Leaf lamina

Although all colobines are to some extent leaf eaters, the proportion of leaves in the diet varies markedly. Table 9.4 gives weighted values for five chemical parameters present in the leaves selected by eight colobine populations (similar weighted values for foliage as a whole are given in Table 9.3 for comparison). Weighting is achieved by considering the total number of leaf feeding records for which biochemical data are available. From this the intake of any biochemical parameter can be gauged using the equation:

$$T = \Sigma(m_s \times p_s)/\Sigma p_s$$

where T is the final weighted value, m is the biochemical measure, p is the percentage of relevant feeding records, and s is the species.

Several trends relating plant biochemistry to food selection are seen from

Table 9.4. *Weighted intake levels (WIL) for total phenolics (TP), condensed tannins (CT), acid detergent fibre (ADF), cellulase digestibility (CDIG) and protein (PROT) and the ratio PROT/ADF for foliage eaten by colobines at different sites*

	Plant part	TP	CT	ADF	CDIG	PROT	PROT/ ADF
Colobus satanas (Douala-Edéa)	All leaves	6.8	5.3	42.0	42.2	18.6	0.61
	Mature leaves	5.3	4.4	43.4	40.0	15.7	0.59
Colobus guereza (Kibale)	All leaves	2.2	0.1	25.2	69.8	24.9	1.06
Colobus polykomos (Tiwai)	All leaves	9.1	3.3	31.6	n/a	20.0	0.82
	Mature leaves	8.4	1.8	36.0	n/a	18.7	0.53
Presbytis melalophos (Kuala-Lompat)	All leaves	2.8	2.0	41.2	45.5	16.0	0.48
Presbytis rubicunda (Sepilok)	All leaves	6.3	9.4	33.7	50.8	18.5	0.67
Procolobus badius tephrosceles (Kibale)	All leaves	5.6	6.2	32.7	51.9	21.1	0.77
	Mature leaves	5.2	7.2	39.2	48.1	16.1	0.46
Trachypithecus auratus (Pangandaran)	All leaves	2.7	0.5	34.7	55.1	11.0	0.35
Trachypithecus johnii (Kakachi)	All leaves	5.7	4.0	30.3	45.3	13.8	0.48
	Mature leaves	4.9	2.4	26.1	45.7	13.5	0.52

Note: n/a, not available.
Source: Feeding records from which WIL values calculated: *C. satanas*: McKey *et al.*, 1978. *C. guereza*: Oates, 1977. *C. polykomos*: Dasilva, personal communication. *Presbytis melalophos* and *P. rubicunda*: Davies *et al.*, 1988. *Procolobus badius*: Struhsaker, 1975. *Trachypithecus auratus*: Kool, 1989. *T. johnii*: Oates *et al.*, 1980.

these calculations. Computed ingestion of fibre (as represented by ADF in all leaves in the diet) varied between 25% and 42.0%. Fibre concentrations showed an inverse correlation with digestibility by cellulase, and it is therefore not surprising that weighted digestibility (42.2–55.1%) was higher for selected foliage than for the average measured at any site. This selectivity was greatest at sites where there had been heaviest investment in structural defences (Douala-Edéa, Sepilok) and in such cases digestibility could become a significant single correlate of leaf selection.

Protein (as estimated from nitrogen × 6.25) was, in most cases, higher in weighted intake computations than in the foliage as a whole. The minimum weighted intake estimate was 11.0%, for *Trachypithecus auratus*, and the highest was 24.9%, for *Colobus guereza*. In comparison, Hladik (1977*b*) reported monthly protein intakes of 10–12% for *Trachypithecus vetulus* (= *T. senex*) in Sri Lanka, with little seasonal variation, and 10–16% for *Semnopithecus entellus*, with considerable seasonal variation.

These data must be treated with caution, as weighted intake computations usually rest heavily on chemical analyses performed on young leaves. All three measures can change dramatically as the leaf matures, often over a relatively short time and without obvious external signs of leaf maturation. In future studies greater efforts must be made to ensure that young leaves to be analysed are of the same age as those observed to be eaten.

Whereas chemical studies in foliage indicate selection for high protein and/or digestibility, and against high fibre, the examination of phenolics and tannins shows no such trends. Weighted total phenol intake values varied four-fold from as low as 2.2% tannic acid equivalents for *Colobus guereza* to as high as 9.1% for *Colobus polykomos polykomos*. Condensed tannin (in quebracho tannin equivalents) varied from as low as 0.1% for *Colobus guereza* to 9.4% for *Presbytis rubicunda*. In some cases weighted values were lower than the average for the site as a whole, in others they were higher (cf. Table 9.3). Intake levels did to some degree reflect levels of investment in phenols and tannins at each site and the inference drawn from this was that tannins and other phenolics are not direct correlates of food selection. However, it is true to say that at each site there did appear to be a trend toward the avoidance of foliage where there was particularly high investment in either condensed tannins or total phenols, suggesting that there are tolerance limits. This was most notable at Douala-Edéa, where leaves rich in condensed tannins were abundant and were avoided as food items, with the result that a simple rank correlation of selection against tannin concentration approached significance.

Milton (1979) for howler monkeys and Oates *et al.* (1980) for colobines both recognized that selection of leaves was primarily dependent on fibre and protein content. McKey *et al.* (1981) and Waterman & Choo (1981) considered the use of various ratios based on the above measures as predictors of suitability as foods for colobines. The simplest of these, the ratio of protein to fibre (or to the sum of fibre and condensed tannin) gave good correlation with leaf selection by *Colobus satanas*. This protein/fibre (PROT/ADF) ratio has generally proved to be a useful predictor of diet at sites where investment in quantitative defences has been high and where plant chemistry would therefore be anticipated to have a prime influence on food selection. Weighted PROT/ADF ratios for eaten foliage are generally, but not always, higher than average values for that site (Table 9.4), usually by a factor of greater than

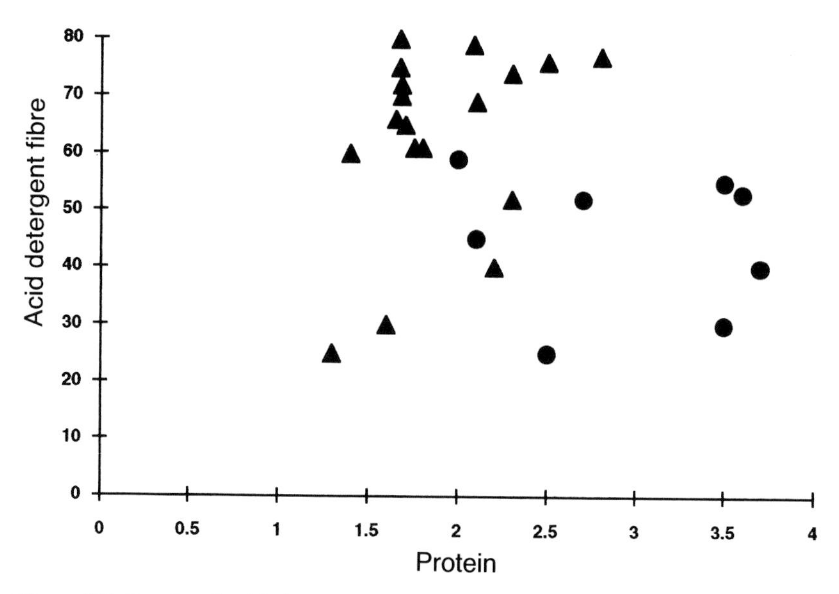

Figure 9.5. Levels of acid detergent fibre and protein (N × 6.25) in leaves selected and rejected by the black colobus monkey (McKey *et al.*, 1981). Triangles indicate 0–5 feeding records and circles over 5 feeding records.

1.5; in the case of *Presbytis rubicunda*, which is, on this criterion, living in the most hostile environment, the ratio is greater by a factor of over 3 (0.67 : 0.15). A graphical representation of this ratio in relation to leaf selection by *Colobus satanas* is shown in Figure 9.5. A similar plot was obtained for *Presbytis rubicunda* but not for colobines at richer sites (e.g. *Procolobus badius tephrosceles* at Kibale), where a far greater proportion of tree species produce leaves with an acceptable PROT/ADF ratio.

Parts of leaf lamina

Most colobines studied have been found to differentiate between apical and basal parts of the lamina of some plant species. The number of biochemical analyses that have differentiated zones within the leaf have been very small; Table 9.5 gives PROT/ADF ratios for four species where separate data have been obtained. In three of these there was a pronounced gradient with the apex having the higher ratio, largely caused by a gradient in nitrogen concentration. This indicates that gross differences can occur within a leaf, but it must not be taken as evidence that the apical portion of the lamina is generally more nutritious and to be preferred; indeed, records for eating basal

Table 9.5. *Ratio of protein to acid detergent fibre (PROT/ADF) for apical and basal parts of the lamina of leaves from four species*

	PROT/ADF	
	Apical	Basal
Calophyllum inophyllum		
Young leaf	0.18	0.15
Mature leaf	0.15	0.19
Terminalia cattapa		
Mature leaf	0.41	0.20
Teijsmanniodendron glabrum		
Mature leaf	1.20	0.45
Teijsmanniodendron pteropodum		
Mature leaf	0.45	0.21

portions are not uncommon. It does point to the need to differentiate between areas of the lamina in biochemical analyses. Where such selective feeding behaviour is observed in future, eaten and uneaten portions should be sampled separately and if possible observations should be made in relation to growth patterns within the expanding leaf.

Petioles

For some colobines leaf petioles form a specific and important dietary item. It does not seem possible to rationalize this on the basis of the biochemical parameters used to define food quality of leaf lamina. Table 9.2 gives the results of a comparison of 22 sets of petiole and corresponding leaf lamina. This reveals that in 18 of these 22, protein levels were at least 20% higher in the lamina than in the petiole. Fibre levels tended to be similar in the lamina and petiole. As a consequence, in 16 out of the 22 the PROT/ADF ratio was greater in the lamina (for 11 the difference was greater than 1.5 : 1) and in only two was the ratio higher for the petiole than the lamina.

Thus, on these criteria petioles represent a less attractive food source, and reasons for their selection must be sought elsewhere. A detailed investigation of petioles and lamina of *Markhamia platycalyx* (Bignoniaceae) at Kibale by Baranga (1983) revealed that over the course of a year petioles had consistently higher moisture content and were richer sources of potassium, calcium and sodium, but not phosphorus. Energy levels were higher in the lamina. One possible explanation for petiole eating would be that it is a response to specific mineral needs. Unusual feeding behaviour by *Colobus guereza* at Kibale has been attributed to the need for sodium (Oates, 1978), an element that is somewhat patchily distributed in foliage (Golley *et al.*, 1980*a*,*b*; Ganzhorn, 1985). Freeland *et al.* (1985*b*) have shown that in mice, sodium intake needs to be supplemented where the diet contains significant amounts of tannin. Another possible explanation is that the petiole, which is the highway into and out of the leaf, might not be as rich in qualitative defences as the leaf.

Flowers

Among colobines studied there is a considerable variability in the use of flowers and flower buds. A few analyses are available from Kibale and Jozani in Africa and Kuala Lompat and Sepilok in Malaysia. The food value of opened flowers, as estimated by the PROT/ADF ratio, is very variable (mean 0.49, range 0.08–1.23, $n = 18$). A flower bud is generally slightly richer in nitrogen than the expanded flower. In 11 cases where ratios for flowers could be compared with those for leaf lamina, computed values lay between those for young leaves and those for mature leaves; so in these terms flowers would appear to represent a better food resource than mature leaves, but a poorer resource than young leaves. It is presumed that the major source of nutrients is in the sexual parts of the flower rather than the petals and sepals, but no study has yet been made to confirm this. Depending on their adaptations for pollination, flowers may possess other nutritionally valuable substances not present to any degree in leaves, such as sugars and amino acids in nectars.

Biochemical parameters of frugivory

Whole fruits

It is less satisfactory to make generalizations about the biochemistry of fruits, because, as noted previously, such chemistry will be a function of the fruit's dispersal mechanism. The approach taken here is to discuss the limited data available from Sepilok, Kuala Lompat and Kibale for fruit chemistry in terms of the items that are eaten (Table 9.6). Some additional data are also available for *Presbytis hosei* observed in the Kutai National Park, Kalimantan, Indonesia (M. Leighton and P. G. Waterman, unpublished). At Sepilok and Kuala Lompat, *Presbytis rubicunda* and *P. melalophos*, respectively, eat a number

Table 9.6. *Chemical analysis for fruits and fruit parts eaten by colobines. CT, condensed tannins; ADF, acid detergent fibre; PROT, protein*

	Source	CT (*n*)	ADF (*n*)	PROT (*n*)	Fats (*n*)
Whole fruit	Sepilok	8.5 (7)	39.3 (7)	8.7 (7)	
Whole fruit	Kuala Lompat	1.2 (2)	42.7 (2)	5.8 (2)	
Whole fruit	Kibale	0.0 (2)	22.3 (2)	17.0 (2)	
Fruit flesh	Kutai[a]	2.8 (7)		7.3 (7)	31.6 (3)
Arils	Kutai	2.9 (11)	22.0 (9)	9.5 (11)	29.2 (16)

Note:
[a]Kutai, Indonesia (M. Leighton and P. G. Waterman, unpublished).

of fruits that are uniformly poor sources of protein and have PROT/ADF ratios of less than 0.20, a value which would normally be regarded as about the lower limit for selection of a leaf by a colobine (Davies *et al.*, 1988). The average protein level of fruit flesh consumed by *Presbytis hosei* is of the same order, but additional analysis of lipids reveals that these fruits are the source of appreciable quantities of fats. In contrast, the fruits of *Teclea nobilis* (Rutaceae), which are eaten by *Procolobus badius tephrosceles* at Kibale, have a PROT/ADF ratio of about 1.0 and contain between 15 and 25% protein. Calorimetric energy measurements on some of these fruits does not suggest that they are selected on this criterion.

It seems clear that the three Malaysian *Presbytis* species do not select fruit for protein content, and it appears possible that *Presbytis hosei* selects for lipids. This colobine also eats lipid-rich arils (Table 9.6), and it is notable that these contain more protein and are low in fibre and tannin, giving a high value for the ratio, even though it will be an underestimate of true nutritional value, as it does not account for the lipids. Interestingly, although arils occur at Sepilok and Kuala Lompat, they are not favoured food items, and analyses of a small sample of uneaten arils collected at Sepilok showed that at this site arils are high in fibre and low in protein (Davies, 1991).

Fruit eating by colobines warrants further study. Mature fruits, likely to contain a high content of simple sugars, generally appear to be avoided by colobines. It has been suggested (Davies *et al.*, 1984) that the reason for their rejection by colobines lies in the inability of the microbial flora of the foregut to metabolize large quantities of simple sugars without impairing the ionic balance of the stomach through the very rapid formation of volatile fatty acids. The chemistry of fruits is known to be capable of rapid change during the maturation process (Swain, 1979; Wrangham & Waterman, 1983) and we

Table 9.7. *Chemical profiles of seeds in relation to acceptability to colobines*

	Category	*n*	TP	CT	ADF	CDIG	PROT
Colobus polykomos	Uneaten	6	22.0	12.3	28.8		13.6
	Eaten	8	3.9	1.7	25.0		21.2
	WIL		4.8	0.2	31.9		29.9
Colobus satanas	Uneaten	12	6.5	3.9			7.0
	Eaten	5	1.7	1.3			12.7
	WIL		2.1	1.4			13.7
Presbytis hosei	Eaten	6	2.4	3.2	22.0	44.2	10.1
Presbytis melalophos	Uneaten	6	2.5	3.4	39.8	53.4	6.5
	Eaten	4	2.3	1.5	32.2	38.8	6.7
	WIL		1.5	0.9	29.7	48.5	9.6
Presbytis rubicunda	Uneaten	6	2.0	3.8	25.0	71.7	8.3
	Eaten	16	1.8	4.7	16.4	59.9	9.3
	WIL		1.8	11.1	22.3	57.4	9.0

Note: TP, total phenolics; CT, condensed tannins; ADF, acid detergent fibre; CDIG, cellulase digestibility; PROT, protein; WIL, weighted intake level (see Table 9.4).

know nothing about this process in relation to acceptability to colobines. If sugar-rich fruit flesh is to be avoided, then unripe fruits of some species could be acceptable, whereas ripe fruits, apparently a richer source of energy, would not.

Seeds

The importance of seeds in colobine diets varies considerably (see Chapters 4 and 5). In total about 70 analyses have been performed relevant to seed eating by *Colobus satanas* (McKey *et al.*, 1981), *C. polykomos* (G. Dasilva, personal communication), *Presbytis melalophos* and *P. rubicunda* (Davies *et al.*, 1988; Davies, 1991) and *P. hosei* (M. Leighton and P. G. Waterman, personal communication). These analyses (Table 9.7) reveal large differences in protein levels. Fibre is often lower than is found in any other plant part except very immature leaves, but this is not uniform. Condensed tannins are often absent or, when present, restricted to the outer layer of the seed (testa). Seeds are commonly the repository of qualitative defences; some examples of toxins known to occur in seeds from Douala-Edéa are shown in Figure 9.6.

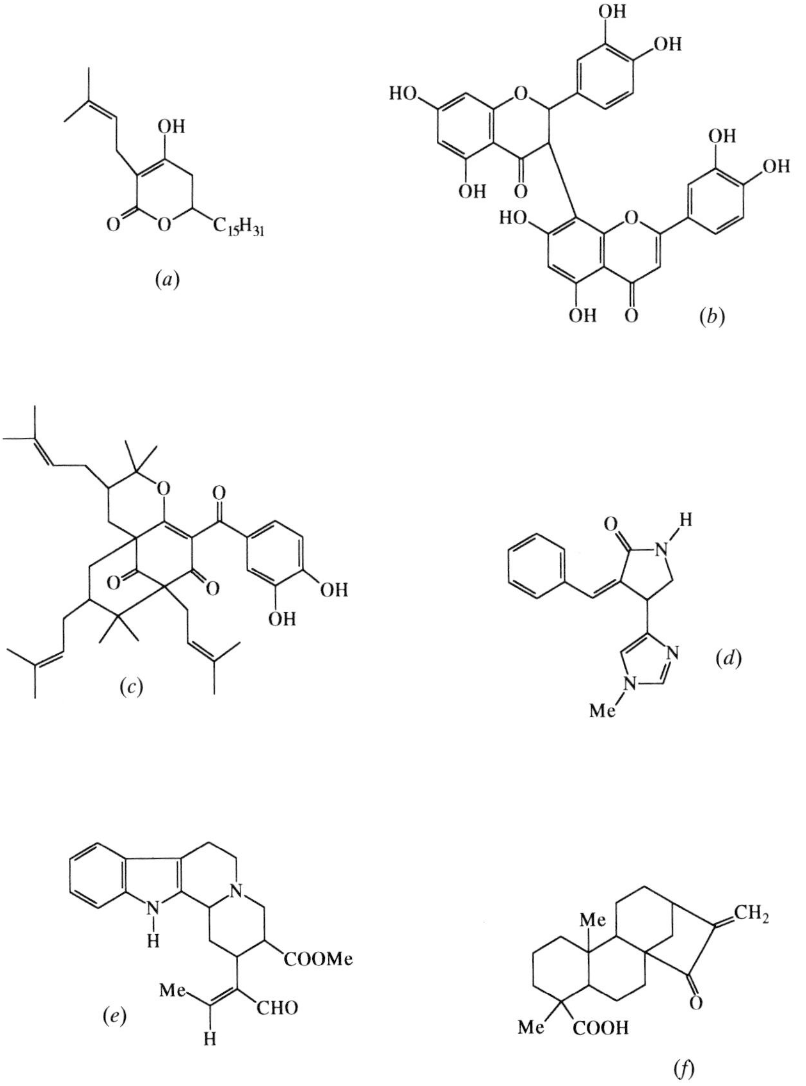

Figure 9.6. Examples of potential qualitative anti-feedants isolated from seeds available to the black colobus monkey in Douala-Edéa Forest Reserve, Cameroon (McKey *et al.*, 1981) (*a*) from *Garcinia conrauana* (Guttiferae); (*b*) from *Garcinia densivenia*, (*c*) from *Garcinia ovalifolia*; (*d*) from *Cynometra hankei* (Leguminosae), (*e*) from *Strychnos tricalysoides* (Loganiaceae), (*f*) from *Xylopia aethiopica* (Annonaceae).

At Douala-Edéa seed selection by *Colobus satanas* appeared to correlate directly with protein concentration (McKey *et al.*, 1981). Eaten seeds for which data are available all contained >10% protein, while all seeds not eaten contained <10%, with two exceptions. These, from *Xylopia quintasii* (Annonaceae) and *Strychnos tricalysoides* (Loganiaceae) are, respectively, sources of diterpenes (Hasan *et al.*, 1982) and alkaloids (Waterman & Zhong, 1982) (see Figure 9.6). The more recent data obtained for the Malaysian *Presbytis* spp. do not show the same correlation of selection with protein, average or weighted values for selected seeds being 9–10% for protein. Unfortunately, there is no obvious correlation with any other biochemical measure.

It seems likely that seed selection will prove more difficult to rationalize on biochemical terms than leaf selection, for a number of reasons. First, seeds represent a more variable nutrient package than leaves and some of the potential combinations, although appearing to be of high nutritional value, may be unacceptable to the rumen microflora. Second, seeds are likely to be more concentrated sources of toxins than are leaves. Third, seeds present greater handling problems, either because of mechanical protection (e.g. within a thick lignified pod, as in many Leguminosae) or because they are surrounded by an unacceptable fruit covering, such as the terpene-rich resins of the pericarp of many dipterocarps.

Conclusions

Where do we stand today?

Using rather crude chemical measurements originally designed to give information on feedstuff quality for ruminants has proved quite successful in rationalizing patterns of folivory observed for a range of colobines from the 4 kg *Procolobus verus* to the 11 kg *Colobus guereza*. This suggests that colobines are really behaving as arboreal browsers in a manner comparable to small artiodactyl browsers such as the dikdik (Hoppe, 1977*a*; Owen-Smith & Novellie, 1982). Their ability to digest fibre was given emphasis by a laboratory study on *C. guereza* (Bauchop & Martucci, 1968; Kay *et al.*, 1976; Watkins *et al.*, 1985) which indicated a fibre digestion capability exceeding that reported for some ruminant species.

Leaf eating by colobines can therefore be expected to show both the benefits and constraints that would be predicted for a terrestrial browser. These include the following:

(1) What is acceptable to any given species will to a considerable extent be

dictated by body size; smaller colobines will need more rapid nutrient capture rates and will be more restricted in the quality of foliage that will maintain them (Freeland, 1991). At the extremes this explains why the very small *Procolobus verus* is associated with fast-growing secondary forest, whereas the very large proboscis monkey (*Nasalis larvatus*) can survive in the seemingly harsh environment of a mangrove swamp. If this is so, then it might be anticipated that the weighted intake level for PROT/ADF would be inversely related to body size. Figure 9.7 shows that the two Malaysian *Presbytis* spp. do not conform to the expected pattern.

(2) The supply of acceptable vegetation will obviously set limits to sustainable biomass. In biochemically harsh sites a large body size with consequent lowering of acceptable leaf quality will be advantageous (*Trachypithecus johnii*). For a small-bodied individual, population density will have to be very low (*Presbytis rubicunda*). Where conditions are less harsh and there is a greater reservoir of edible leaves, a very high biomass can be sustained (*Procolobus badius tephrosceles*), greater than is possible for a sympatric monogastric primate, for which food

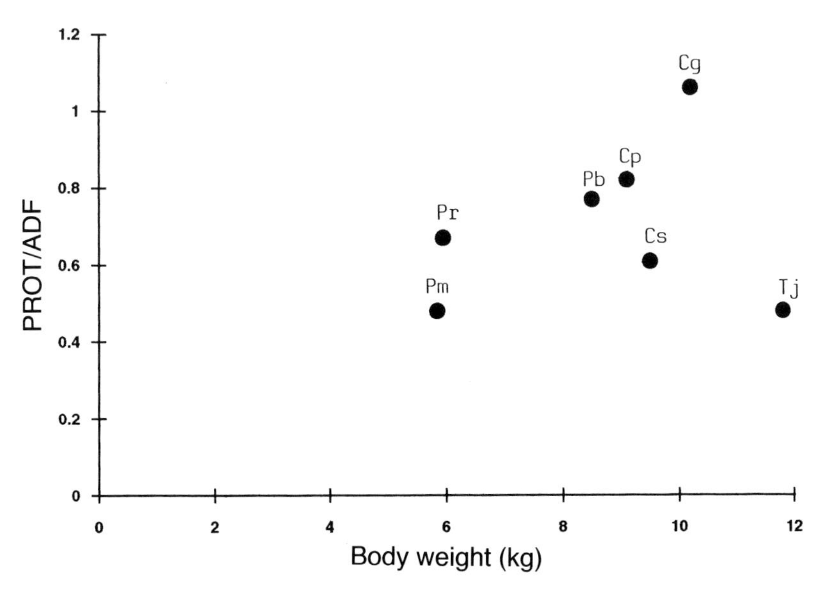

Figure 9.7. A plot of weighted intake levels based on the ratio of protein (PROT = 6.25 N) to acid detergent fibre (ADF) against average adult body weight for the colobines *Colobus guereza* (Cg), *C. polykomos* (Cp), *C. satanas* (Cs), *Procolobus badius tephrosceles* (Pb), *Trachypithecus johnii* (Tj), *Presbytis melalophos* (Pm) and *Presbytis rubicunda* (Pr).

resources will be patchy in both space and time (Struhsaker & Leland, 1979).

(3) Colobines will not react to qualitative defences in the same way as will a monogastric primate, because of the intermediacy of the gut microflora. The potential of the rumen to act as a detoxification chamber has been stressed by Freeland & Janzen (1974), and is well illustrated by data from domestic ruminants (Keeler *et al.*, 1978). Although no direct evidence of detoxification is available for colobines, the capacity of *Colobus satanas* to exploit the leaves of *Rauvolfia vomitoria* (Apocynaceae) despite the presence of a number of potentially toxic indole alkaloids (Amer & Court, 1980) can be explained by rumen detoxification. Other examples where qualitative defences appear to be detoxified by colobines include seeds of *Erythrophleum* sp. (Leguminosae) eaten by *Procolobus badius* and *Colobus satanas*; leaves of *Funtumia latifolia* (Apocynaceae) eaten by *Procolobus badius* and the fruits of *Strychnos potatorum* (Loganiaceae) eaten by *Semnopithecus entellus* (Waterman, 1984). *Colobus angolensis* (Moreno-Black & Bent, 1982), *C. guereza* (Oates *et al.*, 1977) and *Trachypithecus johnii* (Oates *et al.*, 1980) are all reported to eat foliage containing alkaloids.

(4) Reliance on the fermentation processes of a gut microflora does also have some disadvantages. Food choice will be constrained by compounds that, although not toxic to the physiology of the primate, are anti-microbial. No studies have yet been made of antibiotic activity in relation to food selection by colobines, but these would certainly be worthwhile. As noted above, colobines may also be forced to avoid rich energy sources based on simple carbohydrates because of the effect these may have on the environment in the fermentation chamber.

The above comments consider colobines as leaf eaters, but this is only half the story. Some, perhaps a majority, are equally fruit and seed eaters (Davies *et al.*, 1988). An interesting question arises as to whether in such cases they are preferentially frugivores or semivores that are able to maintain themselves by folivory in times of shortage in their favoured resource, or whether they are folivores that turn to other foodstuffs at seasonal bottlenecks, when young leaves are absent and acceptable mature leaves are rare (Harrison & Hladik, 1986; Oates *et al.*, 1990). The evidence is conflicting. *Procolobus verus* certainly fits the latter pattern, but *Colobus satanas*, *C. polykomos, Presbytis rubicunda, P. melalophos* and *Trachypithecus auratus* seem to fit the former. A recent report on the diet of *Trachypithecus pileatus* (Stanford, 1991) suggests it prefers young leaves and fruit when these are available. At present we know insufficient about criteria governing fruit and seed selection, other than that they generally differ from those governing leaf selection.

What are the priorities for future research?

It seems doubtful that much will be gained from further studies of the correlation between plant chemistry and colobine food selection using the very general chemical analyses employed to date. These have been valuable in identifying major selection factors and in highlighting how the phytochemical environment in which the arboreal colobines exist can differ greatly with consequences for the animal. The following areas of study are worthy of consideration for future research.

First, in the field there has to be much greater attention to linking more closely what is observed to be eaten with what is actually analysed (Glander, 1982). The considerable changes in plant chemistry that can occur for a single plant part not only within individuals of a species but within a single tree are only now being realized. This is of particular importance if we are going to understand the biochemical basis of young-leaf and fruit selection.

A second imperative of future field work is to concern ourselves more with fluctuations in daily intake and with positive or negative associations between the selection of food items. This is necessary to develop our understanding of how the short-term nutrient and energy budget is maintained and to ascertain if food selection is in any way designed to take account of the intake of specific nutrients and/or defence compounds relative to one another.

Other areas need to be investigated in the laboratory. First, we need much more data on the fermentation efficiency and throughput rates of colobines. The paper by Watkins *et al.* (1985) to some degree points the way, but what we want are investigations on newly captive animals with a microflora used to handling the complex diet of a wild colobine and presented with natural foliage of known composition, rather than a study of long-term captive animals fed high-fibre biscuit. An understanding of the mechanisms of microfloral detoxification of qualitative defences is perhaps one of the greatest requirements. We need to know the range of structures that the microflora can modify, what type(s) of modification occur and how quickly the microflora can respond to the introduction of new toxins. Laboratory studies can also add much to our present very confused picture of the impact of tannins on colobine digestion. Do tannins interact with salivary mucoproteins? What concentration of tannins can be tolerated without unacceptable protein loss? Are there certain tannin : protein ratios that lead, through formation of soluble complexes, to enhanced rates of protein digestion in the gut? Do tannins alter the profile of nitrogen metabolism between the foregut and the remainder of the digestive tract?

Answers to any of the questions posed in this section would promote our understanding of the role of plant chemistry in shaping colobine food selection. In the field, it requires detailed observation and collection to a level of refinement so far not achieved. For this, and for needed laboratory studies, a research centre will be required that is not only capable of undertaking the necessary biochemical investigations, but that also has access to animals that are confined but maintained in as near natural environments as possible.

Acknowledgements

K. M. K. thanks the Indonesian Institute of Sciences (LIPI) and the Nature Protection and Wildlife Management Division (PHPA); financial support to K. M. K. was provided by a Commonwealth Postgraduate Research Award and the Joyce W. Vickery Scientific Research Fund (Linnean Society of New South Wales). Biochemical studies performed by P. G. W. were supported by Natural Environment Research Council grants GR3/3455 and GR3/5554.

10

Colobine populations

A. GLYN DAVIES

Colobine population densities have been recorded to range from about 15 individuals/km^2 for *Presbytis rubicunda* in north Borneo, to over 315 individuals/km^2 for *Colobus guereza* at Bole in Ethiopia. The causes of this variation are of interest from both ecological and conservation viewpoints but they are still poorly understood. In part this is because most colobine monkeys occupy forests that are rich in plant species and complex in their ecology, but also because there have been few long-term studies to document patterns of reproduction and population growth in these long-lived mammals. Despite this shortcoming, studies have been carried out at over 20 sites in Asia and Africa where colobine population densities and biomass have been recorded (Figure 10.1), and these provide a reasonable data-set for preliminary investigation of the factors that limit colobine populations.

In this chapter efforts have been made to use the most recent measures of body weight (see Chapter 3, Table 3.2) and reported bisexual group composition to improve the accuracy of biomass calculations (Table 10.1). The biomass estimates (kg/km^2) are calculated using published figures for population densities (individuals/km^2), but use of new body-weight figures means that they do not correspond exactly with previously published biomass figures (Tables 10.2 and 10.3).

Colobine population density and biomass

In Africa, one of the highest population densities recorded is in the Kibale Forest. Both *Procolobus badius* (red colobus) and *Colobus guereza* occur, but the former contributes a much greater proportion of the numbers (298 individuals/km^2) in the primary forest of compartment 30 in the Kanyawara study area (Struhsaker, 1975). In nearby forest, with a mosaic of lightly felled and unfelled patches, the *P. badius* numbers declined by one-third, but the

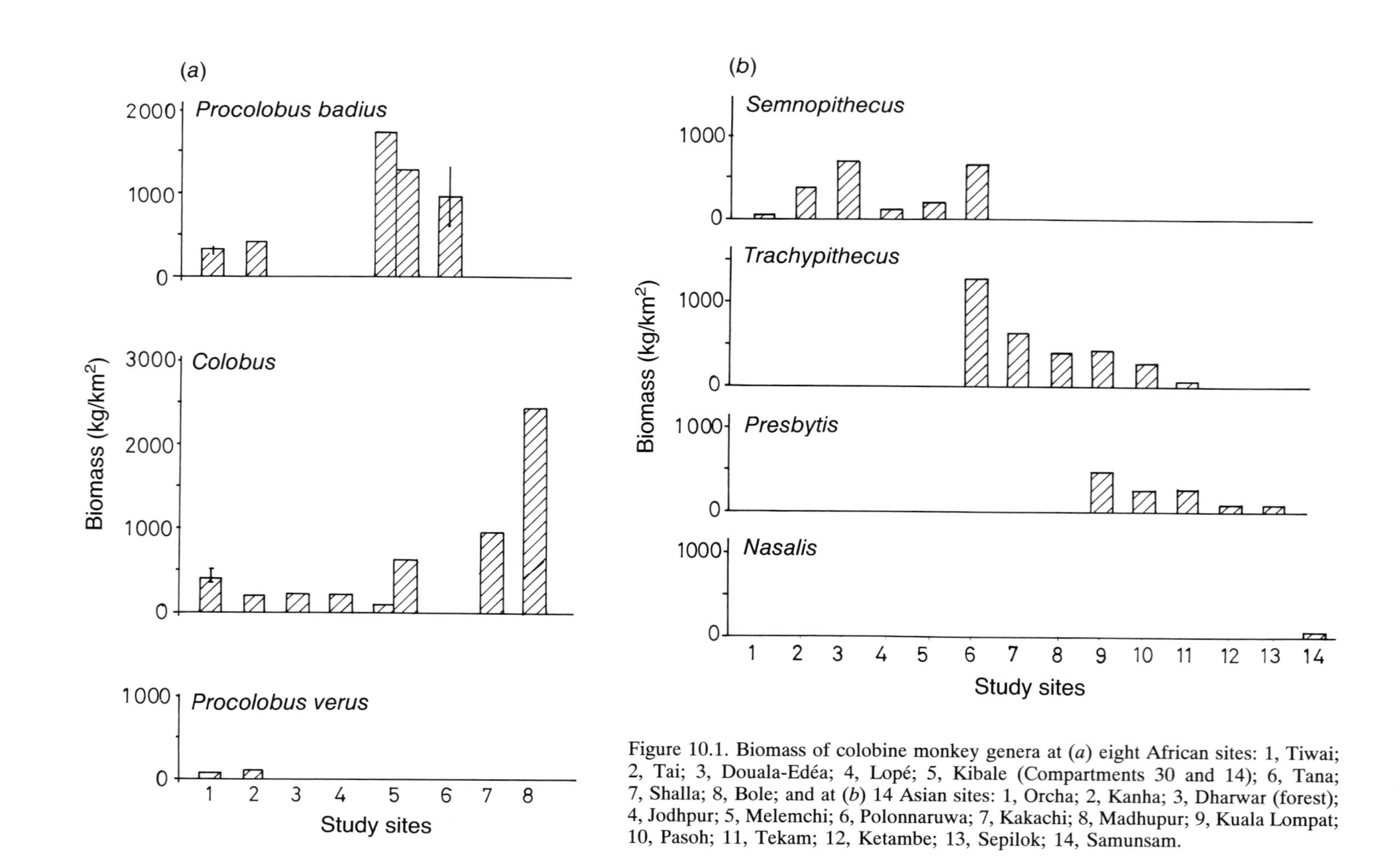

Figure 10.1. Biomass of colobine monkey genera at (*a*) eight African sites: 1, Tiwai; 2, Tai; 3, Douala-Edéa; 4, Lopé; 5, Kibale (Compartments 30 and 14); 6, Tana; 7, Shalla; 8, Bole; and at (*b*) 14 Asian sites: 1, Orcha; 2, Kanha; 3, Dharwar (forest); 4, Jodhpur; 5, Melemchi; 6, Polonnaruwa; 7, Kakachi; 8, Madhupur; 9, Kuala Lompat; 10, Pasoh; 11, Tekam; 12, Ketambe; 13, Sepilok; 14, Samunsam.

Table 10.1. *Group comparison (used to calculate the average weight of individuals in bisexual groups)*

	Groups (*n*)	Adult males	Adult females	Non-adults	Source
P. rubicunda (Tanjung Puting)	4	4	9	11	Supriatna *et al.* (1986)
P. melalophos (Kuala Lompat)	3	5	18	19	Curtin (1980)
	3	3	23	19	Bennett (1983)
T. obscurus (Kuala Lompat)	2	5	12	17	Curtin (1980)
T. pileatus (Madhupur)	11	13	32	34	Stanford (1991)
T. vetulus (Polonnaruwa)	>15	52	132	31	Rudran (1973*b*)
S. entellus					
(Polonnaruwa)	1	6	10	9	Newton (1984, 1988*a*)
(Kanha)	14	15	132	157	Newton (1984, 1988*a*)
P. badius badius (Tiwai)	1	6	12	17	Oates *et al.* (1990)
P. badius tephrosceles (Kibale)	1	7.5	17.5	25	Struhsaker (1975)
P. badius rufomitratus (Tana)	13	19	131	95	Marsh (1978*a*)
P. verus (Tiwai)	1	1	2	3	Oates *et al.* (1990)
C. polykomos (Tiwai)	1	2	4	3	Oates *et al.* (1990)
C. guereza occidentalis (Kibale)	7	12	24	40	Oates (1977*b*)
C. guereza guereza					
(Bole)	12	19	25	40	Dunbar (1987)
(Shalla)	6	6	12	29	Dunbar (1987)
C. satanas (Douala-Edéa)	1	2.2	7	6.2	McKey (1978*a*)

C. guereza showed a nearly five-fold increase – from 22 to about 100 individuals/km^2 (Oates, 1977*c*).

This is interesting from two perspectives. First, it shows the great range of population densities a single species can achieve within a small area, and second, it confirms the different habitat preferences of the two species. This preference of red colobus for primary forest should not be overstated, however, since red colobus populations also occur in the gallery forests of Kenya, Tanzania and Senegambia. Indeed, the biomass of red colobus in the gallery forests of Tana river is lower than recorded at Kibale (Marsh, 1978*a*), but higher than in the rain forest at Tiwai (Oates *et al.*, 1990).

Table 10.2. *Comparison of population densities and biomasses of colobine species at selected Asian sites*

Study site	Body weights (kg)				Biomass (kg/km^2)		
	Adult male	Adult female	Average (including non-adults)	individuals/km^2	EGA	BSG	Source
Orcha							
S. entellus	18.3	11.2	9e	4.4		40	Newton (1984)
Kanha							
S. entellus	18.3	11.2	8.6	46.2	55	370	Newton (1984, 1985)
Dharwar (forest/grassland)							
S. entellus	18.3	11.2	9e	85/16[a]		765/144[a]	Sugiyama (1964)
Jodhpur							
S. entellus	18.3	11.2	8.6e	18		162	Vogel (1973); Mohnot *et al.* (1981)
Melemchi							
S. entellus schistacea	19.8	15.6	12.5	15.2		178	Bishop (1979)
Polonnaruwa							
S. entellus thersites	10.6	6.7	6.4	100–200		640–1280	Rudran (1973*a*); Hladik, (1977*b*); Dittus (1985, personal communication)
T. vetulus philbricki	9.9	7.1	7.2	150–200		1080–1440	
Kakachi							
T. johnii	12.7	10.9	8.7e	71		617	Oates *et al.* (1980)
Madhupur							
T. pileatus	12.8	10.0	8.3	52		423	Stanford (1991)

Kuala Lompat						
P. melalophos	5.9	5.8	4.5	108	486	Marsh and Wilson (1981)
T. obscurus	7.3	6.6	5.0	87	435	
Pasoh						
P. melalophos	5.9	5.8	4.5	57	256	Marsh and Wilson (1981)
T. obscurus	7.3	6.6	5.0	57	285	
Tekam						
P. melalophos	5.9	5.8	4.5	47.6	214	Johns (1983)
T. obscurus	7.3	6.6	5.0	+		
Ketambe						
P. thomasi	6.0e	6.0e	4.5e	27	121	Rijksen (1978)
Sepilok						
P. rubicunda	6.2	5.7	4.5	16.2	73	Davies (1984)
P. hosei	6.2	5.7e	4.5e	+	15e	
Kutai						
P. hosei	6.2	5.7e	4.5e	21	94	Rodman (1978)
P. rubicunda/P. frontata				+		
Paitan						
P. potenziani	6.5	6.4	4.7e	14.2	67	Anon. (1980)
S. concolor	9.5e	7.1	5.3e	12.4	66	
Samunsam (riverine/all forest)						
N. larvatus	21.2	10.0	8.5	5.9/2.6[a]	50/22[a]	Bennett and Sebastian (1988)

Note: Body weights from Table 3.2; subspecies for which there are no data are given weights of nearest subspecies in size (from Napier, 1985). Average individual weight calculated from known group sizes, extrapolated from the same (sub)species at other sites, or estimated from average individual value for similar-sized species. e, estimate; EGA, extra-group animals; BSG, bisexual groups; +, present, but very rare

[a]In two types of vegetation at the site.

Table 10.3. *Comparison of population densities and biomasses of colobine species at selected African sites*

	Body weight (kg)				Biomass (kg/km²)		
	Adult male	Adult female	Average (including non-adults)	individuals/km²	EGA	BSG	Source
Tiwai							
P. badius badius	8.9	8.2	6.2	45–55	18	279–341	Oates *et al.* (1990)
C. polykomos	9.9	8.3	7.3	49–70	10	356–511	
P. verus	4.7	4.2	3.2	66–78		21–25	
Tai							
P. badius badius	8.3	8.2	6.2e	66		409	Galat-Luong (1983);
C. polykomos	9.9	8.3	7.3e	23.5		171	Bourlière (1985)
P. verus	4.7	4.2	3.2e	21		67	
Douala-Edéa							
C. satanas	10.9	–	6.5e	30		225	McKey (1978*b*)
Lopé							
C. satanas	10.9	–	6.5e	25–30		187–225	Harrison (1986)
Kibale Compartment 30							
P. badius tephrosceles	10.5	7	5.8	298	25	1728	Struhsaker (1975); Oates
C. guereza	10.1	7.9	6.2	22	25	136	(1977*c*)
Compartment 14/30							
P. badius tephrosceles	10.5	7	5.8	223	25	1293	
C. guereza	10.1	7.9	6.2	100	25	620	
Tana							
P. badius rufomitratus	8e	7.5e	6.1	100–216		610–1318	Marsh (1986);
				56		342	Decker (1989)
Shalla							
C. guereza guereza	13.5	9.2	6.9	138		952	Dunbar (1987)
Bole							
C. guereza guereza	13.5	9.2	7.9	315		2488	Dunbar (1987)

Note: Body weights from Table 3.2; subspecies for which there are no data are given weights of nearest subspecies in size (from Napier, 1985). Average individual weight calculated from known group sizes, extrapolated form the same (sub)species at other sites, or estimated from average individual value for similar-sized species.
e, estimate; EGA, extra-group animals; BSG, bisexual groups.

The guereza biomass in the gallery forest at Bole in Ethiopia (Dunbar, 1987) is three times higher than that achieved by the species in the logged and swampy forests at Kibale, and ten times higher than that recorded for *C. satanas* in the rain forests at Douala-Edéa (McKey, 1978*b*) and Lopé (Harrison, 1986, personal communication). The Bole biomass is the highest recorded for colobines at any site and this begs the question: Why do the lush, evergreen, apparently productive rain forests of west and west–central Africa not support more colobine monkeys than the species-poor, broken-canopy gallery forests in the savanna regions of Kenya and Ethiopia? This question is addressed later in this chapter.

In South Asia, the combined population density of *Semnopithecus entellus* and *Trachypithecus vetulus* in the semi-evergreen forests of Polonnaruwa exceeds the combined abundance of *P. badius* and *C. guereza* at Kibale, and the population density of *C. guereza* at Bole. However, the smaller body size of the Asian langurs results in a biomass lower than at Bole (2490 kg/km^2). Other South Asian sites have much lower population densities of langurs; for example, *S. entellus* in the forest areas at Dharwar have densities of 85 individuals/km^2, but only 16 individuals/km^2 in the adjacent grasslands (Sugiyama, 1964); *S. entellus* in the Sal forests of Kanha have 46 individuals/km^2 (Newton, 1984); and Nilgiri langurs in the evergreen Shola forests at Kakachi have 71 individuals/km^2 (Oates *et al.*, 1980). Very sparse populations occur in the desert areas of Rajasthan and in the cold areas of the Himalayas (See Chapter 5).

In South-east Asian evergreen forest at Kuala Lompat, there is an abundance of *Presbytis melalophos* and *Trachypithecus obscurus* (combined, there are 195 individuals/km^2), but their small body size leads to only a moderate biomass (920 kg/km^2). Colobine population densities are generally even lower at other South-east Asian sites, especially those in the dipterocarp-rich forests of Borneo. At Sepilok, for example, colobine biomass is a meagre 90 kg/km^2 (Davies & Payne, 1982). Even the large proboscis monkeys, in the riverine forests and mangrove swamps of Samunsam, north-western Borneo, live at such low density that their biomass is low despite their large body size (Bennett, 1986*a*).

What limits primates?

There is therefore a wide range of population densities on both continents. One of the environmental factors which has most frequently been shown to limit herbivore populations is food supply, which is discussed in detail later in the chapter. Other limiting factors considered here are disease, predation and competition.

Disease

The classic example of disease being inferred to have reduced a primate population is a yellow-fever epidemic which swept through Barro Colorado Island, Panama, in the 1940s, drastically reducing the population of howler monkeys (Collias & Southwick, 1952). However, no such epidemic has been described for any population of colobine monkeys, and there are few instances of diseased animals even being seen. This is in marked contrast to the descriptions of heavy bot-fly and screw-worm infestations of howler monkeys which probably kill off animals that are in poor nutritional condition (Milton, 1982).

Predation

There have been many anecdotal descriptions of predator attacks on colobines: domestic dogs killing sub-adult male *T. vetulus* langurs that descend to the ground at Polonnaruwa (Rudran, 1973*a*), jackals killing *T. pileatus* that travel along the ground at Madhupur (Stanford, 1989), and reports of rare attacks by golden jackals on immature *S. entellus* travelling along the ground at Kanha (Newton, 1985). However, these are all infrequent events which cannot be expected to lower overall population densities, although they may affect group structure (Anderson, 1986).

Predation of primates which do not descend to the ground has been considered at Kibale, in terms of the mortality of primates over a 13-year period from 49 instances of losses to, or attacks by, crowned hawk-eagles (*Stephanoaetus coronatus*), and another 36 records of mortality (Struhsaker & Leakey, 1990). The eagles seemed to show a preference for adult males of *C. guereza*, perhaps because they were solitary extra-group males, or animals which gave loud calls that attracted the eagle, or were simply less cautious of the eagles than other age–sex classes. Red colobus which lived in large groups but did not give loud calls were taken less frequently than expected from their population density.

The overall predation by eagles remained small, approximately 0.3% of individuals from the five commonest primate species in the forest, which cannot be considered a major influence on the colobine population densities, although there is a possibility that this predation had some influence on the sub-adult and adult males of *C. guereza*.

More commonly reported, and probably a more significant pressure, is chimpanzee predation on African colobines and other primates (Busse, 1977; Boesch & Boesch, 1989; Takahata *et al.*, 1984; Wrangham & Van Zinnicq Bergmann Riss, 1990). In Gombe, the population density of red colobus was very high (Clutton-Brock, 1975*b*), and chimps showed a preference for hunting infant and juvenile red colobus whenever they chanced upon a group (Busse, 1977). No adult male red colobus were caught, and hunting success was lowered whenever adult male red colobus intervened. Despite their fierce defence, however, about 8–13% of the population of 300–500 red colobus was lost in a 2-year period (Busse, 1977). This loss did not appear to depress the population, probably because the loss of individuals in the lower age classes was offset relatively quickly by new births.

A subsequent survey of chimpanzee hunting at Gombe indicated an exceptionally high predation rate, with an estimated 20–40% loss of red colobus annually (Wrangham & Van Zinnicq Bergmann Riss, 1990). This was based on estimated chimpanzee hunting frequencies in different periods; if mortality had continued at this rate for protracted periods, the red colobus would be unlikely to survive; twenty years later, however, they are still common at Gombe. Our understanding of the effects of chimpanzee predation on red colobus population densities therefore needs to be studied through regular surveys of red colobus group demography at Gombe, in relation to chimpanzee attacks.

In the Mahale mountains (Tanzania) chimpanzee hunting behaviour was studied over a 34-month period (Takahata *et al.*, 1984). Here, red colobus were also caught, most often by male chimpanzees, and preliminary analysis of the results indicates that young animals were caught most frequently. The number of observations from a population of about 100 chimpanzees is quite low; it seems unlikely that this predation limited the red colobus population density, and this applies also at all study sites other than Gombe (Wrangham and Van Zinnicq Bergmann Riss, 1990).

Today, human hunting and trapping is probably the most pervasive form of predation, severely reducing primate densities in many areas. The effects of human hunting have been analysed in most detail for South American primates, and body size has been shown as an important factor in determining the likelihood of a species being hunted (Freese *et al.*, 1982). All colobines, with the exception of *Procolobus verus*, fall within the class of the largest, most vulnerable size.

Throughout West Africa 'bushmeat', including primates, is consumed by rural communities and traded in urban areas (Asibey, 1974; Anadu *et al.*, 1988; Davies, 1987*a*). In the Gola Forests, hunters showed a preference for

P. badius, which has a good taste when smoke-dried, and can be transported to urban centres for sale. As a consequence, in the more accessible forest areas, *P. badius* has been greatly reduced by commercial hunting, but the less conspicuous *Colobus polykomos* has not shown such marked population declines (Davies, 1987*a*).

In Asia, preliminary surveys in the Krau Game Reserve attributed a depression of primate population densities to hunting by aboriginal peoples (Chivers & Davies, 1979), which is well documented at other Asian sites (Bennett, 1992; Whitten, 1980).

Competition

At sites where two colobine species are sympatric, for example Kibale with *P. badius* and *C. guereza*, Polonnaruwa with *T. vetulus* and *S. entellus*, and Kuala Lompat with *P. melalophos* and *T. obscurus*, there is the possibility that the two species may compete for food, or some other limiting resource.

At Kibale, for instance, the top two tree species in the guereza diet, *Markhamia platycalyx* and *Celtis durandii* (which together contributed 63% of the annual diet), were ranked 1 and 3 in the diet of the red colobus, although contributing only 25% of the annual intake (Struhsaker, 1975; Struhsaker & Oates, 1975). Thus there is considerable scope for food competition between the two species, but different plant parts were eaten by each species at different times, and *C. guereza* numbers were both high and low at two sites where *M. platycalyx* and *C. durandii* were abundant. No conclusion can therefore be drawn about competition between these two species of monkeys.

At Kuala Lompat, *P. melalophos* are sympatric with similar numbers of *T. obscurus*. The former have a large intake of young leaf parts and seeds (Bennett, 1983; Davies *et al.*, 1988) while the latter have a high mature leaf intake (Curtin, 1980; MacKinnon & MacKinnon, 1980), so there is little overlap in their diets. However, during the most difficult time of year, when young leaves and seeds are scarce, both leaf-monkeys rely on mature leaf parts and might compete with one other at this time. Competition may not be intense, though, because digestible mature leaf parts are moderately abundant at Kuala Lompat throughout the year (Waterman *et al.*, 1988).

The inland dipterocarp forests of Borneo are the only place in Asia where two or three species in the same genus are sympatric. *Presbytis rubicunda* and *P. hosei* can hardly be distinguished from one another on skeletal features and both have similar diets (Rodman, 1978; Davies, 1984) and social systems (Chapter 5), giving ample scope for competition. In Sabah, surveys of 20 different forest sites showed that although both species were found throughout the state, only one species predominated at any given site (Davies & Payne, 1982). Which conditions favoured one species over the other were not clear, but the predominance of a single species at every site is strongly indicative of competitive exclusion. In West Africa, *Procolobus badius* and *P. verus* are sympatric, but differ greatly in their body size, foraging behaviour and diet (A. G. Davies, G. L. Dasilva and J. F. Oates, unpublished data).

Simple-stomached cercopithecine monkeys and apes are unlikely to be strong competitors for colobine foods because of their major dietary differences, although colobines share some succulent fruit resources with these primates (see e.g. Davies, 1991) and frugivorous birds. Furthermore, seed resources may have to be shared with squirrels (see e.g. Payne, 1980; Emmons, 1980).

Since the only way to test conclusively whether competition is limiting population size is to remove one competitor while keeping all other ecological variables constant, and look for an increase in the numbers of remaining species (Simberloff & Connor, 1981), there is little chance of reaching convincing conclusions about the effects of competition between colobine monkeys, except under very unusual circumstances.

Crowding

Available space will only become a limiting factor to population growth when populations are crammed into small areas at high densities. This may apply to the guereza population at Bole, where Dunbar (1987) contends that space may be a limiting factor. Bisexual group territory size was very small (1.6–2.5 ha), but territories were stable and supported an equivalent of 1 guereza/10 trees (Dunbar, 1987; Figure 10.2). Newly formed groups grew rapidly and reproductive rate only declined when a large group was compressed within a small territory; Dunbar reported that aggression between group members often interfered with reproduction under these circumstances.

The ratio of monkeys:trees was maintained by periodic splitting of groups and emigration of sub-groups from the forest area to the surrounding savanna. Since mature leaf parts were eaten throughout the year, and over 75% of the diet comes from just seven tree/shrub species, it is possible that there was no shortage of food. Female reproductive rates, group fission and sub-group emigration may therefore occur in response to a lack of space (Dunbar, 1987). However, precise information on food availability in different seasons of the year was not given, so it is not possible to reject the hypothesis that food supply diminishes with increasing group size and that it is this that limits population growth rather than lack of space.

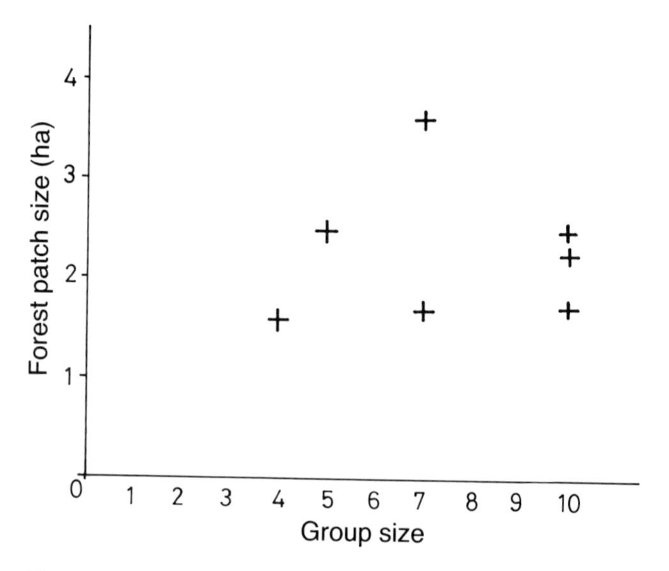

Figure 10.2. The relationship of *C. guereza* group size to forest patch size in the gallery forests at Bole, indicating that larger groups occur in larger forest patches (from Dunbar, 1987).

Does food supply limit colobine populations?

It is widely accepted that populations of non-migratory herbivorous mammals, such as colobines, will be largely limited by the food resources within their home ranges, and this tenet is not contradicted by the small or infrequent effects of disease, predation and competition in limiting populations. It is therefore important to examine the available evidence that food supply limits colobine populations.

Ecological crunches

There is considerable seasonal variation in plant-part production, even in the most equable of rain-forests, and there are also marked fluctuations in fruit production from year to year (Struhsaker, 1978; Milton, 1982). These variations make it very difficult to decide what the long-term carrying capacity of an area might be, or when it should be measured. Cant (1980) therefore proposed that primate populations were limited by infrequent but catastrophic dearths of food, termed 'ecological crunches' (after Wiens, 1977).

This hypothesis received support from Barro Colorado Island, where both spider monkeys and howler monkeys rely heavily on fruits that are scarce at the end of the wet season each year (Milton, 1982). Furthermore, if the wet season is prolonged or early, which has been reported five times between 1929 and 1979, flowering may be disrupted, with resultant failure of the fruit crop (Foster, 1982). Fruit-crop failure in 1970 led to the widespread death of many frugivores, including the primates (Mittermeier, 1973; Foster, 1982). These infrequent events might limit populations irrespective of food abundance at other times of the year, or in years with good fruiting.

Milton (1982) has reviewed population change of howler monkeys on BCI over five decades, however, and her results do not support the hypothesis that 'ecological crunches' limit their population densities. The number of howlers on BCI has steadily increased, following the inferred yellow-fever epidemic in the 1940s (Collias & Southwick, 1952), to occupy all of the suitable forest. The population stabilized at 1250–1350 animals between 1970 and 1979 (Milton, 1982), and the severe fruit-crop failure in 1970 did not lead to any obvious long-term population crash, despite animals being found dead on the forest floor.

Since BCI is the only site where relevant long-term, detailed data have been collected on the effects of ecological crunches, and since no other reports of such phenomena have been received from other sites of long-term primate study, the hypothesis that very erratic super-annual changes in food supply limit primate populations is poorly supported at this stage. At least three factors help to explain why this might be so:

(1) Variation in fruit availability within a year is often greater than differ-differences in the same month or season between years.
(2) Consequently, primates have a considerable capacity to switch diets and eat low-quality less-preferred foods such as mature leaves to get through difficult periods (Chapters 4 and 5).
(3) Even these relatively large-bodied, long-lived, slow-breeding mammals, whose populations have been shown not to fluctuate markedly for periods of over ten years (e.g. Kibale – Struhsaker & Oates, 1975; Struhsaker & Leland, 1979; Kuala Lompat – Chivers, 1974; Marsh & Wilson, 1981) have a considerable capacity for rapid recovery of population losses by immediately increasing reproductive output (see e.g. Southwood *et al.*, 1974).

In this chapter, therefore, two approaches are taken to consider whether colobine monkey populations are limited by food supply. First, single populations are examined which have experienced sudden and permanent changes in food supply, whether through natural catastrophes or human perturbation. Second, a single species (or two closely related species) occupying different forests are compared to see if variations in food supplies give corresponding differences in population density.

Population change as a result of natural catastrophes

The gallery forests of the Tana River provide some evidence that food supply limits populations of the Tana red colobus, *Procolobus badius rufomitratus*. A decade after his intensive ecological study of the red colobus, Marsh recensused the same populations in 1985 (Marsh, 1981*a*, 1986). Marsh recorded a major population crash, with numbers falling from 1210–1800 to 200–300; a devastating loss considering that this is the only area in which this subspecies occurs (Marsh, 1986).

There was no obvious disappearance of gallery forest to account for the crash, since only two of 24 forest patches were lost in a decade. The population decline was a result of far fewer forest patches being occupied by monkeys: down from 17 out of 24 patches to 9 out of 22. Furthermore, the size of groups that still survived had declined from an average of 18.8 monkeys (12–30, $n = 13$ groups) to 9.9 monkeys (6–13, $n = 6$). The reproductive success of females also appeared to have declined, falling from 34% of females carrying young animals to 22% (Marsh, 1986).

There was no evidence that hunting of the monkeys had occurred, and indications that a disease epidemic had been responsible were difficult to reconcile with the widespread loss of monkeys from forest patches that were isolated from each other. The loss of food trees seemed to be the most likely cause of the fall in numbers of red colobus, and botanical evidence from the Mchelelo study area supported this hypothesis.

At Mchelelo, 28 animals in two groups declined to seven animals in one group, and there was a conspicuous decline in food-tree abundance. There was a small but noticeable loss (12%) in the number of all trees, but this would not be expected to account for the major loss of monkeys. In a more detailed breakdown, it appeared that some major food trees suffered high losses: *Albizia gummifera* (–58%), *Acacia robusta* (–23%) and *Ficus sycomorus* (–40%), although the absolute numbers of trees involved was not great (Figure 10.3). The loss of the last species may be telling, because it produced fruits and young leaves at different times throughout the year, while most other foods were produced seasonally.

In a subsequent study of two groups from the same red colobus population in Baomo South forest (5 km south of Mchelelo), Decker confirmed the overall population crash but concluded that there had not been a significant loss of food trees, certainly insufficient to explain the population crash (Decker,

Figure 10.3. The loss of food-tree species of Tana red colobus (*Procolobus badius rufomitratus*) at Mchelelo, Tana River, over a 10-year period. 1, *Ficus sycomorus* (29.3% annual diet); 2, *Sorindeia obtusifoliolata* (19.6%); 3, *Acacia robusta* (15%); 4, *Albizia gummifera* (9.6%); 5, *Majidea zanguebarica* (4.4%) (from Marsh, 1986). Abundance = no. of trees > 10 m tall in 9-ha study area.

1989). Medley (1990) had shown that there was a 56% loss of forest area in 1960–75, and fragmentation of the forest from five to 15 patches. On the basis of this information, and the fact that both Baomo and Mchelelo colobus had a much lower mature leaf intake in 1986–88 than the Mchelelo group in 1975 (1–2% versus 11%), Decker concluded that past forest loss and fragmentation had concentrated the red colobus into small forest patches at the time of Marsh's first study, in excess of the carrying capacity, and they consequently had been forced to eat less preferred foods (i.e. mature leaves). This led to an inevitable if delayed population crash.

If this conclusion is correct, then the population decline has been caused by loss of forest, rather than natural degeneration of forest as part of a long-term cyclical pattern of forest change as suggested by Marsh (1986). But irrespective of the details of the cause, the conclusion from both studies is that food resources were limiting population size.

At Polonnaruwa, direct measures of food abundance in the semi-evergreen forest were related to food consumption by *S. entellus* and *T. vetulus* (Hladik & Hladik, 1972). In addition, Dittus made detailed measurements of the tree flora and observed the effects of a cyclone on both the vegetation and the langurs (Dittus, 1977, 1985).

A single individual of *T. vetulus* consumes about 400 kg fresh weight of foliage per annum, which was only 10% of the estimated plant-part production in the study area. The difference between production and consumption was consistent in all home ranges. The food available for groups of *S. entellus* was more difficult to assess, because it was distributed more heterogeneously, over a larger area. However, important food trees like *Drypetes sepiara* and *Ficus* spp. occurred in all three ranges and the overall numbers of fruits from several species were similar in each home range. An average of 500 kg of fruit was available per individual per year, about half of which was consumed (Hladik, 1977*a*).

These figures give the impression of a superabundance of food. Food supplies were dramatically reduced, however, when a cyclone tore through the Polonnaruwa forest in 1978. Over 45% of all trees in the emergent and upper canopy were killed, and 29% of all sub-canopy trees were lost. In all, 40% of all trees over 10 cm in girth were lost, as were 21% of all shrubs. Many of the remaining trees suffered substantial damage to their crowns, with an immediate loss of wind-blown flowers and fruits and an estimated 50% reduction in foliage-bearing crown area. There was also a marked drop in production of flowers and fruit after the cyclone (Dittus, 1985; Table 10.4).

Inevitably, this meant a loss of food trees, and over-browsing exacerbated some of these losses. Tall *Walsura piscidia* trees were the commonest food source for *S. entellus* before the cyclone, but 55% of the trees were destroyed. In response, one *S. entellus* group abandoned its home range after killing off the few remaining *Walsura* trees by over-browsing, and moved to a new area in which some of these trees still survived. *T. vetulus* killed off *Adina* trees which could not cope with browsing on top of the damage trees suffered during the cyclone (Dittus, 1985), and even before the cyclone Hladik (1977*b*) reported *T. vetulus* killing some *Alangium salvifolium* trees by over-browing the young leaves, a situation also reported for *Mallotus* trees browsed by *S. entellus* at Kahna (Newton, 1984).

So there is evidence that at least some of the preferred food sources were in short supply, certainly unable to meet the langurs' needs in some areas even before the cyclone, and pressure on these trees after the cyclone led to their elimination through over-browsing.

Similarly, if food availability limits colobine populations, then a large reduction in trees and shrubs, compounded by a reduction in production of flowers and young leaves would be expected to lead to a loss of langurs. Surprisingly, this did not immediately come about; both species suffered mortality of only about 5–10%, which supports the contention that the langurs had a super-abundance of food.

Table 10.4. *Loss of important food trees (those accounting for 70% of the diet of* T. vetulus *and approximately 30% of the diet of* S. entellus*) at Polonnaruwa after a cyclone (data from Dittus, 1985). Abundance before the cyclone measured as Importance Value Index, a combination of size, frequency and dispersion*

	Abundance	Percentage/loss
Adina cordifolia	12.8	23
Schleichera oleosa	21.1	6
Drypetes sepiara	55.5	0
Walsura piscidia	8.5	55

Before it is possible to accept that food supply is not limiting these langurs' populations, however, the following data are needed:

(1) Long-term data to show whether the langur populations do eventually collapse, because any population decline through adult mortality is likely to be a slow process, as indicated for the red colobus in Tana.
(2) Data on the abundance of foods at the worst time of year, likely to limit population size; for example, there may be a dearth of foods at the end of the wet season (December–January) when very few fruits or young leaves are produced.
(3) Data on nutritional value; for example, the mature leaves of *Adina cordifolia*, which are a major food of *T. vetulus*, are remarkably lacking in nutrients (Hladik, 1977*b*). It is probable that they are a dietary staple that must be supplemented with rarer, more nutritious items; a strategy that has been suggested for black colobus (McKey, 1978*b*).

Given the gaps in our knowledge, no firm conclusions can be drawn, but the poor quality of some of the mature foliage, and the fact that langurs deserted their home range when the preferred trees were destroyed, both indicate that food supply may still be the main constraint on populations of both langur species at Polonnaruwa (Dittus, 1985).

The effects of logging

Throughout the humid tropics, forests are being altered by timber extraction, which can have obvious effects on the food supply for arboreal colobines.

Two detailed studies of the effects of timber extraction on colobine monkeys have been conducted, one in Africa and the other in South-east Asia.

In the forests at Kibale, Uganda, Skorupa surveyed three habitats: unlogged, lightly logged and heavily logged forest, using line-transect methods to get statistically acceptable measures of relative density (groups/km surveyed) of *P. badius tephrosceles* and *C. guereza*. He then related these measures to the composition of tree species at each site (Skorupa, 1986).

The effects of logging are clearest for red colobus, which suffered a noticeable, but statistically insignificant, fall in populations following light logging of *Parinari* forest (Figure 10.4). However, there was a major decline in sightings of red colobus (38% drop) in the heavily logged forest, where 49% of all trees were lost and the basal area of food trees supplying more than 80%

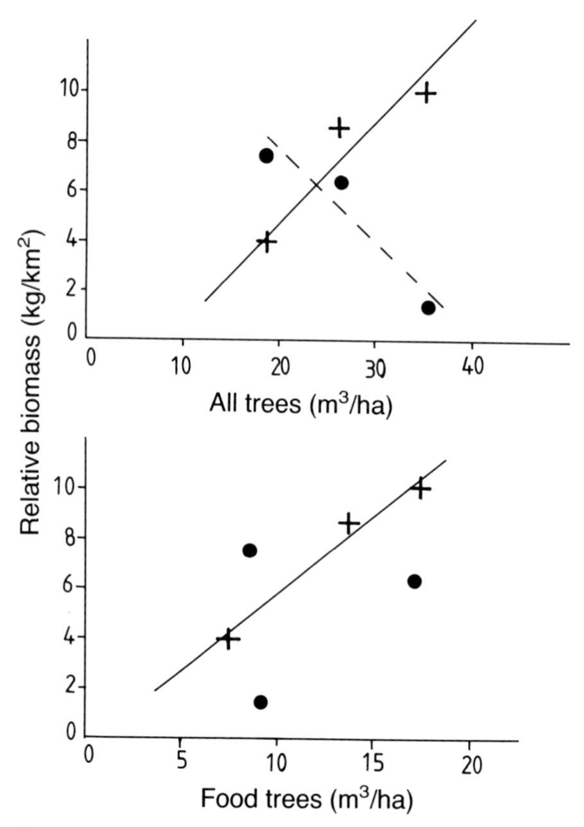

Figure 10.4. Different biomasses of *Procolobus badius tephrosceles* (cross) and *Colobus guereza* (circle) in unlogged, lightly logged and heavily logged forest at Kibale, showing decline in *P. badius* populations as a result of logging (after Skorupa, 1986).

of the red colobus' diet declined by 42%. This gave a clear indication that food-tree abundance probably limits population density of red colobus.

A similar situation has been reported from Ghana (Martin & Asibey 1979) and Sierra Leone (Davies, 1987*a*), where the numbers of red colobus dropped after logging in closed-canopy semi-evergreen forests. It remains a mystery, however, why red colobus consistently appear to be so sensitive to this type of habitat alteration in rain forests, when populations have adapted to live in gallery forests in Tana to the east and Abuko to the west (Chapter 4).

Skorupa's findings reconfirmed that guereza numbers increased in heavily and lightly logged forests, but showed no correlation between the abundance of food trees and guereza biomass. What features of the secondary forest favour guerezas remains unclear, but plants common in 'growth-phase' forest (Whitmore, 1984*b*), such as lianes and colonizing tree species, are important food sources when young leaves of *Celtis durandii* are unavailable (Oates, 1977*a*). Unfortunately, these important food sources were not included in the enumerations and have yet to be related to guereza biomass.

In South-east Asia, Johns looked at populations of banded leaf-monkeys (*Presbytis melalophos*) at Tekam, both before and after timber was extracted from the forest. Dusky leaf-monkeys (*Trachypithecus obscurus*) were also encountered in the area, but at such low densities that reliable data on population size or change could not be collected.

Dipterocarp trees accounted for 7.7% of the stems sampled in the uncut forest, but a much larger proportion of the tree-crown area, since they were the largest trees. In the course of road-building, felling and extracting of large timber trees there was a 50% loss of all trees, and a similar reduction in the stem basal area from 25 m^2/ha to 10–18 m^2/ha. These reductions were randomly distributed across all tree species, and so affected leaf-monkey food trees and non-food trees alike (Johns, 1985).

This might be expected to have led to a 50% reduction in overall food supplies. However, data on the production of fruit, flowers and young leaves indicated that loss of trees was buffered, at least in part, by in increased productivity by the remaining trees (Johns, 1983). Fruiting and flowering were not seen to increase immediately after logging, but young-leaf production almost doubled, leaving the overall abundance of this important leaf-monkey food supply approximately constant, despite a loss of half the trees. Johns (1983) considered that increased illumination of tree crowns following logging led to greater foliage production, and noted that fruit production by roadside trees that were not damaged may also have increased.

The *P. melalophos* groups shifted their home ranges in direct response to disturbance by chainsaw crews and tractors, but largely reoccupied them after logging, as might be expected for this highly territorial population (Johns, 1985). Patterns of ranging and sub-group formation did change after logging, in response to the patchier distribution of food resources, and there was probably an increase in foliage consumption (Figure 10.5).

Leaf-monkey mortality during the logging operations was high, but restricted to infants and juveniles. There was also an absence of new infants in the birth season following logging (A. D. Johns, 1986), probably because mating behaviour coincided with logging activities, and was disrupted. However, there was no indication that any animals died of starvation. Furthermore, estimates of group densities in logged forest immediately following timber extraction were low (*c.* 2.3 groups/km^2), but returned to original levels (*c.* 3.4 groups/km^2) after at most 5–6 years (A. D. Johns, 1986).

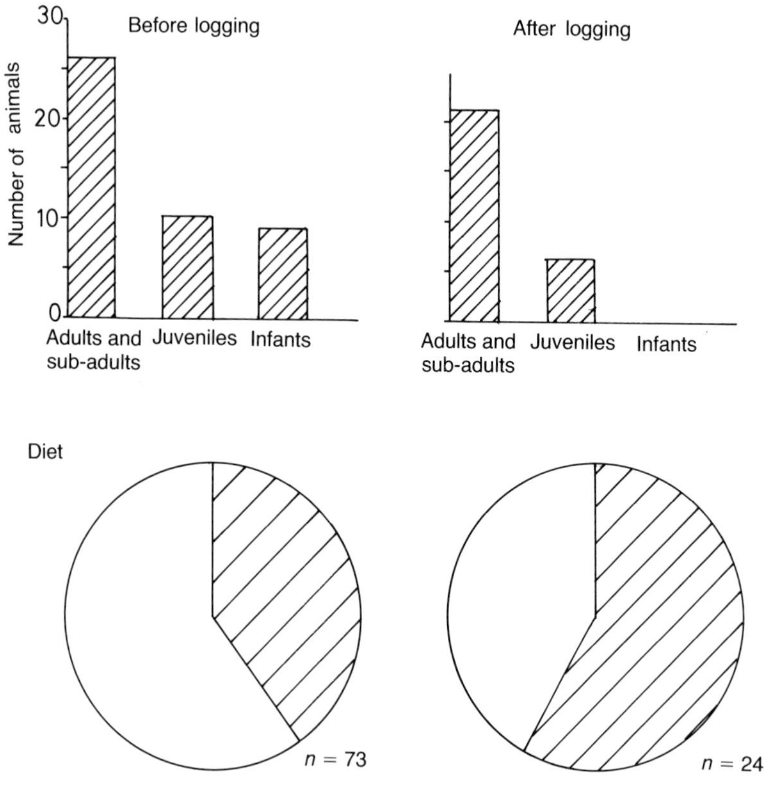

Figure 10.5. Loss of *P. melalophos* from the study area at Tekam following logging, and indications that diet had also changed: unshaded sector, fruit and flowers; hatched sector, foliage (after Johns, 1985).

The initial indication from this thorough study is that the banded leaf-monkeys were not being limited by food supply, since half of the trees were removed for minimal reduction of the monkey population. In the absence of any evidence that alternative factors, such as disease or predation, were limiting populations, Johns (1983) concluded that infrequent, super-annual 'ecological crunches' might be responsible.

This conclusion is equivocal, however, since the leaf-monkeys clearly can recover their numbers quickly, and the food supply may not have been seriously depleted, given that tree losses were offset by increased production of young leaves, and possibly fruit. A further complication in such a study is the very high species-diversity of trees and lianes in forests of South-east Asia, which allows monkeys to switch diet easily, including lianes if necessary, and makes estimation of changes in food supply before and after logging very difficult. Finally, estimation of food availability in forest 5–6 years after logging, when populations had returned to their original levels, have not been reported.

Relationships between colobine biomass and tree-family abundance

Within the forests of the Malesian floristic realm, west of Wallace's line, it is possible to compare the abundance of tree families shared between different sites at which colobine monkeys have been studied. The highest biomass of colobines recorded in non-coastal forests of the region is at Kuala Lompat, less than 200 km from Tekam, in an area of forest which is characterized by an abundance of leguminous trees and a scarcity of dipterocarps (Davies *et al*., 1988). The reverse holds true for Sepilok, across the South China Sea, where there are few legumes and very few colobine monkeys (Chapter 5).

Leguminous trees supply substantial portions of South-east Asian leaf-monkey diets (e.g. Curtin, 1980; Davies *et al*., 1988), a preference also shown by *S. entellus* in India (Oppenheimer, 1977), and colobine population density and biomass showed a positive correlation with legume biomass at inland sites in Peninsular Malaysia (Marsh & Wilson, 1981), and between Peninsular Malaysia and Sabah (Davies *et al*., 1988; Waterman *et al*., 1988). A comparison of the abundance of legumes at nine sites across Sundaland, from Siberut in the India Ocean to Sepilok beside the Sulu sea (Figure 10.6) shows a highly significant positive correlation (r_s=0.741, n=9, P<=0.02) between *Presbytis* biomass and abundance of legumes (Figure 10.7; Davies, 1984).

This pattern over such a wide geographical area, involving different climatic conditions, different tree species and different *Presbytis* species, lends strong support to the hypothesis that food-tree abundance determines

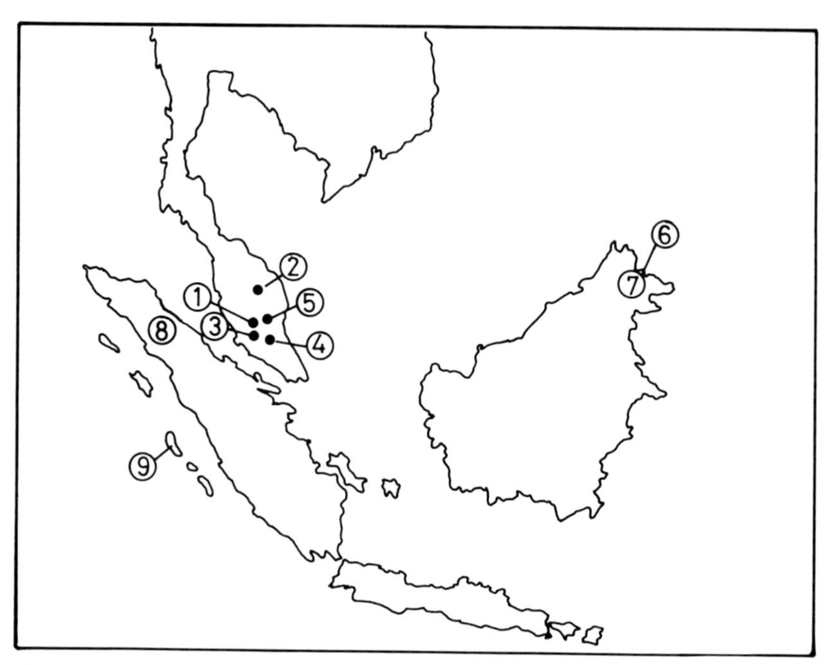

Figure 10.6. Study sites across South-east Asia from which data on tree-family abundance and colobine biomass were compared: 1, Kuala Lompat; 2, Kenyam; 3, Pasoh; 4, Lesong; 5, Tekam; 6, Sepilok; 7, Silabukan; 8, Ketambe; 9, Paitan (from Davies, 1984).

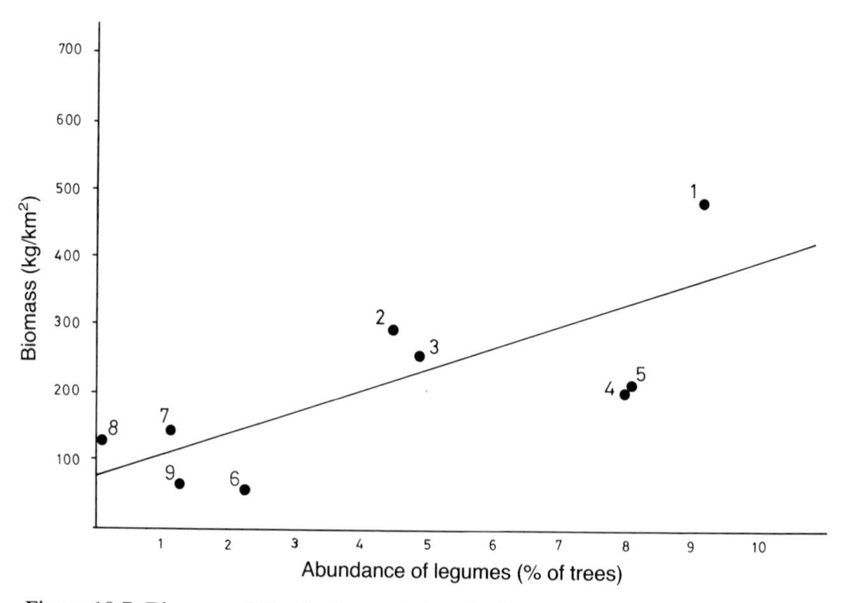

Figure 10.7. Biomass of *Presbytis* populations in South-east Asian dipterocarp forest shown increasing with abundance of leguminous trees. Numbers refer to study sites on Figure 10.6.

colobine abundance in the inland forests of Sundaland. This is further supported by the positive correlation between legume abundance and the combined biomass of colobines at the same sites where *Trachypithecus* and *Presbytis* are sympatric.

The conclusion is further underscored by the lack of such a relationship between legume abundance and biomass of gibbon species that are sympatric with the *Presbytis* species, but do not respond to this measure of forest quality as the colobines do. Gibbons have fairly constant territory sizes and population densities across a wide geographical range (Chivers, 1980).

To prove the positive relationship between legumes and population densities of leaf-monkeys, site by site, more detailed studies will be needed. First, no account has yet been taken of the seasonal nature of the production of fruit and young leaves (the main colobine foods); second, legume-tree abundance is negatively correlated with dipterocarp-tree abundance (Davies, 1984) and it may be the abundance of the latter non-food trees which influenced colobine biomass by suppressing all other potential colobine food trees; and third, lianes are often exploited, especially if trees supply few foods, but liane abundance has not been considered in the above comparison.

Analysing the relationship between leguminous-tree abundance and colobine biomass in African forests failed to show the same trend, despite the strong preference for seeds and young leaves from this family by some colobus species (McKey, 1978*a*; Dasilva, 1989; Maisels *et al.*, 1994). At some sites where leguminous trees are common, such as on Tiwai Island, colobine numbers can be very high (Oates *et al.*, 1990), but even higher colobine biomass has been recorded in the Afro-montane forests of Ethiopia, Uganda and Kenya, where leguminous species are relatively uncommon.

Influence of tree-leaf chemistry

As a result of an early comparison between the sparse population of black colobus at Douala-Edéa (Cameroon) and the very high population densities of guereza and red colobus at Kibale, McKey (1978*b*) proposed that the year-round availability of digestible mature leaves, upon which colobine monkeys can rely when higher-quality foods are unavailable, limits the size of colobine populations. This proposition does not require that mature leaves are seldom eaten where mature-leaf quality is low, but rather that if digestible mature-leaf sources are few and/or widely scattered, the forest cannot support high densities of colobines.

Measures of overall mature-leaf acceptability, in terms of the ratio of protein: fibre (Chapter 9), allow inter-continental comparison between leaf quality and colobine-monkey communities that are very different. The colobine population densities show a direct positive relationship to average mature-leaf acceptability across their entire longitudinal range (Davies, 1984; Waterman *et al.*, 1988; Oates *et al.*, 1990); as mature-leaf quality declines, the colobine population densities also decline (Figure 10.8).

Once again it is remarkable that this relationship holds between such different forests and such different colobines (in diet, body size and taxonomic relationship). This broad relationship strongly supports the hypothesis that mature-leaf chemistry influences colobine biomass. Irrespective of seasonal variation in food supply, all species face similar ecological constraints on their population size. This relationship may become less close at sites where there is no seasonal variation in the preferred, high-quality food supplies, such that monkeys do not need to fall back on mature leaves to sustain them at the worst times of year.

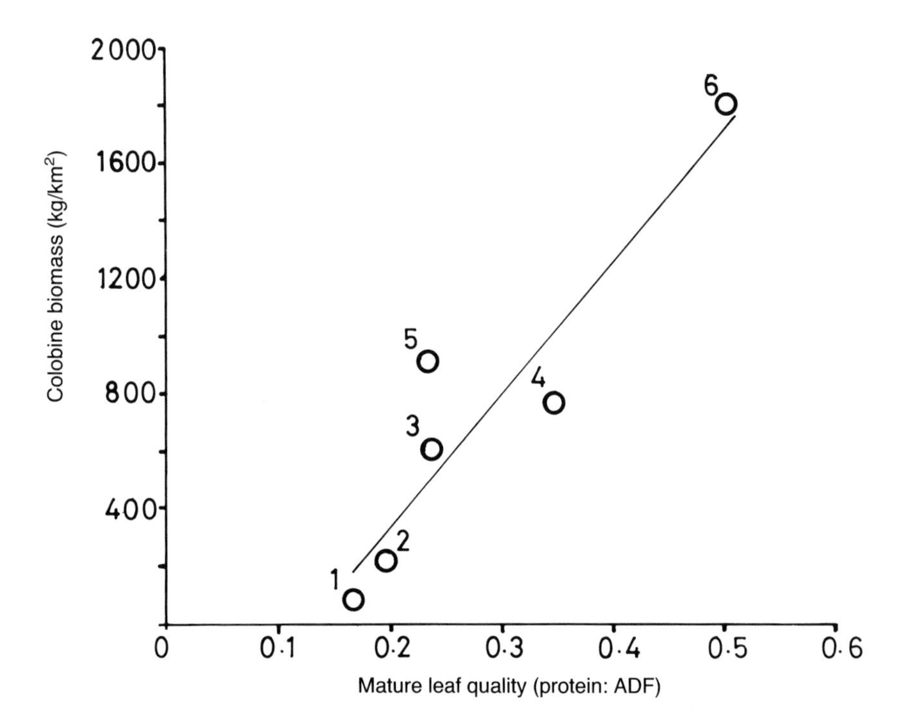

Figure 10.8. Colobine study sites in Africa and Asia at which mature leaf quality (protein/acid detergent fibre) has been compared with colobine biomass: 1, Sepilok; 2, Douala-Edéa; 3, Kakachi; 4, Tiwai; 5, Kuala Lompat; 6, Kibale. See Tables 9.3, 10.2 and 10.3.

Why trees should have an almost three-fold difference in mature-leaf quality between the worst site (Sepilok) and the best site (Kibale) is not clear. Janzen (1975) identified three 'ecological gradients' which he considered would influence the extent to which plants should invest in chemical protection of their leaves:

(1) Plants with a deciduous pattern of leaf production should invest less in chemical defence of their leaves than evergreen species, because their energy is concentrated on initial rapid growth – any investment in chemical defences would be lost when leaves were dropped, a theory supported by comparison of leaves from deciduous and evergreen trees in Lopé (M. Harrison, personal communication).
(2) Colonizing plants should invest less in chemical defence of their foliage than climax species, for similar reasons – their strategy is to grow quickly and be relatively short-lived. This corresponds well with selection of foods by *C. guereza* (Oates, 1977*a*), *C. satanas* (McKey *et al.*, 1981) and *C. polykomos* (Dasilva, 1989) from areas where the forest canopy is broken, and colonizing shrubs, trees and lianes grow in greater abundance compared with the dark understorey of mature, closed-canopy patches of forest.
(3) Plants growing on poor soils should invest more heavily in chemical defence because of the high costs of replacing lost leaves in an oligotrophic environment.

An early comparison of the poor soils at Douala-Edéa and fertile soils at Kibale supported the hypothesis that trees growing on poor soil had less-digestible foliage (McKey, 1978*b*), and a similar comparison of the soils at Kuala Lompat and Sepilok continued to support the hypothesis – very poor soils at Sepilok had trees with mature foliage that was extremely indigestible (Davies & Baillie, 1988; Waterman *et al.*, 1988).

However, Harrison (1986) has pointed out that the soils at Lopé are not as poor as those at Douala-Edéa, yet the population densities of black colobus are similar at both sites. Furthermore, the soils on Tiwai Island are poorer than those at either Douala-Edéa or Lopé, but the mature-leaf foliage is moderately digestible and colobine numbers are moderately high (Oates *et al.*, 1990). The relationship between mature-leaf chemistry and soil quality is therefore not as simple as had been proposed, but the relationship between mature-leaf chemistry and colobine biomass still stands.

What limits colobines?

Based on these discussions, it is reasonable to conclude that colobine populations are commonly limited by food supply, although precise measures of food availability are difficult to obtain, especially in forests with a species-rich tree flora and seasonal variation in plant-part production. The relationships between mature-leaf digestibility and colobine biomass across the Old World tropics, and between legume abundance and biomass across South-east Asia are both robust links in a chain of logic which supports the hypothesis that the major influence on colobine population densities is food availability. Another link is the decline in red colobus populations at Kibale following logging, when many food trees were lost.

There are many missing links, however. The guereza population density at Kibale varied erratically in response to logging, and it is clear that food availability for this species is still poorly measured at any site. The lack of any population crash after logging at Tekam was also unexpected, but many more details on the diets of *P. melalophos* are needed before it is possible to know whether the monkeys' food supplies declined at this site. Even more important is to learn whether the langur numbers at Polonnoruwa declined if food supplies indeed remained low after the cyclone.

Accepting these gaps in our knowledge, there is even less evidence to show that factors other than food supply, except human hunting and trapping, limit colobine numbers. Disease, competition, crowding and non-human predation, even where they have been reported, have not been shown to cause permanent population declines. There is therefore ample scope for more thorough research into the factors which control colobine populations, since it is all too easy to over-state that food supply limits colobine populations, on the basis of the currently available but far-from-complete information.

Acknowledgements

Drafts of this chapter were read by Robin Dunbar, John Oates and Barbara Decker; many of their helpful comments have been included. Lydia Abiero and Dennis Milewa helped with typing and figures, respectively.

11

Colobine monkey society

PAUL N. NEWTON and ROBIN I. M. DUNBAR

Introduction

Although the basic features of social organization and ecology are known for at least one population of most species of colobine (see chapters 4 and 5), few populations have been studied for long periods and intraspecific variation has been poorly documented. The field investigation of colobine ecology and social behaviour has, like much mammalian zoology, tended to remain in separate compartments. Therefore, our understanding of the relationships between environment, ecology and behaviour in the sub-family has remained limited. Theoretical considerations suggest that ecological factors are likely to influence the social organization of colobines and other primates, but it has proved difficult to quantify the variables, such as predation rates and food-patch distribution, presumed to be important (Wrangham, 1987). None the less, the data available for colobines allow us to undertake some preliminary analyses.

Colobines show great diversity in the organization of their societies; this variation is summarized in Figure 11.1 and Tables 11.1 and 11.2. The extremes of colobine troop size range from the Mentawai leaf-monkeys (with troops of fewer than four animals) to the golden snub-nosed monkeys (whose troops coalesce into aggregations of more than 400). Like most Old World monkeys, the majority of colobine species live in matrilineal or female-bonded societies. Examples include the Hanuman langur and Nilgiri langur, the guereza, proboscis monkey and red leaf-monkey. Most of these live in troops of 7–20 animals, including only one immigrant adult male. Some populations of Hanuman langurs and proboscis monkeys live in multi-male, matrilineal troops, but no colobine species is found in only groups of this type. In matrilineal societies, males leave their natal troops at puberty to spend time as solitaries or in all-male bands, before acquiring a breeding group of their own.

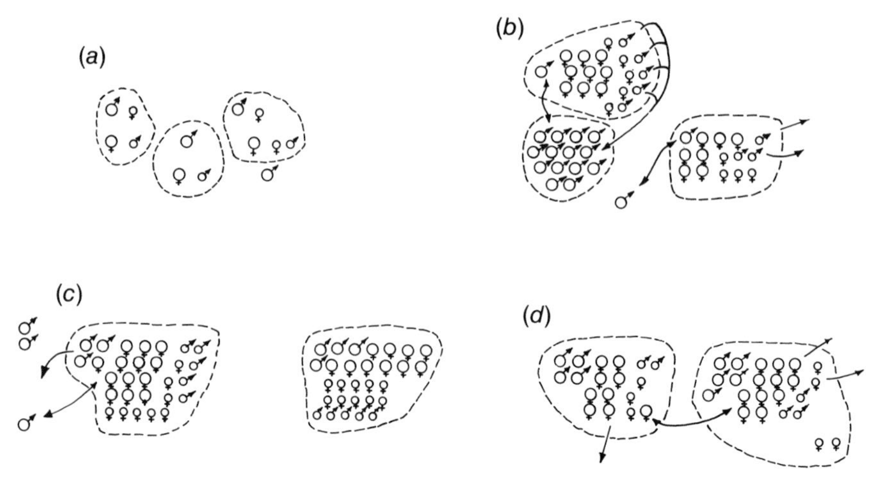

Figure 11.1. Summary of colobine societies: (*a*) Monogamy, e.g. *Presbytis potenziani, P. comata, Simias concolor*; (*b*) Matrilineal – harem, e.g. *Colobus guereza, Presbytis melalophos, P. comata, P. rubicunda, P. thomasi, Trachypithecus vetulus, T. johnii, T. geei, T. cristatus, T. pileatus, T. obscurus, Nasalis larvatus, Semnopithecus entellus*; (*c*) Matrilineal – multi-male, e.g. *Colobus satanas, Nasalis larvatus, Semnopithecus entellus*; (*d*) Patrilineal – multi-male, e.g. *Procolobus badius, P. verus, Colobus polykomos.*

There are only a few clear exceptions to this generalization. The Kibale red colobus, and perhaps the olive colobus and western black-and-white colobus, live in multi-male, patrilineal troops in which, idiosyncratically, the males tend to remain in their natal troop whilst the females emigrate. Although monogamy is rare amongst mammals, three colobine species have been found living in monogamous pairs. The Mentawai leaf-monkey and simakobu both live on the Mentawai Islands off Sumatra; the former is consistently monogamous, whilst multi-male troops of the latter species have been recorded (Watanabe, 1981). One monogamous population of Javan leaf-monkey (*Presbytis comata*) has been reported (Ruhiyat, 1983). Gambian red colobus and proboscis monkeys, although living in patrilineal and matrilineal societies, respectively, are variants in that the smaller social units in which they habitually reside frequently coalesce and divide. Variation in other aspects of colobine social life, such as grooming patterns, sexual swellings, flamboyant natal coats, caretaking and infanticide are probably linked to this variation in social organization (Clutton-Brock & Harvey, 1977; Chalmers, 1979).

The other subfamily of Old World monkeys, the cercopithecines, shows a similar breadth of variation, but differs in the content of the variability (see

Andelman, 1986; Cords, 1987; Dunbar, 1988). Because they are both easier to observe in the wild and important in biomedical research, more is known about cercopithecine social systems, both in the wild and in captivity. Field research on colobines has concentrated on a few species, notably Hanuman langurs in South Asia and red colobus in East Africa. This may well give us an unbalanced view of colobine social systems since the more neglected species found in western and central Africa and Indo-China may behave differently. Interestingly, the two best studied species have the widest range of intraspecific variation in social behaviour known in the Colobinae.

Reproduction

Reproductive behaviour

Female colobines characteristically solicit copulations. Red colobus (*Procolobus badius*), olive colobus (*Procolobus verus*) and western black-and-white colobus (*Colobus polykomos*) are the only species known in which females do not usually solicit matings (Table 11.2). In the Hanuman langur (*Semnopithecus entellus*), all copulations in one series were preceded by female solicitation (drooping of tail, shaking head and presenting genitals towards the male), although only 11.5% of solicits were followed by mating within 5 min (Newton, 1987).

Harassment of matings is a characteristic but puzzling feature of the society of both Hanuman langurs and red colobus, but it is not found in western black-and-white colobus (Dasilva, 1989). In Hanuman langurs up to 83% of copulatory attempts may be harassed, mostly by adult males and high-ranking females (Hrdy, 1977; Newton, 1987; Sommer 1989*a*). Sommer (1989*a*) provides evidence that four rather different advantages may accrue to the harasser depending on its status. Thus, adult males gain by limiting their competitors' mating success. Adult females, on the other hand, gain by imposing contraception on their peers so as to limit future resource competition for their offspring. Females may harass copulations to prevent ejaculation, to improve the quality and quantity of sperm available to themselves. Amongst immatures, 33% of harassment is by offspring harassing their mothers; this could be aimed at reducing the probability of future resource competition from siblings.

In the multi-male societies of red colobus and Himalayan Hanuman langurs, the dominant adult males achieved most matings, and presumably therefore sired most offspring (Struhsaker, 1975; Boggess, 1980). However, Laws & Vonder Haar Laws (1984) found that, in the Himalayan foothills,

Table 11.1. *Summary of colobine social organization*

Species	Site	Habitat	Density/ km^2	Size of troop	Size of band	Mean no. adult males/ troop	Mean no. adult females/ troop	Percentage of troops one-male	Reference
Colobus polykomos	Tiwai	Rain forest	50	8	0	2–3	3	0	Dasilva (1989)
C. guereza	Kibale	Evergreen forest	100	10.9	0	1.7	3.1	71.4	Oates (1977*b*)
	Ethiopia	Riverine forest	50–140	6.7	0	1.2	1.8	83.3	Dunbar & Dunbar (1974)
	Ethiopia	Riverine forest	–	7.3	0	1.4	2.1	77.8	Dunbar (1987)
C. satanas	Gabon	Rain forest	–	15	0	2.3	6.0	–	McKey & Waterman (1982)
	Gabon	Rain forest	25–30	9–13	0	–	–	–	Harrison (personal communication)
Procolobus badius	Kibale	Evergreen forest	300	50	0	3.5	9.1	0	Struhsaker (1975)
	Tana River	Riverine forest	–	18	0	1.5	9.6	53.9	Marsh (1979*b*)
	Senegal	Riverine forest	–	29	0	6.0	12.0	0	Gatinot (1975)
	Gambia	Riverine forest	–	23	0	2	9	0	Starin (1981)
	Gombe	Riverine forest	–	55	0	11	27.5	0	Clutton-Brock (1972)
	Tiwai	Riverine forest	60	35	0	6	–	0	Oates *et al.* (1990)
Procolobus verus	Tiwai	Rain forest	7.2	6.3	0	1–2	2.1	0	Oates *et al.* (1990)
Pygathrix nemaeus	Vietnam	Moist deciduous forest	–	9.3	1	1.3	3.7	66	Lippold (1977)
P. roxellana	China	Coniferous forest	–	600	–	–	–	–	Happel & Cheek (1986)
P. bieti	China	Coniferous forest	–	30–100	–	>1	–	–	Happel & Cheek (1986)
P. brelichi	China	Evergreen and deciduous forest	–	6.1	5–7	1.0	2.2	100	Bleisch *et al.* (1993)
Nasalis larvatus	Sarawak	Riverine forest	5.9	9	1–12	1	3.6	100	Bennett & Sebastian (1988)
	Sabah	Mangrove and nipa	–	19.7	0	5	9	0	Kawabe & Mano (1972)
Simias concolor	Siberut	Evergreen rain forest	9.5	3.5	0	1.0	1.0	–	Tilson (1977)
	Siberut	Evergreen rain forest	7.0	3.2	0	1.0	1.0	–	Watanabe (1981)
	Siberut	Secondary forest	220	6.3	1	1.0	1.9	–	Watanabe (1981)

Presbytis melalophos	Kuala Lompat	Rain forest	108	15.0	5.5	1.0	7.7	100	Bennett (1983)
	Sungai Tekam	Hill forest	–	14	–	–	–	–	Johns (1983)
P. rubicunda	Sabah	Rain forest	–	6.1	0	1.0	2.6	100	Supriatna *et al.* (1986)
	Sabah	Rain forest	19	7.0	3.0	1.0	2.0	100	Davies (1984)
P. comata	Borneo	Rain forest	20.4	8	0	–	–	–	Rodman (1978)
	West Java	Montane forest 1	35	6	0	1.2	1.4	80	Ruhiyat (1983)
		2	11.5	7.5	0	1.0	3.0	100	Ruhiyat (1983)
P. potenziani	Siberut	Rain forest	–	2–6	1.0	1.0	1.0	–	Tilson & Tenaza (1976)
		Rain forest	13.5	3.0	0	1.0	1.0	–	Watanabe (1981)
P. hosei	Sabah	Rain forest	–	7.0	0	1.0	–	–	Davies & Payne (1982)
P. thomasi	Sumatra	Rubber plantation	11.8–29.1	8.0	6.0	1.2	3.6	82.6	Kunkin (1986)
Trachypithecus vetulus	Polonnaruwa	Dry deciduous forest	215	8.6	7.5	1.1	4.4	93.1	Rudran (1973*a*)
	Horton	Cloud forest	92.6	9.0	6.5	1.0	3.7	100	Rudran (1973*a*)
T. cristatus	Malaya	Parkland	–	28.1	0	1.2	14.4	83.3	Bernstein (1968)
	Kalimantan	Rain forest	23–61	9	0	1.0	–	–	Supriatna *et al.* (1986)
T. auratus	Java	Secondary forest and tree plantation	345	6–21	0	1.2	6.6	77.8	Kool (1989)
T. johnii	Nilgiris	Evergreen forest	–	8.9	1.8	1.6	–	71.4	Poirier (1970*b*)
	Periyar	Evergreen forest	–	14.3	1.0	1.0	8.0	100	Tanaka (1965)
	Mundanthurai	Deciduous forest	9.8	7.0	4.0	1.0	3.3	100	Ali *et al.* (1985)
T. obscurus	Malaya	Rain forest	93	17.0	0	–	–	–	Curtin (1980)
T. pileatus	Madhupur	Moist deciduous forest	92	7.5	3.3	1.1	3.0	100	Green (1981)
			52	8.5	4.0	1.0	3.6	100	Stanford (1991)
T. phayrei	Thailand	Rain forest	2.1–6.8	12.9	0	–	–	–	Fooden (1971)
T. geei	Manas	Moist deciduous forest	–	10.3	5.3	1.1	4.1	87.5	Mukherjee (1978)
Semnopithecus entellus	Orcha	Moist deciduous forest	4.4	19	1	3.7	6.0	0	Newton (1988)
	Kanha		46.2	21.7	14	1.1	9.4	93.0	Newton (1988)
	Rajaji		90	46.3	5.5	3.4	13.5	25.0	Newton (1988)
	Dharwar	Dry deciduous forest	85	15.5	15	1.5	7.8	77.0	Newton (1988)
	Abu	Village and farm	72	21	–	1.0	8.3	87.5	Newton (1988)
	Jodhpur		18	35	15	1.0	20.0	95.0	Newton (1988)
	Junbesi	Himalayan forest	1.5	11	–	2.0	3.4	35.0	Newton (1988)
	Melemchi		15.2	32	1	2.5	8.0	0	Newton (1988)

Table 11.2. *A summary of colobine social systems and reproduction*

Species	Female solicitation of copulation	External sign of oestrus	Natal coat type	Infant handling	Social system	Territorial	Intergroup loud calls	Infanticide observed	All-male bands
Colobus polykomos	–	–	F	+	M–M, P–M		+	–	–
C. guereza	+	–	F	+	M–M, M–H	+	+	–	–
C. satanas		–			M–M?			–	–
Procolobus badius	–	+	D	–	P–M, F–F	–	–	+	–
P. verus	–	+	D/A	–	P–M?			–	–
Nasalis larvatus	+	–	A	+	M–H, F–F	–	+	–	+
Simias concolor		+			Mo, M–H		+	–	–
Presbytis melalophos		–	F	+	M–H	+	+	–	+
P. comata	+	–	F		M–H, Mo	+	+	–	–
P. rubicunda	+	–	F	+	M–H	+	+	–	+
P. potenziani		–			Mo	+	+	–	–
P. thomasi					M–H, M–M			–	+
Trachypithecus vetulus	+	–	F	+	M–H	+	+	–	+
T. cristatus	+	–	F	+	M–H	+	+	+	+
T. johnii		–	D	+	M–H	+	+	–	+
T. obscurus		–	F	+	M–H			–	+
T. pileatus	+	–	F	+	M–H	–	–	–	+
T. geei	+	–	F/A	+	M–H			–	+
Semnopithecus entellus	+	–	D	+	M–H, M–M	+/–	+	+	+

Note: A, resembles adult coat; F, flamboyant and differs from adult coat; D, dark and differs from adult coat; M–H, matrilineal–uni-male; M–M, matrilineal–multi-male; P–M, patrilineal–multi-male; F–F, fusion and fission; Mo, monogamous.
Source: Cheney, 1987; Dasilva, 1989; Hrdy, 1976, 1977; Hrdy and Whitten, 1987, Lee, 1983; McKenna, 1979; Napier and Napier, 1967; J. Oates (Chapter 4, this volume); Stanford, 1990; Struhsaker, 1975; Tenaza, 1989.

Hanuman langur males, temporarily immigrating into bisexual troops, were as successful at mating as the resident.

The colobine for which the most detailed data on reproduction are available is the Hanuman langur (Table 11.3). Menarche is reached at 26–42 months, with first conceptions at about 33–45 months and the first live birth at 54 months (Winkler *et al.*, 1984; Harley, 1988). In monkeys in captivity, David & Ramaswami (1969) found a menstrual cycle of 21–26 days, with solicitations from day 7 and ovulation on days 9–11. Winkler *et al.* (1984) found a menstrual cycle length of 24.1±3.8 days in the wild. For those few species for which estimates are available, the gestation period is of the order of 175–200 days (Table 11.3). The rate of conception per oestrus period declines with adult female age, with evidence for a menopause and significant post-reproductive survival (Borries 1986; Sommer *et al.* 1992). Hanuman langur males appear to reach reproductive maturity at about 5–6 years of age, some 2 years later than females (Vogel & Loch, 1984; Roonwal & Mohnot, 1977). Sommer & Rajpurohit (1989) calculated, for the Jodhpur langur population, that the mean lifetime reproductive success for a female was 5.6 infants, but 27.6 infants for a troop male.

Sexual swellings

Prominent perineal 'sexual swellings' have been noted in 17 out of 31 cercopithecine species. In contrast, they have been recorded in only three out of 24 colobine species: the red colobus, olive colobus and simakobu (Table 11.2; and see Chapter 4).

The distribution of sexual swellings amongst primates remains puzzling and appears to have evolved independently in three lineages (the great apes, colobines and cercopithecines). There is an association between multi-male troop societies and sexual swellings that is strong within the colobines – two of the species with sexual swellings do consistently live in multi-male troops (probably patrilineal) and usually show low levels of female solicitation (Clutton-Brock & Harvey, 1976; Hrdy & Whitten, 1987). It is possible that the swellings increase a female's opportunities to mate with different males, thereby decreasing an individual male's confidence in the paternity of his offspring and hence reducing the risk that he will kill them. However, it has also been suggested that the swellings allow a male to pinpoint the time of ovulation, thus increasing the male's confidence in the paternity of the subsequent infants and encouraging greater paternal investment. A third explanation is that sexual swellings evolved to incite intermale competition for females, so increasing the probability that a female would conceive with the

Table 11.3. *Colobine quantitative reproductive parameters*

Species	Site	Age at maturity (months)		Gestation (days)	Interbirth interval (months)	Birth timing
		Male	Female			
Colobus polykomos	Tiwai	–	–	186	24	Season
C. guereza	Kibale	–	–	–	25.2	–
Procolobus badius	Kibale	38–46	35–58	–	25.5	2 peaks/year
P. verus	Tiwai	–	–	155–186	–	Peak
Trachypithecus vetulus	Polonnaruwa	–	–	200–226	23	Peak
	Horton	–	–	200–226	16.5	No regular pattern
Semnopithecus entellus	Jodhpur	34–47	60	200	15.3	Peak
	Kanha	–	–	174–226	26.5	Season
	captive	–	–	200	15.4	–

Source: Dasilva, 1989; Harley, 1985; Hrdy, 1977; Newton, 1987; J. Oates, personal communication; Rudran, 1973*b*; Struhsaker and Leland, 1987; Vogel and Loch, 1984; Winkler *et al.*, 1984.

fittest male (Hrdy & Whitten, 1987). The available data, however, do not allow us to distinguish between these hypotheses at present.

Birth seasons and intervals

Most colobines that have been studied breed throughout the year with a variable pattern of birth peaks (Table 11.3, and Struhsaker & Leland, 1987). This is in contrast to the cercopithecines, most of which breed with strict seasonality (Andelman, 1986). In red colobus in Kibale, there is possibly a link between the temporal pattern of birth peaks and climate, with peaks during the rainy months (Struhsaker & Leland, 1987). In contrast, the birth peaks in the Polonnaruwa population of purple-faced langurs appeared to occur within the dry season (Rudran, 1973*b*).

Strict birth seasons have been found in some Hanuman langur populations (Newton, 1987), western black-and-white colobus (Dasilva, 1989), capped langurs, *Trachypithecus pileatus* (Stanford, 1990) and also possibly the high-altitude *Rhinopithecus* species (Struhsaker & Leland, 1987). All the Himalayan Hanuman langur populations studied have a pronounced December-to-May birth season (Figure 11.2), presumably because of the pronounced seasonality of mountain climates and food resources (Bishop, 1979). Of the nine peninsular populations of Hanuman langurs studied, birth peaks were found in five and birth seasons in four, with no apparent distinction between the two groups in terms of climate, population structure and patterns of social change. Harley (1985) suggested that access to crops and human provisioning during seasonal dry periods allows some populations to breed throughout the year. Indeed, all peninsular populations with strict birth seasons inhabit forested game reserves without access to human cultivation or food provisioning, while those populations showing birth peaks either crop-raid or can take advantage of provisioning, which is provided for them as they are held sacred by Hindus (Newton, 1987).

There is also a significant negative correlation between evenness with which births are distributed across the year and latitude in Hanuman langur populations (Figures 11.2 and 11.3; $r = -0.709$, $n = 13$, $P < 0.02$ two-tailed), as might be expected on the grounds that climate becomes more seasonal with increasing latitude. This effect is independent of altitude and proximity to villages: removing high-altitude Himalayan populations and those living near villages has little effect on the relationship between latitude and birth seasonality ($r = -0.613$, $n = 7$).

Detailed interbirth interval data are available for red colobus and Hanuman langurs (Table 11.3). The mean interbirth interval in the Kibale red colobus

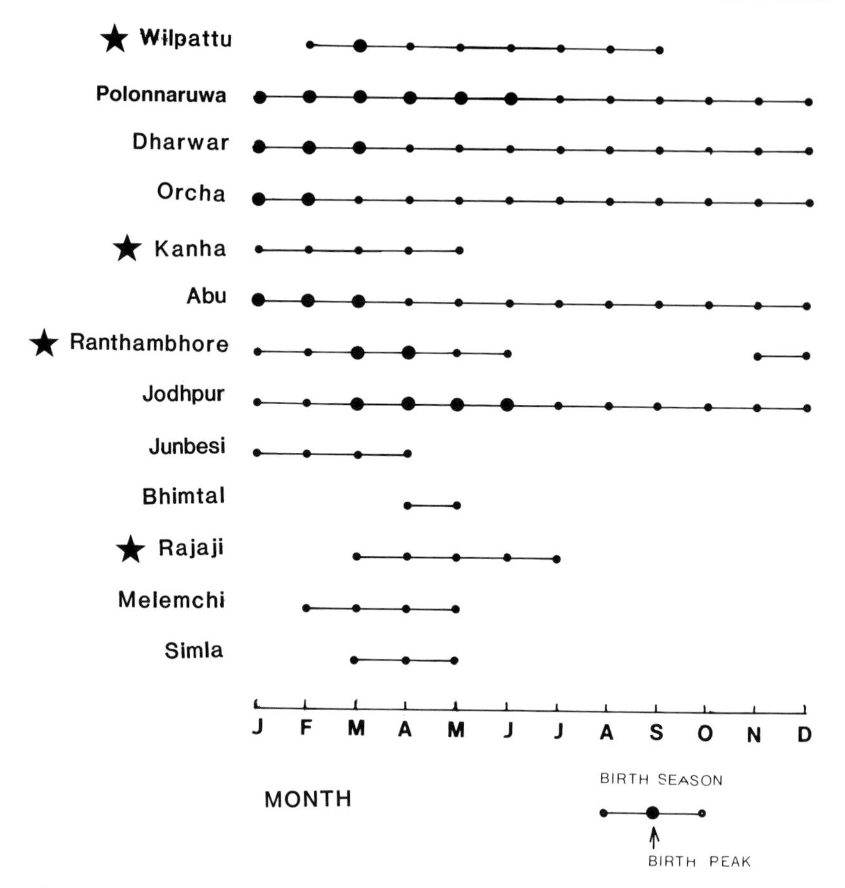

Figure 11.2 Summary of Hanuman langur birth seasonality, in decreasing order of latitude. Star, game reserve population. Adapted from Newton (1987).

was 24 months (Struhsaker & Leland, 1987) whilst values of 15.3 months (Winkler *et al.*, 1984) and 26.5 months (Newton, 1987) were found in two populations of Hanuman langurs. Infant death has been shown to reduce interbirth intervals in red colobus to 12.1 months and in Hanuman langurs to 11.6 and 11.7 months (Winkler *et al.*, 1984; Struhsaker & Leland, 1987; Newton, 1987). Indeed, there is a significant, positive correlation between age of Hanuman langur infant at death or disappearance and the mother's subsequent interbirth interval (Figure 11.4). As colobine litter size is almost always one, interbirth intervals will be the major determinant of fecundity (Andelman, 1986). Rudran (1973*b*) suggested that variation in the interbirth intervals of purple-faced langurs (*Trachypithecus vetulus)* might be a con-

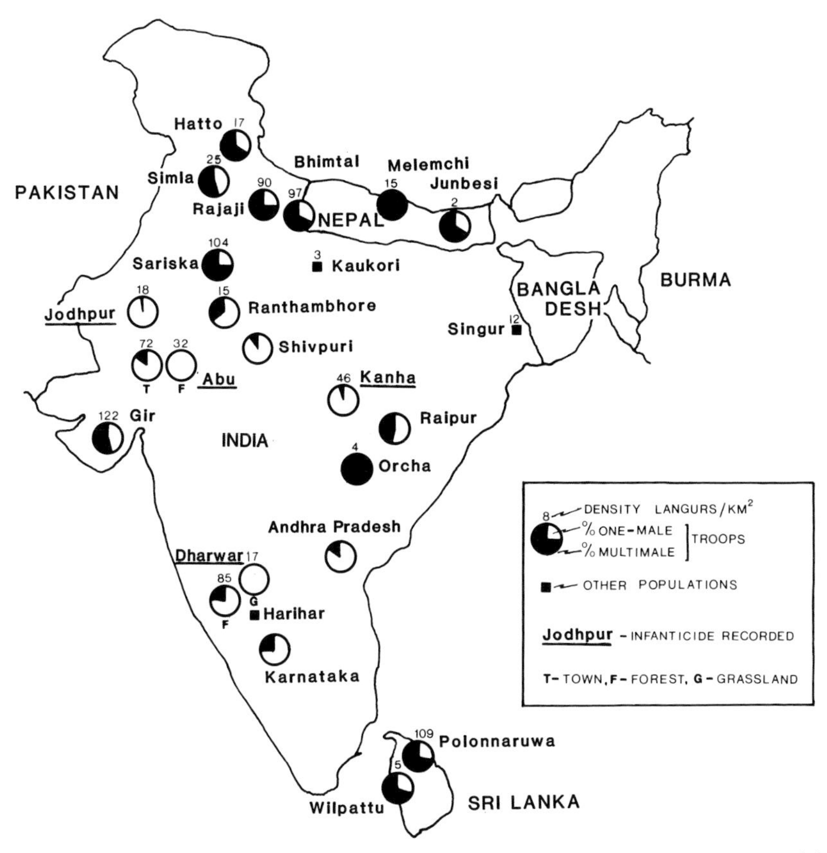

Figure 11.3 Map of South Asia showing variation in Hanuman langur social organization. From Newton (1988*a*).

sequence of nutrition, the interval being shorter in habitats with less climatic variability and a more abundant food supply. Similar effects have been suggested for cercopithecines (Andelman, 1986; Struhsaker & Leland, 1987). Winkler (1988) showed that Hanuman interbirth intervals were significantly shorter after male than after female infants were born and speculated that mothers were investing more in female offspring. Rajpurohit & Sommer (1991) found that the Jodhpur infant and adult mortality was predominantly male whilst immature mortality was predominantly female, suggesting that the higher investment in females was an attempt to reduce this immature female loss. The reason for the high male infant death rate was that more males than females were killed by infanticidal males. Interbirth intervals were

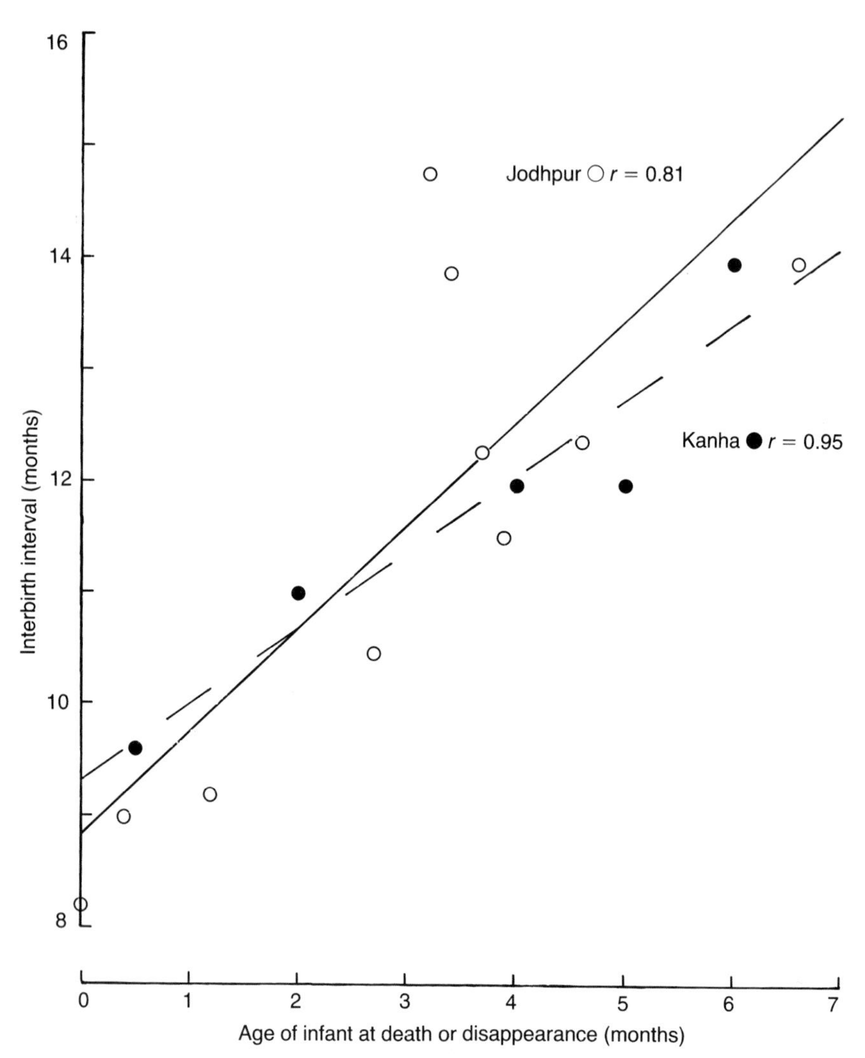

Figure 11.4 Relationship between Hanuman langur infant's age at death or disappearance and mother's subsequent interbirth interval. From Newton (1987).

significantly shorter in small harems than in large harems, perhaps because of more intense female sperm competition in larger troops. As larger troops are more frequently provisioned, food scarcity is unlikely to be the cause (Sommer & Rajpurohit 1989).

Social behaviour

As in other social mammals, colobine social interactions within groups tend to be affiliative and relatively frequent, while those that occur between groups tend to be rarer and more aggressive. In comparison to cercopithecine society, interactions within colobine groups appear to be relatively benign, relaxed and less obvious to the human observer.

In matrilineal societies grooming is most frequent between females and whilst males receive more grooming than expected, they give less than predicted if grooming was distributed at random (Hrdy, 1977; Oates, 1977*b*; McKenna, 1978; Struhsaker & Leland, 1987). The one known exception is the patrilineal red colobus, in which adult males groom each other more, and adult females less, than expected. Colobine male–female interactions are largely confined to grooming, mating and infanticidal attacks and there is no evidence for the existence of long-term 'special relationships' between the sexes as has been found in *Papio* baboons (Smuts, 1983).

Amongst males

In matrilineal, multi-male colobine societies, there is little affiliative social contact between adult males. Agonistic vocal interactions occur between males much more frequently than between females. Within multi-male troops of Himalayan langurs, Boggess (1980) found that only 0.3% of adult grooming interactions were between males. Dunbar (1987) suggested that in Ethiopian *Colobus guereza* the higher frequency of agonistic interactions in multi-male troops than in uni-male troops resulted in female reproductive suppression (see below). Dominance hierarchies among males in multi-male troops have also been noted in a number of Hanuman langur populations and in western black-and-white colobus (Boggess, 1980; Laws & Vonder Haar Laws, 1984; Dasilva, 1989).

Male Hanuman langurs leave uni-male troops before reaching adulthood (Sugiyama, 1967; Hrdy, 1977). Although little is known of the details, the male infants are harassed by resident adult males and approach nearby all-male bands, which they later join (Hrdy, 1977; Mohnot, 1978, 1984). In the Jaipur predominantly uni-male population, Mathur & Manohar (1991) found that male juveniles left their natal troop only after takeover by a new male and not when they reached puberty. Male relationships within bands have received little attention. Moore (1984, 1985*a*, 1986) provided evidence that bands at Ranthambhore consisted of sub-groups which were cohorts of paternal half-siblings expelled with their father from a troop (see also Rajpurohit,

1991). However, presumed paternal cohorts did split up and males probably migrated between these cohort sub-groups. Sommer (1988) detected a linear dominance hierarchy amongst males within bands at Jodhpur.

A correlation between male dominance rank and the number of females mated has been noted in the multi-male troops of some seasonally breeding langur populations (e.g. Junbesi: Boggess, 1980) but not in others (e.g. Rajaji: Laws & Vonder Harr Laws, 1984). This can be explained by the monopolization hypothesis (Emlen & Oring, 1977; Dunbar, 1988; Cowlishaw & Dunbar, 1991; and see below). At Junbesi, which has a mean of 3.4 adult females per troop, the adult male is able to monopolize reproductive access. At Rajaji, with 13.5 adult female per troop, the adult male is unable to monopolize the females and many males are able to mate at high frequency.

In the patrilineal, multi-male society of red colobus, Struhsaker (1975) described a pronounced dominance hierarchy amongst troop males that was expressed through priority of access to space, food, copulations and grooming position.

Adult males of most species of colobine studied to date make distinctive loud calls. However, such calls are absent in capped langurs and red colobus and are of unusually low amplitude in olive colobus (Table 11.2). Male calls of the Hanuman langur can be heard over 1 km from the source and occur predominantly in the first few hours after dawn (Newton, 1984), perhaps because thermal effects aid long-distance transmission (Waser & Waser, 1977). It is generally assumed that loud male calls are directed towards other groups and mediate intergroup spacing. Although there is direct experimental evidence for this in cercopithecines, cebids and hylobatids, such data are lacking for colobines (Cheney, 1987). Loud calls may also assist in maintaining intratroop cohesion and attracting females (Bennett, 1983). The only colobine with equivalent female calls is the monogamous Mentawai leaf-monkey (*Presbytis potenziani*) in which the adult female adds a coda and display to the male's call and display (Tilson & Tenaza, 1976). Such female calls appear to be characteristic of monogamous primates (Cheney, 1987).

Amongst females

Relationships amongst colobine females are subtle, infrequent and hard to record in comparison to those amongst cercopithecine females. As suggested by McKenna (1979) and Oates (1987) this distinction might be related to different foraging strategies in the two subfamilies. As a broad generalization, cercopithecines are 'foragers', consuming a diet derived from small dispersed patches, whilst colobines are 'banqueters', consuming food from large abundant patches. It is suggested that, as feeding interference is likely to be less frequent amongst banqueting animals, dominance relationships will also be less frequent and important in their interactions. Intraspecific variation in female relationships amongst Hanuman langurs tentatively supports this hypothesis. Rank does not appear to be important amongst wild forest Hanuman langur females (Jay, 1965; Sugiyama, 1976; Newton, unpublished). However, female rank amongst the Mount Abu langurs, which are more likely than forest animals to be 'foragers', is important and is correlated with female age (and therefore reproductive value); young females rise in rank at the expense of their elders, so that the lowest-ranking animals are the very young and the very old (Hrdy & Hrdy, 1976; Hrdy, 1977). Hrdy & Hrdy (1976) suggest that old females have little or no remaining reproductive value, but may be able to gain genetic representation in future generations via kin selection (for example, by assisting close relatives in troop defence). Young females, however, stand to gain in fitness by attempting to outcompete other females for available resources and by selfishly leaving troop defence to other individuals. Dolhinow *et al.* (1979) have disputed this hypothesis with evidence from captive studies; however, since their troops were artificially constructed, it is doubtful whether the observed relationships are comparable with those observed in wild matrilineal troops.

Amongst female *Colobus guereza, Procolobus badius* and *Presbytis melalophos*, social interactions are rare; the only colobines with evidence of a linear female hierarchy in the field are *Trachypithecus johnii* and *Colobus polykomos* (Poirier, 1970*a*; Oates, 1977*b*; Bennett, 1983; Dasilva, 1989). Within the patrilineal society of red colobus, in which females are unlikely to be closely related to each other, adult females groomed each other less than would be expected if grooming was distributed at random (Struhsaker, 1975). There is no evidence to suggest that colobine daughters acquire ranks similar to those of their mothers, unlike the case in most cercopithecines (Melnick & Pearl, 1987).

Hanuman langur adult females frequently mount each other. Srivastava *et al.* (1991), working at Jodhpur, found that a female participated in these pseudocopulations approximately every five days and that they were usually initiated by the mounter. This behaviour occurred predominantly among ovulating females, with mounters tending to be of higher rank than mountees. Srivastava *et al.* (1991) suggest that homosexual mountings are a reproductive strategy of the mounter female to try to reduce the number of solicitations the mountee will make to males and hence her probability of insemination. Similarly, Sommer *et al.* (1992) suggest that post-conception oestrus is a female strategy to remove sperm from males, thereby depriving competing females of conception and increasing the availability of resources for the expected infant. Evidence for sperm competition is further strengthened with the finding that the probability of conception for a particular female increases with an increased number of copulations but with a decrease in the number of females copulating on the same day (Sommer *et al.* 1992).

Adult females and infants

A prominent feature of female social structure in most colobines, but only a few cercopithecines, is the early transient transfer of neonates from the mother to other females, especially juvenile and pregnant adult females (Table 11.2). These females are variously known as 'caretakers', 'aunts' or 'allomothers' but are termed 'infant-handlers' here as a more neutral label (see Wasser & Barash, 1981). Compared to cercopithecines, infant-handling in colobines is found in more species, involves younger infants, is not dependent on the social status of the handler and rarely involves kidnapping or male handling (McKenna, 1979). A Hanuman langur infant can spend up to 50% of its first few days of life on handlers (Hrdy, 1977).

Five major factors have been proposed to explain the evolution of colobine infant-handling. First, it is suggested that infant-handling will benefit the mother by releasing her to forage unhampered ('baby-sitting'). Poirier (1968), for example, found that 50% of transfers in Nilgiri langurs (*Trachypithecus johnii*) were followed by the mother moving away to feed, while Vogel (1984) found that the time a Hanuman langur mother spent feeding increased three-fold when her infant was being handled.

Second, it has been hypothesized that handling may improve the development of maternal skills by the handler, thereby enhancing her own future infants' survival ('learning to mother'). Hrdy (1977) suggested that Mount Abu nulliparous Hanuman langurs did improve maternal skills and competence by handling. However, Vogel (1984) found that young juvenile females were fully competent and careful handlers. In addition, the 'learning to mother' hypothesis cannot explain all langur infant-handling, as some 50% of handling events were undertaken by multiparous females. Furthermore, Poirier (1968) records that, in the Nilgiri langur, only adult females were handlers. There is no clear evidence that handling infants benefits both mother and handler through reciprocal altruism or kin selection.

Third, the infant might benefit from handling if, by being quickly integrated into the troop, it increases the probability that it would be adopted should the mother die or become disabled (Lancaster, 1971; Hrdy, 1976;

Quiatt, 1979). Although adoption has been reported in colobines, it occurs only rarely (Dolhinow & DeMay, 1982); moreover Vogel (1984) found no clear relationship between the individual patterns of handling and subsequent adoption. However, there is also great potential cost to the infant, and therefore to the mother, in terms of abuse, wounding, malnourishment, dehydration and death (so-called 'aunting to death'; Hrdy, 1977; Vogel, 1984). Indeed, a fourth hypothesis suggests that the handling has evolved as a competitive adaptation: the abusive handler reduces resource competition for her own offspring (Wasser & Barash, 1981). This idea was not supported by data from the Jodhpur langur population (Sommer, 1989*b*). Although allomaternal nursing has been recorded in both Hanuman and Nilgiri langurs, it does not appear to be frequent or widespread (Jay, 1965; Poirier, 1968). In any case, nocturnal suckling may, at least in part, compensate for any malnutrition cost (A. G. Davies, personal communication). Fifth, Quiatt (1979) and Scollay & DeBold (1980) suggest that infant-handling is a fortuitous result of maternal behaviour, without evolutionary significance. The field data do not allow any clear conclusion as to the adaptive significance of infant-handling amongst colobines; it is unclear who benefits.

The absence of infant-handling among red colobus might be attributable to the relative lack of cohesive grooming and affiliative behaviour amongst females (the most sociably mobile sex in this patrilineal species; Struhsaker & Leland, 1979). This explanation is supported by the absence of infant-handling in the olive colobus, which is also probably patrilineal (J. F. Oates, personal communication). Infants of this species are idiosyncratically carried in the mother's mouth, perhaps as an adaptation to extremely thick forest or because the adult pelage is too short for the neonate to grip (Booth 1957).

Kidnapping of infants by sub-adult and adult females from different troops has been noted in Hanuman langurs. In these instances, the kidnappers returned to their own troops, where other females also handled the infant. Although the majority of infants were retrieved by the mother, infants did disappear from the kidnapper's troop and are presumed to have died. The explanation for such dramatic behaviour is unclear; in the majority of cases, neonates were present in the kidnapper's own troop (Sugiyama, 1965; Hrdy, 1977; Mohnot, 1980). Starin (1981) mentions attempted kidnapping of Gambian red colobus by vervets (*Cercopithecus aethiops*).

A further interesting feature of social structure, common within the Colobinae but rare in other primate taxa, is the prevalence of flamboyant natal coats (Table 11.2). Analysis of the literature is complicated by variable definitions. Here, natal coats are classified as either resembling adult pelage, flamboyant (striking, highly visible) or dark (differing from adult coats but dark and not highly visible). Only one colobine, the proboscis monkey (*Nasalis larvatus*), has a coat that closely resembles the adult pattern, but the neonates have bright blue faces. Hrdy (1976) suggests that flamboyant coats have evolved to facilitate infant-handling by attracting other females, implying a benefit to infants from being handled (see above). Only three species do not have flamboyant natal coats. Assuming that bright coats increase vulnerability to predation, the dark coat of the Hanuman langur infant might be a cryptic adaptation to their unusual partially terrestrial habit and concomitant increased predation risk. The dark natal coats of red and olive colobus might in contrast be related to the absence of infant-handling (Hrdy, 1976).

Infanticide

Since the first report in 1965 of the killing and disappearance of Hanuman langur infants following takeover by adult males, infanticide has been seen some 13 times in this species with additional observations in red colobus and silvered langur (*Trachypithecus cristatus*) (Table 11.4). These observations have dominated research on Hanuman langurs and have generated considerable controversy. Some have regarded infanticide as typical for the species (Hrdy, 1977) whilst others have dismissed it as a myth (Schubert, 1982); indeed, Darwin (1871, p. 52) wrote that 'our semi-human progenitors would not have practiced infanticide . . . for the instincts of the lower animals are never so perverted as to lead them regularly to destroy their own offspring'.

Hrdy (1977) suggested that langur infanticide is an evolved male reproductive strategy. This sexual selection hypothesis suggests that, because an adult male may have only 2 years in which to breed within a troop, there will be considerable reproductive advantage in concentrating as much breeding as possible into his tenure. However, many of the females will be unable to breed after takeover because of lactational amenorrhoea; consequently, a new resident male would accelerate his breeding by killing unrelated infants. Secondarily, the elimination of these young would also remove potential food and reproductive competitors with his own offspring. The essential components of this hypothesis are that the invading male should not kill his own offspring but should accelerate and sire post-takeover conceptions (Hrdy, 1977; Hrdy & Hausfater, 1984).

Up to 1985 some 36 Hanuman infants were recorded as disappearing during or soon after takeover in three populations (Dharwar, Jodhpur and Abu, see Figure 11.3); actual infanticide was observed only three times and wounding three times (Hrdy, 1977; Boggess, 1984). In contrast, at other long-term study sites such as Rajaji, Junbesi, Orcha and Melemchi (Figure 11.3),

no evidence suggesting infanticide was found (Jay, 1965; Boggess, 1980). This dichotomy fueled a controversy. Some observers argued that the cases of infant disappearance following takeover could have resulted from disease, predation or accidents (Curtin, 1977; Curtin & Dolhinow, 1978*a*; Boggess, 1979, 1984). They also claimed that key components of the sexual selection hypothesis were violated: namely, that interbirth intervals were not shortened by infant death and that there was insufficient evidence to confirm that infanticidal males did not kill their own offspring but did sire subsequent infants (Curtin, 1977; Boggess, 1979). An alternative explanation, the social pathology hypothesis, was therefore proposed. This suggests that infanticide is not an evolved male strategy but rather an aberrant, incidental result of escalated aggression in populations where human encroachment and provisioning has produced artificially high densities; infanticidal males gain no advantage over non-infanticidal males and infant killing would thus not be expected in 'undisturbed habitats at low or moderate density' (Curtin, 1977; Curtin & Dolhinow, 1978*b*, p. 699; Schubert, 1982; Boggess, 1984).

The data published up to the early 1980s were equivocal: the field evidence neither categorically disproved nor supported either hypothesis (Chalmers, 1979). The social pathology hypothesis could be most easily refuted by the observation of infanticide in an undisturbed low-density population (Curtin & Dolhinow, 1978*b*), whilst the sexual-selection hypothesis could be most easily disproved by demonstrating that invading males do frequently kill their own offspring and do not accelerate or sire subsequent births. A direct test between these two alternative hypotheses comes down unequivocally in favour of the sexual selection hypothesis (Figure 11.5; Newton, 1986). Infanticide has now been seen in two undisturbed forest populations, at Kanha and Mundanthurai, both with moderate langur population density (Newton, 1986, 1987; Ross, 1993). Both infanticidal and non-infanticidal populations are equally distributed across all population densities, but infanticide has only been seen in populations with a high proportion of one-male troops, whereas non-infanticidal populations typically occur where multimale troops predominate. However, the one exception does support the hypothesis of an association between one-male troop structure and infanticide, in that the invading band at Mundanthurai killed an infant in the only uni-male troop in a predominantly multi-male population (Ross, 1993).

Agoramoorthy *et al.* (1988) have recently described seven cases of abortion seen in 10 years of observing Hanuman langurs at Jodhpur; five of these cases were in association with takeover or infanticide. They suggest that, by inducing abortions, males may return a female into breeding condition. Given a male's ability to kill infants, it may be advantageous to the female to abort

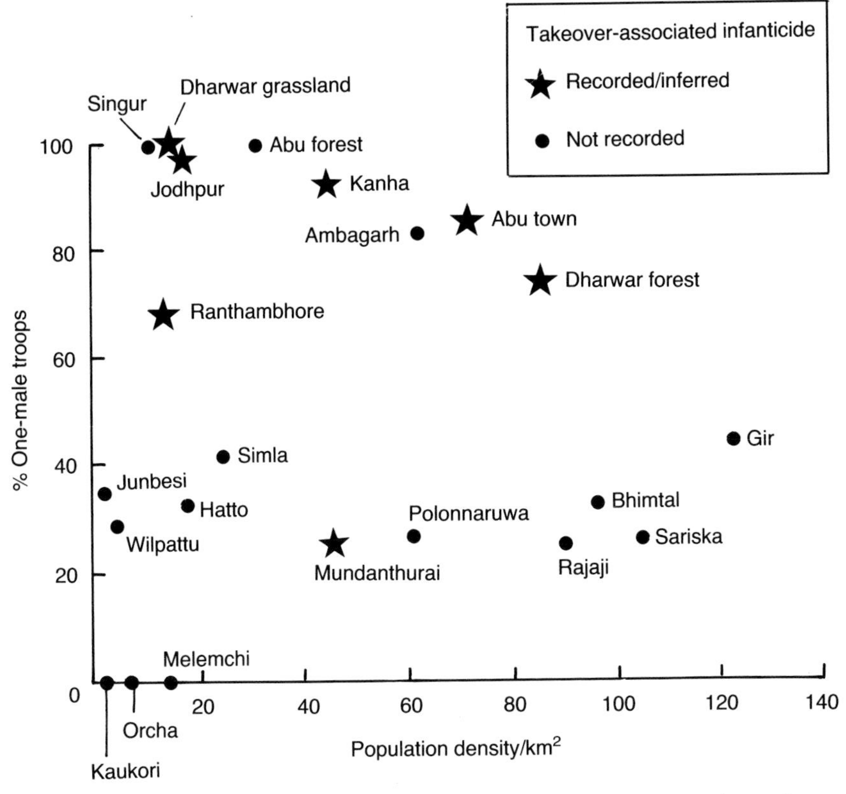

Figure 11.5 Relationship between Hanuman langur population density and prevalence of one-male troops as a percentage of total troops in 22 populations. From Newton (1986), Reena & Ram (1992) and Ross (1993).

when faced with an infanticidal male, in order to terminate investment in an infant that would in any case be killed later (Bruce, 1960).

Although most of the evidence for infanticide in colobines derives from Hanuman langurs, it has also been observed in red colobus and the silvered langur. Struhsaker & Leland (1985) and Leland *et al.* (1984) have documented one observed and two suspected infanticides in red colobus at Kibale. At first glance the sexual selection hypothesis would not seem to be supported as an explanation for infanticide in a patrilineal society, since an infanticidal male is likely to be born in the troop in which it killed. However, intriguingly, Struhsaker & Leland (1985) provide evidence that the infanticidal male was conceived outside the troop, as his mother immigrated whilst pregnant (i.e. he was unlikely to have been closely related to the victim). The interbirth interval of the victim's mother was reduced and the infanticidal male might have sired subsequent births; therefore, the essential postulates of the sexual-selection hypothesis were not violated.

The sexual-selection hypothesis appears to explain infanticide in both matrilineal, harem society and patrilineal, multi-male society. Why has infanticide not been found in colobine matrilineal multi-male troops? Leland *et al.* (1984) suggest that a uni-male troop structure predisposes to infanticide because intermale reproductive competition and variance in mating success in such populations are greater than in multi-male society. Infanticide will be facilitated as incoming males are unlikely to be closely related to troop infants and there are no other males present who might defend infants and contest post-takeover reproductive access. In contrast, in multi-male societies, competition for mating will occur within the troop, rather than between troop and band. Promiscuity will confuse paternity, increasing the probability that an infanticidal male would kill kin and that other males would defend the victim. In addition, the chances that the infanticidal male would breed with the victim's mother are reduced (Hrdy, 1979; Leland *et al.*, 1984).

Chapman & Hausfater (1979), Hausfater *et al.* (1982) and Hausfater (1984) have modelled alternative male langur reproductive strategies using assumptions and data derived from field studies. These suggest that an infanticidal male in an otherwise non-infanticidal population would gain a substantial reproductive advantage. However, if an infanticidal male replaced an infanticidal male, infant killing would be advantageous only at certain tenure lengths. This is a consequence of the fact that at certain durations of tenure the male's infants would be unweaned at the end of his residency and hence susceptible to infanticide by the next incoming male. Although Chapman & Hausfater (1979) and Newton (1988*a*) found that estimates of male tenure from the wild were consistent with the model, the determinants of tenure length are unknown. The evidence is inconclusive in that, in uni-male troop societies of Hanuman langurs, male tenure is shorter at higher population density. Tenure length is independent of habitat disturbance (Reena & Ram, 1992). Male tenure at Jodhpur ranged from 3 days to > 74 months (Sommer & Rajpurohit, 1989) whilst the maximum at Kanha was > 9 years, 9 months (Newton, 1994). At Jodhpur no male achieved residency in more than one troop and it was estimated that 25% of adult males never acquired a troop. These excluded males were likely to have extremely low reproductive success. To achieve high reproductive success a male must take risks, take over troops and kill infants (Sommer & Rajpurohit, 1989). The preponderance of male infant infanticide found by Rajpurohit & Sommer (1991) may have arisen if invading males opt to kill future competitors of their sons but spare female infants which they may be able to inseminate if their tenure as new residents is sufficiently long (Newton, 1994).

Hrdy (1979) and Leland *et al.* (1984) suggested that an infanticidal male would not be expected to gain reproductive advantage in a strictly seasonally breeding population, as interbirth intervals would not be shortened: infanticidal males would have to 'wait' until the next breeding season whether or not they killed infants. However, observations suggest that infanticide does occur in populations with a short breeding season (Newton, 1986). If takeovers are an extra-troop male strategy to increase reproductive success, it is predicted that the optimal time for invading troops would be between the birth and mating seasons in seasonally breeding populations. If invasion occurred after the mating season, the females would probably already have been inseminated. As new residents do not usually kill post-takeover births (Hrdy, 1977; Hausfater, 1984; but see Agoramoorthy & Mohnot (1988) for four examples of males doing so), takeover before the birth season would result in a cohort of infants appearing after the onset of the male's tenancy, with a consequent delay in the male's access to oestrous females. The temporal distribution of the onset of Hanuman langur takeovers, relative to the median date of the birth season, agrees with this prediction (Figure 11.6). Therefore the environment, in the form of climate seasonality and human

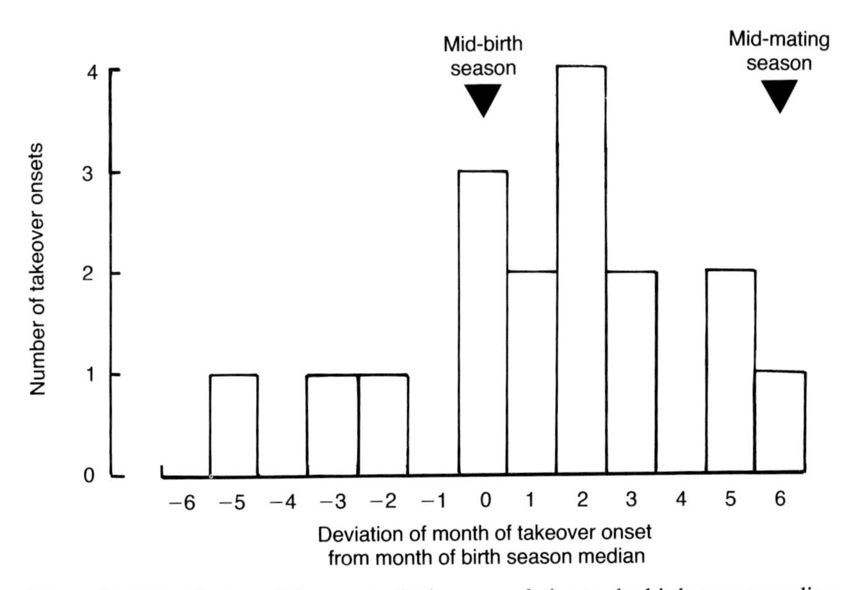

Figure 11.6 Distribution of the onset of takeovers relative to the birth season median in Hanuman langurs. Data from Newton (1986) and Sommer & Mohnot (1985).

cultivation and provisioning, may, through its effects on birth seasonality (see above), influence the timing of langur infanticide and takeover.

Irrespective of whether an infanticidal male does gain a reproductive advantage, infant killing is clearly disadvantageous to the infant and its parents. Mothers may therefore be expected to have evolved counter-strategies for the protection of their young; these may be classified as 'defence' or 'penalty' counter-strategies (Hrdy, 1977, 1979; Chapman & Hausfater, 1979; Hausfater, 1984).

A defensive counter-strategy involves the defence of the infant with an attempt to reduce the reproductive success of the infanticidal male to that of a non-infanticidal male. Possible defensive strategies include the active defence of young, emigration, premature weaning of infants and the manipulation of the time of takeover. Although active defence is the only counter-strategy for which there is direct field evidence in langurs, it is generally ineffective (Hrdy, 1977; Chapman & Hausfater, 1979; Newton, 1986). This is not surprising, considering the pronounced sexual dimorphism in body weight and canine length in Hanuman langurs. A notable feature of many takeovers is the increase in female sexual behaviour during and after male replacement (Sugiyama, 1965; Hrdy, 1977; Vogel & Loch, 1984; Newton, 1986). Hrdy (1977, 1979) suggested that these females were pregnant and that by soliciting incoming males they were inducing the male to tolerate future offspring. Indeed, there have been few reports of the killing of infants born after takeover (Sommer, 1987; but see Agoramoorthy & Mohnot, 1988). However, field data do not support the hypothesis that this pseudo-oestrus is a counter-strategy (Newton, 1987; Sommer, 1987). A simpler explanation is that females are exhibiting mate choice; unencumbered females choose successful, infanticidal males whilst females with infants remain distant from the dangerous band males. Penalty counter-strategies, on the other hand, involve tactics which penalize infanticidal males so that their reproductive success is reduced to less than that of a non-infanticidal male; there is no evidence for their existence in colobines (Hausfater, 1984).

Social organization and ecology

Spacing and social change

Although the majority of social interactions between monkeys occur within groups, groups do not live in isolation, but interact with other troops and bands to produce varying patterns of spacing and social change. The majority of colobines have range sizes of less than 1 km^2. However, the variation spanned is enormous, from 0.03 km^2 for purple-faced langurs to 25–30 km^2 for the Sichuan snub-nosed monkey (see Chapter 5, this volume). The general colobine trait appears to be the almost exclusive occupancy of at least the central part of the range, with some overlap with adjacent troops at the periphery (Cheney, 1987; Struhsaker & Leland, 1987). For example, a Hanuman langur troop at Kanha occupied a 74-ha range with exclusive, intensive use of the central 50% and overlap with adjacent troops in the outer 50%. Aggressive encounters with other troops at the periphery, involving both sexes, occurred with a frequency of 0.68/day (Newton, 1992). As suggested by Wrangham (1987), females living in a matrilineal society tend to respond aggressively to other troops.

There are only three clear exceptions to this pattern. In the patrilineal red colobus, three or more troops showed almost complete overlap with aggressive intertroop interactions expressed as intergroup dominance; bands have not been recorded (Struhsaker & Leland, 1979, 1987). The only matrilineal species known to have completely overlapping ranges are the proboscis monkey and capped langur (Bennett & Sebastian, 1988; Stanford, 1990, 1991). The latter species is unusual in that it lives in uni-male matrilineal troops with complete range overlap. The frequency of male displays and male 'herding' of females is higher in encounters with 'familiar' troops than with unknown groups (Stanford, 1990, 1991).

Amongst Hanuman langur populations, range size declines with increasing population density in a classic Huxley (1934) compression curve (Figure 11.7, log transformation $r = 0.735$, $n = 15$, $P < 0.01$; Newton, 1984). The data suggest that these colobine territories behave like rubber discs that become compressed as groups compete for space at high population densities. As shown by Waser (1977) for a cercopithecine, the grey-cheeked mangabey (*Cercocebus albigena*), an overdispersed range pattern could result from either site-independent avoidance or site-specific territorial defence. The field evidence suggests that the pattern of Hanuman langur troops results from site-specific territorial defence, mediated by adult male loud calls, perhaps as a means of defending food resources (Newton, 1984). Amongst primates in general, and also within the Colobinae, there is a negative relationship between the proportion of the day spent moving and the proportion of foliage in the diet. Predominantly folivorous species, such as the guereza, spend relatively little of each day moving in comparison to more frugivorous *Presbytis* species, presumably because of smaller patch size and wider patch dispersion of fruit relative to leaves (Oates, 1987).

Relationships between troops and all-male bands differ profoundly from those between bisexual troops. In Hanuman langurs, all-male bands, but not

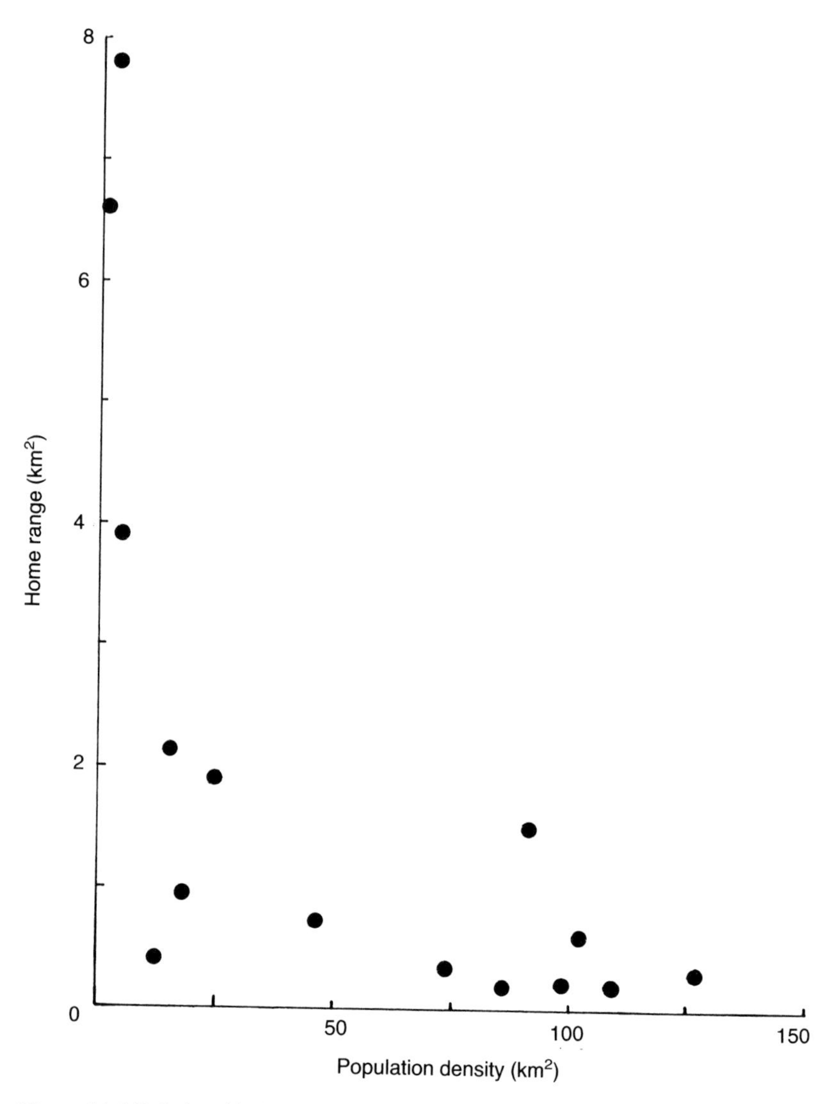

Figure 11.7 Relationship between Hanuman langur home-range size and population density for 15 populations. Data from Newton (1984).

neighboring bisexual groups, may be found in the central part of a troop's range. However, the band males rarely feed there, being engaged in conflict with the resident male. Bands have a much larger range size than troops, encompassing some 3–7 troop ranges (Hrdy, 1977; Mohnot, 1984; Sommer, 1988; Newton, 1992, 1994). The often made suggestion that langur bands occupy sub-optimal habitat has not yet been substantiated (Pusey & Packer, 1987).

The existence of extra-troop males in many colobine populations suggests that males move from their natal troops. Indeed, long-term monitoring of Hanuman langurs, Nilgiri langurs, banded leaf-monkeys, purple-faced langurs, proboscis monkeys, red leaf-monkeys (*Presbytis rubicunda*) and guerezas suggests that females do usually remain in their natal troops, whilst immature males disperse. Whilst in some species these males remain solitary, in others they join bands from where they attempt to gain access to troop females (Table 11.2; Rudran, 1973*a*; Hrdy, 1977; Bennett, 1983; Davies, 1984; Newton, 1987). Interestingly, extra-troop males tend to exist as solitaries among African colobines but in bands amongst Asian colobines (Table 11.2). The only colobines in which females are known to emigrate more frequently than immature males are the red colobus and olive colobus (Starin, 1981; Struhsaker & Leland, 1987; Oates, Chapter 4, this volume). Starin (1981) found that, within a Gambian population of red colobus, immature females voluntarily left their natal troops. Those immature males who left were harassed into leaving and vigorous attempts were made by residents elsewhere to prevent their subsequent immigration. Indeed, resident males and females killed one and possibly two other adult males attempting to enter the troop. Most males returned to their natal troop after a period of adolescent exile. Despite predominant female emigration, affiliative relationships between related females remained important as they tended to transfer together (Starin, 1991), analogously to the cohorts of immature Hanuman langur males migrating into all-male bands.

Aside from the migration of immatures, a great variety of social changes have been recorded. In Hanuman langurs, change from uni-male to multi-male troops and vice versa, troop formation from fragments of other troops and bands, gradual resident-male replacement and rapid resident-male replacement (or takeover), with or without infanticide, have been noted (Sugiyama, 1967; Roonwal & Mohnot, 1977; Hrdy, 1977; Newton, 1987). Takeovers have also been recorded or inferred in the uni-male societies of purple-faced langurs, guerezas and red leaf-monkeys (Rudran, 1973*a*; Oates, 1977*b*; Davies, 1987*b*). Sommer (1988) noted, within the Jodhpur Hanuman langur population, that it was the highest-ranking male within the attacking band that became the new resident. Low-ranking males appeared to be important in assisting with takeovers and may have gained through sexual contact during the confusion of the social change before being expelled by the new resident. Takeovers may occasionally be lethal to adults as well as to infants, as witnessed by the killing of two adult males of Thomas' leaf-monkey in Sumatra by a young adult male who subsequently became the troop resident (Kunkin, 1986) and the suspected killing of a Hanuman resident male by an invading band (Rajpurohit *et al.*, 1986). Interestingly, a 15-month study of five capped langur troops, two bands and solitary males, revealed no takeovers in this non-territorial species. The extra-troop adult males may have obtained the few females who left troops, forming new uni-male troops (Stanford, 1991).

Preliminary reports on the enigmatic Chinese snub-nosed langurs suggest that the uni-male troops and all-male bands coalesce into enormous aggregations of over 400 animals (Happel & Cheek, 1986; Bleisch *et al.*, 1993). More is known of the fusion–fission society of proboscis monkeys in which specific uni-male troops and male bands consistently form large clusters at riverine sleeping sites (Bennett & Sebastian, 1988; Yeager, 1991). This society resembles the fusion–fission systems of the gelada (*Theropithecus*) and hamadryas (*P. hamadryas*) baboons. Why such distantly related cercopithecids have converged is not clear. However, the baboons and proboscis monkeys are unusual in occupying linear sleeping sites, with restricted escape from predators, the proboscis monkey sleeping along river banks and the baboons on cliff edges.

Hrdy (1977), Bishop (1979) and Laws & Vonder Haar Laws (1984), have described a pattern of social change in some langur populations in which band males associated with otherwise uni-male troops. In the Rajaji population these influxes resulted in the formation of temporary multi-male troops during the mating season (Laws & Vonder Haar Laws, 1984). Although the lack of objective definitions of takeover and gradual male replacement hampers the analysis, both gradual and rapid patterns have been reported in a few populations (Newton, 1988*a*). In general, however, uni-male troop populations of Hanuman langurs tend to show rapid, aggressive adult-male replacement with evidence of infanticide. In contrast, multi-male troop populations tend to show gradual male replacement with a staggered pattern of male introductions and exclusions (Boggess, 1984; Newton, 1986). In other cases, new troops may arise through the fission of a larger group and the subsequent emigration of part of the original group. Dunbar (1987), for example, has documented troop fission in Ethiopian guereza when groups exceeded a size that gave each individual less than ten trees within the troop range. In contrast, despite 13 years of study, few social changes have been recorded within the Kibale population of red colobus; in this case, social troops appear to be very stable, perhaps as a result of intermale competition for females occurring within rather than between troops (Struhsaker & Leland, 1987).

Group size and composition

The diversity of colobine societies is summarized in the Introduction; their interpretation is bedevilled by two problems. First, although cercopithecids have traditionally been classified into matrilineal, patrilineal, uni-male, multi-male and monogamous societies, fieldwork suggests some species do not fall neatly into such compartments, but change through space and time (Tsingalia & Rowell, 1984; Chism & Rowell, 1986). Second, a primate's social organization does not necessarily equate with its mating system: for example, females in harem troops sometimes mate with extra-troop males.

In order to explain the evolution of such a diversity of social systems, we need to be able to answer at least four key questions. First, why do animals live in groups? Second, what determines the size of the group that females live in? Third, when females do live in groups, why do they live with related females in some cases (matrilineal societies) but not in others (patrilineal societies)? Finally, what determines the number of males in a group?

Why live in groups?

Two main hypotheses have been advanced to explain why animals live in groups. One is that the risk of predation is proportionately reduced as group size increases; the other is that larger groups are more effective at defending exclusive access to food resources (Alexander, 1974). In a series of tests between these hypotheses, Van Schaik (1983) argued that birth rates would be inversely related to group size in the first case (since the primary benefit from the increased group size comes from a reduced risk of predation to juveniles and adults), but would be an inverted U-shaped function of group size under the second hypothesis (since larger groups would enable females to acquire more food and thus breed faster, but only up to the point where the costs of group living began to outweigh the benefits). The data for most primates tend to support the predation-risk hypothesis (Van Schaik, 1983; Dunbar, 1988), though some species are anomalous in this respect.

Among colobines, data with which to test these two predictions are available for only two species. In one population of Hanuman langurs, there is a significant positive correlation between adult female group size and the number of infants per adult female at the end of the birth season (Figure 11.8; $r = 0.67$, $n = 18$, $P < 0.01$; Newton, 1987), thus supporting the resource-defence hypothesis. In contrast, data from four populations of East African guereza yield a negative correlation (correlations for individual populations: $r = -0.94$, $n = 6$; $r = -0.45$, $n = 7$; $r = -0.35$, $n = 10$; $r = -0.23$, $n = 6$; Fisher's (1958) pooled significance test $P < 0.01$; Dunbar, 1987), thus supporting the predation-risk hypothesis. In the case of one of these four populations, Dunbar (1987) provides evidence that the negative relationship is a consequence of higher birth rates in smaller troops and not differential post-natal mortality. The evidence for colobines is thus equivocal, possibly suggesting that the primary factor promoting group-living may vary between species, depending on dietary niche and the level of predation risk.

What determines group size?

It is usually assumed that the size of groups in which females live is determined by the interaction between individual foraging strategies, the spatial and temporal dispersion of food resources and the risk of predation

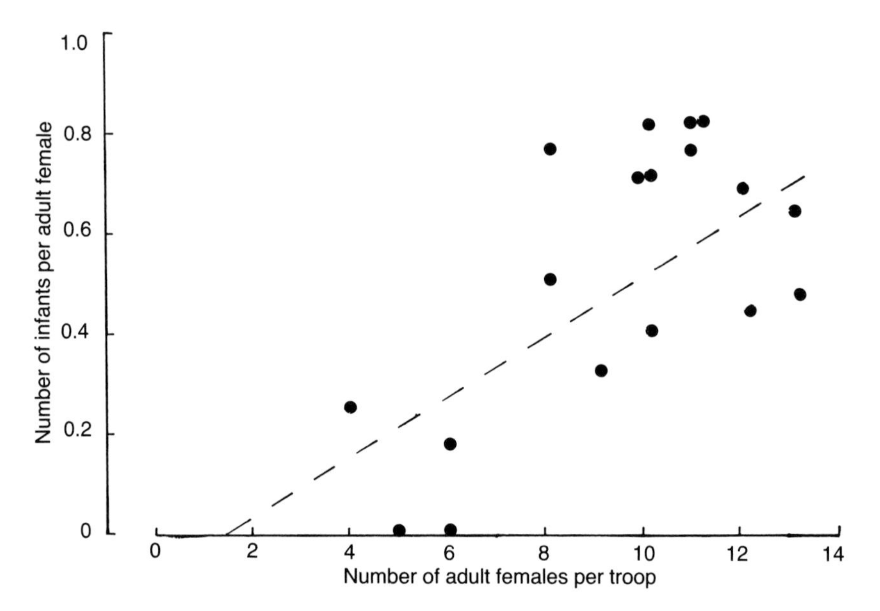

Figure 11.8 Relationship between number of Hanuman langur adult females per troop and number of infants per adult female. Data for 18 Hanuman langur troops on the Kanha meadows (From Newton 1987).

(Wittenberger, 1980; Van Schaik & Van Hooff, 1983; Dunbar, 1988). Group sizes will be smaller when predation risk is low and when food-patch size is small and/or abundant; when food patches are large but ephemeral, animals will be forced to live in large groups and range widely. Indeed, red colobus which feed on a diverse diet of flowers, fruits, shoots and leaves live in large groups in large ranges, whereas guereza, which have a more restricted diet based on the leaves of a few key tree species, live in small groups in small defended ranges (Clutton-Brock, 1974*b*). Hladik (1975) developed a similar argument to explain differences between sympatric Hanuman and purple-faced langurs in Sri Lanka.

There is some evidence to suggest that, at very high densities, group size is limited by the quality of the territory. In one guereza population, for example, territory sizes were compressed to the minimum, with no room for expansion as groups grew in size; groups were then forced to undergo fission once their size exceeded a territory-specific threshold (Dunbar, 1987). As a result of this, group sizes were typically smaller in smaller forest blocks (e.g. scrub and secondary forest) than in larger forest blocks (e.g. climax forest, Figure 11.9).

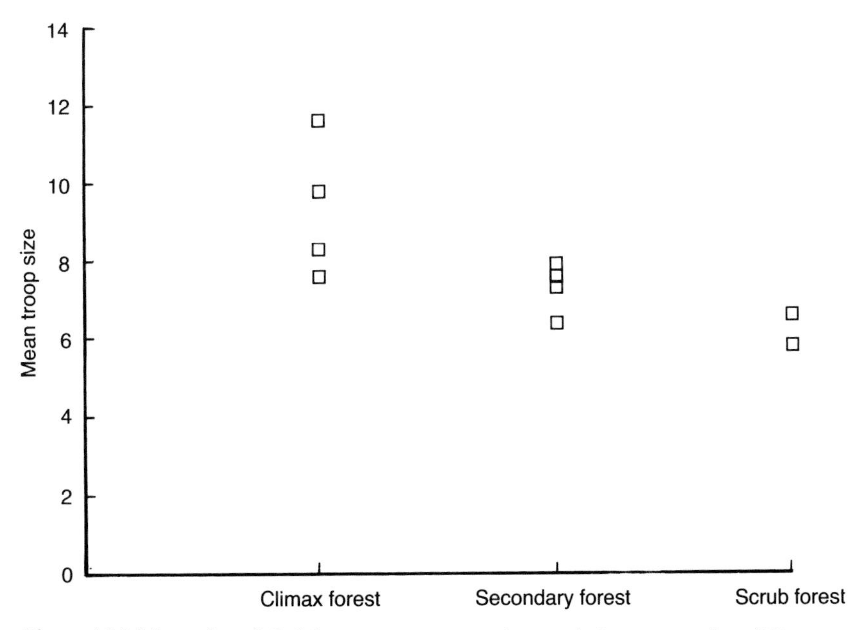

Figure 11.9 Mean size of *Colobus guereza* groups in populations occupying different kinds of forest. From Dunbar (1987).

Monogamy will result when the size of groups in which females live is reduced to the minimum of one. Group size will generally be at the minimum when predation pressure is negligible and feeding competition makes it advantageous for females to space out. The monogamous colobines live on islands where large predators are rare or even absent (Cheney & Wrangham, 1987; Tilson & Tenaza, 1976).

Why live with relatives?

It is generally assumed that living with close relatives yields benefits via kin selection: in other words, given that it is advantageous to live in a group with other individuals (e.g. for reasons of protection against predators or for resource defence), additional gains may be obtained by doing so with relatives rather than with unrelated individuals, since relatives share a higher proportion of their genes in common. Wrangham (1980) argued that defence of food resources by groups would lead to the evolution of related groups of females staying together to act as coalitions. The same is true if females band together for protection against their predators. This claim is difficult to test, because we have no alternative hypotheses against which to compare it.

A more puzzling phenomenon is the evolution of patrilineal societies in some African colobines in which females within the group appear not to be closely related to each other. The evidence for this is based largely on the observation that females commonly migrate between groups in red colobus (Marsh, 1979*a*; Struhsaker & Leland, 1987). In some populations of red colobus (e.g. at Kibale), males tend to remain in their natal groups, whereas in others (e.g. at Tana River), males also migrate. Considerable uncertainty surrounds the extent to which these species really do differ from other 'matrilineal' colobines, since female migration is more common among colobines in general than it is among cercopithecines (Moore, 1984). Busse (1977) has argued that a multi-male group structure may have evolved in the red colobus as a way of minimizing predation by chimpanzees (*Pan troglodytes*). At Gombe, where it is estimated that 8–13% of red colobus are killed each year by chimps, individuals are significantly less likely to be preyed on by chimpanzees if an adult male is nearby. Female red colobus may therefore select groups of males best able to defend themselves, their females and their offspring. This explanation is consistent with the observation that red colobus live in multi-male troops in habitats where chimpanzees are sympatric (Gombe, Kibale) and in uni-male troops where they are not (Tana, Senegal) (Struhsaker & Leland, 1987). The difference may, however, also reflect variation in group size, in that groups are also larger in those habitats where chimpanzees are sympatric; in this case, multi-male groups would be expected to occur when group size exceeds a certain size (see below).

What determines the number of males?

The reasons for variation in the occurrence of uni-male and multi-male troops have remained 'one of the most striking enigmas in primate social behaviour' (Clutton-Brock & Harvey, 1976). The field data do not support the hypotheses that variation in the number of males per troop is a function of population density, climate, predation risk or differential male mortality (Clutton-Brock & Harvey, 1977; Marsh, 1979*b*; Van Schaik & Van Hooff, 1983; Dunbar, 1984, 1988; Newton, 1988*a*).

However, there is increasing evidence that the number of resident males is a function of the number of adult females. Emlen & Oring (1977) argued that a male's breeding strategy would depend on his ability to monopolize access to receptive females and that this, in turn, would depend on female group size and the degree of synchrony in the females' reproductive cycles (see Wrangham, 1980). Whereas the reproductive success of males is limited by the availability of females, that of females is limited by the availability of food resources. We would therefore predict a positive correlation between troop size (or female group size) and the number of males in a troop (Dunbar, 1988). There is growing evidence from a wide range of primates to support this prediction (Alexander, 1974; Van Schaik & Van Hooff, 1983; Terborgh, 1986; Andelman, 1986; Crockett & Eisenberg, 1987; Dunbar, 1988).

Among the colobines, Marsh (1979*b*) demonstrated a positive relationship between group size and the number of adult males in Senegalese red colobus, and Dunbar (1987) found a similar relationship in East African guereza. Newton (1988*a*) also found a significant positive correlation between the number of adult males in troops of Hanuman langurs and both group size and the number of females in a troop. The Hanuman langur data suggest that a single male is able to monopolize up to 6–12 breeding females, but that a larger number of females is not economically defendable and a multi-male troop forms. The relationship between population mean group size and the mean number of adult males per troop for the various colobines is shown in Figure 11.10, with separate regression lines for Hanuman langurs and red colobus. The alternative explanation, that females aggregate around groups of males in proportion to the number of males (and hence their likely ability to defend food resources), is unlikely, given that, with the exception of red colobus, female troop membership is relatively static.

Three other factors may affect a male's ability to monopolize a troop.

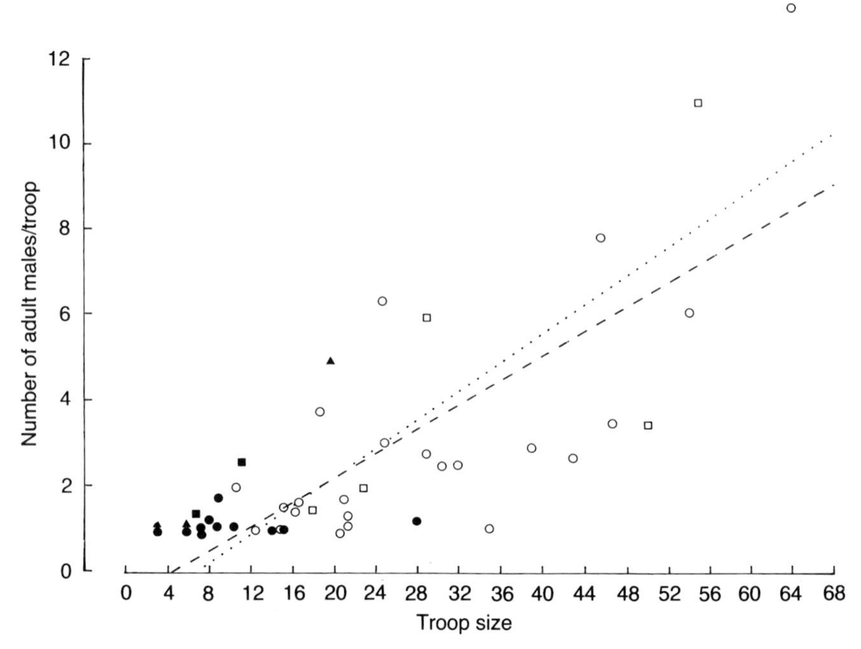

Figure 11.10 Relationship between mean troop size and number of adult males per troop for colobine populations (from Table 11.1). Open circles, *Semnopithecus entellus*; open squares, *Procolobus badius*; filled circles, other *Presbytis* and *Trachypithecus* spp.; filled squares, other *Procolobus* and *Colobus* spp.; filled triangles, other Colobinae. Regression lines: dotted line, red colobus, $r = 0.728$; dashed line, Hanuman langurs, $r = 0.752$.

First, wide female dispersion within a troop will reduce a male's ability to monopolize them (Van Schaik & Van Hooff, 1983). Second, Terborgh & Janson (1986) suggest that males will be better able to monopolize females if they have ample time free from foraging. However, there is no clear association between the social organization and the proportion of fruit in a species' diet; frugivores tend to spend less time resting and would therefore be predicted to be predominately uni-male (Dunbar, 1988; P. N. Newton, unpublished).

Third, Emlen & Oring (1977) suggest that extreme synchrony or asynchrony in female reproductive cycles would reduce the potential for male monopolization, whilst moderate asynchrony would increase it. There is evidence in support of this hypothesis amongst a sample of 33 primate species (Ridley, 1986; but see Altmann, 1990) and amongst cercopithecines (Andelman, 1986). Newton (1988*a*) used an intraspecific comparison to test the prediction that the monopolization of Hanuman langur troops would be easier in nonseasonally reproducing populations; although the results were in the predicted direction, the difference was not significant.

In Figure 11.11 we give a tentative summary of colobine mating systems and reproductive behaviour. Although there are exceptions in each of the clusters and some species (e.g. olive colobus, black colobus) may have been misplaced, some clear-cut patterns do emerge. Matrilineal species have frequent female–female interactions and live in overdispersed ranges; the females typically solicit matings, have no sexual swellings and handle infants

Social association	Species
PATRILINEAL Intergroup dominance Sexual swellings Dark natal coats No infant handling Rare female mating solicitation Male–male grooming common **African rain forest**	*Procolobus badius, Procolobus verus? Colobus polykomos??*
MONOGAMOUS Female loud calls **Mentawai Islands** **Sumatran rain forest**	*Presbytis potenziani, Simias concolor, Presbytis comata*
MATRILINEAL uni-male **multi-male** Over-dispersed ranges No sexual swellings Flamboyant natal coats Infant handling Female solicitation of mating Female–female grooming common **African rain forest, Asian rain forest, mangrove, deciduous forest, Himalayan forest, scrub and villages**	*Colobus satanas?, C. guereza, Presbytis melalophos, P. comata, P. thomasi, P. rubicunda, Trachypithecus vetulus, T. geei, T. johnii, T. cristatus, T. obscurus, T. pileatus, Semnopithecus entellus, Nasalis larvatus.*

Figure 11.11 Social associations and colobine species.

that frequently have flamboyant coats. Patrilineal species are confined to equatorial Africa; in these species, there are infrequent female–female interactions and females rarely solicit matings, typically have sexual swellings and rarely handle infants (which in turn seldom have flamboyant coats).

Explanations for the evolution of colobine social systems are, at best, tentative. The data summarized here suggest that they have been subject to the same evolutionary forces that have given rise to the social systems of other catarrhine primates. In the final analysis, however, the relationships between colobine ecology and behaviour will only be resolved by more fieldwork, in particular fieldwork that is designed to yield data that are directly comparable with those from other studies. Comparative analyses from populations differing widely in habitat characteristics, as well as studies of the neglected colobine species, will go a long way to resolving many of the uncertainties apparent here.

Acknowledgements

PNN thanks the Government of India, Madhya Pradesh Forest Department, SRC, Mungal & Mohan for their assistance in working in Kanha and to M. J. Coe and Sir Richard Southwood for permission to use the facilities of the Department of Zoology, Oxford. He also thanks the Sasakawa Fund, Royal Society, Professor Y. Sugiyama and Professor T. Nishida for facilitating a visit to the Primate Research Institute and Department of Zoology, University of Kyoto, where this chapter was partly written. RIMD thanks the Wildlife Conservation Organization of the Government of Ethiopia and the Institute of Ethiopian Studies of the University of Addis Ababa for their support and sponsorship during fieldwork in Ethiopia. We are greatly indebted to the editors and Sarah Hrdy, Georgina Dasilva, Lysa Leland and Tom Struhsaker for suggesting improvements to the manuscript.

12

Conclusions: the past, present and future of the colobines

JOHN F. OATES and A. GLYN DAVIES

The first colobines

Today's 31 colobine species are the remnants of a long series of adaptive radiations. The variety of niches occupied by these species reflects the constraints and opportunities provided by the original colobine strategy and particular events in the history of each lineage. As described in Chapter 2, the colobines probably diverged from other cercopithecid monkeys in Africa in the middle Miocene. Early in their history, they invaded Eurasia, and subsequently there were a number of separate radiation events in Eurasia and Africa. The oldest currently known fossils that are definitely colobine come from south-eastern Europe and western Asia and date from 8–9 million years ago, the late Miocene. The colobines appear to have fared differently in each continent they have occupied. They are not known in Europe after the middle Pliocene, and in Africa their generic diversity has declined from a Pliocene peak. Asian colobines, on the other hand, seem to have reached their greatest diversity only recently.

A phylogenetic analysis of the cercopithecid foot by Strasser (1988, 1989) suggests that this foot was originally adapted to a semi-terrestrial way of life, and that it was remodelled by the earliest colobines for a more arboreal existence in rain-forest, with particular adaptations for leaping. In Chapter 2, Delson presents evidence that the colobines of the past, like those alive today, were predominantly inhabitants of forest and woodland. Although some fossil colobines (such as *Dolichopithecus*) may have been quite highly terrestrial, it seems that colobines have no significant history of open-country dwelling, unlike their cercopithecine sister group.

It has long been assumed that the earliest colobines were folivorous, and that this feeding strategy was their key original difference from the frugivorous cercopithecines. This view has been derived from several lines of evidence: the presence in living colobines of a divided stomach with a fermentation chamber, resembling that of ruminants; the findings of early field studies, which reported colobines to be eating large quantities of leaves; and the high relief on the cheek teeth of both living and fossil colobines. But as several chapters in this book have shown, recent findings and analyses have raised doubts about the essentially folivorous nature of these monkeys.

For instance, several extant colobine populations have been found to have diets dominated by seeds, especially immature seeds, at least for part of the year (Chapters 4 and 5), while some of the anatomical peculiarities of colobines may not be so closely tied to folivory as has been assumed. Lucas and Teaford (Chapter 6) have shown that the colobine dentition, and particularly molar structure, may be more useful for processing immature seeds than for processing foliage. McKey (1978*b*) has argued that foregut fermentation by colobines could be primarily an adaptation not for fibre processing but for detoxifying plant secondary compounds such as alkaloids (and see Chapter 9); seeds could be just as important a target for detoxification as leaves.

Similar arguments have been developed to explain the evolution of forestomach fermentation in the Bovidae. Demment & Van Soest (1985) suggested that the earliest ruminants, which appear to have been small-bodied, evolved fermentation as an adaptation for detoxification or nutrient synthesis, rather than for cellulose digestion. This argument is based on the observation that foregut fermentation of fibre is not a viable primary digestive strategy in a mammal smaller than 5 kg. Countering this argument, Cork & Foley (1991) found the actual evidence for detoxification of plant secondary compounds in the mammalian foregut equivocal. They argued that the chief benefit of a foregut fermenting system in arboreal mammals, particularly primates, is that it allows a mix of fruits, seeds and foliage to be consumed, with the most highly digestible components passing rapidly through the tubiform midstomach and the less digestible fibre being retained in the sacciform part. Kay and Davies (Chapter 8) also emphasize the adaptability of the colobine digestive system, in which complementary forestomach and caeco–colic fermentation facilitate digestion of foliage and seeds.

A possible evolutionary scenario, therefore, is that the first colobine was a relatively small, forest-living and predominantly fruit-eating African monkey that had developed adaptations for the more efficient processing of immature seeds and/or leaves. Such adaptations would have opened up new adaptive zones, allowing the exploitation of food resources not readily handled by other anthropoids, and leading to increased population densities and/or the occupation of new habitats (for instance, habitats having major seasonal fluctuations in the availability of ripe fruit). From this point on, as colobines diversified, different advantages of their food-processing system could have come under selection according to circumstance.

For instance, selection is likely to have operated on some lineages to increase body size, since a larger size would be an advantage in situations (such as strongly seasonal environments) where foods other than mature foliage were scarce or absent for at least part of the year. This is consistent with the observation that today's largest colobines are those living either in the most highly seasonal environments (the Chinese snub-nosed monkeys, and Himalayan populations of Hanuman langurs) or in environments offering a limited range of highly digestible food (proboscis monkeys in Bornean mangrove swamps).

Socioecological differences from cercopithecines

While the details of the diet of the earliest colobines are conjectural, it is apparent that key features of the original colobine strategy were adaptations to arboreality in forest habitats and to the processing of plant foods, rather than to life in open country and searching for highly dispersed food – the likely strategy of the early cercopithecines. It is common to see colobines described as ecological specialists compared to cercopithecines, but this is an inaccurate characterization, for the colobine food-processing system can cope with a wider range of plant foods (and with more commonly available foods) than can that of cercopithecines. In consequence, populations of colobine monkeys can achieve biomasses higher than those of any other primate (Chapter 10). At the same time, colobines have not come under selection for abilities to search out and gather hard-to-find foods like ripe fruits and insects (like guenons), or to dig in grassland soils for rhizomes (like baboons).

There is no one typical colobine social system, and as field studies have multiplied it has become clear that colobine social systems are as varied as those of cercopithecines. However, one general difference between cercopithecine and colobine societies (discussed in Chapter 11) is that colobine females tend to have lower frequencies of overt intragroup agonism. This may be because colobine foods occur on average in larger, denser patches than cercopithecine foods, with the consequence that contest competition for food has historically been of less value to colobines. An example of how such low-competition feeding strategies play out in the lives of colobines is provided by *Presbytis rubicunda* in Borneo. Groups of these leaf-monkeys split up into smaller foraging parties when young leaves are taken from small understorey trees (Davies, 1984).

Another ecological consequence of the evolutionary path followed by the colobines is that they typically have a low species-richness in any one community. At least in Africa, forests usually contain many more cercopithecine than colobine species. Nine cercopithecine species co-occur in the central part of the Ituri Forest of Zaire (Hart *et al.*, 1986), and in parts of mainland Equatorial Guinea (Sabater-Pi & Jones, 1967); yet the Ituri Forest contains just three colobus species, and in most of Equatorial Guinea there is only one. Several factors probably interact to produce this pattern. First, one or more cercopithecine species typically occupy semi-terrestrial niches unoccupied by colobines; for instance, four of the Ituri cercopithecines are partly or predominantly terrestrial. Second, because cercopithecines pursue strategies that involve acquiring foods that are often hard to find and/or require special skills to capture or gather, a tropical forest may provide niches for several sympatric cercopithecine species, each foraging for a different (if overlapping) set of of foods distributed in a different fashion or requiring different capture or handling skills. A third factor promoting species diversity is habitat specialization within a general area. For instance, *Cercopithecus neglectus* and *Cercocebus galeritus* (present in communities in both Ituri and Equatorial Guinea) are cercopithecines especially associated with riverine forest. Emmons *et al.* (1983) recognize these two monkeys as members of a distinct group of African forest 'watercourse species', animals that have an obligatory requirement to include a watercourse in their home range.

Colobine communities

It is rare for any forest, African or Asian, to contain more than two sympatric colobines. Where more than two occur, the extra species often have the properties of watercourse species, showing some association with riverine or swamp vegetation. In the areas of the Upper Guinea forest in West Africa where three colobus species co-occur, the small olive colobus seems to be strongly associated with riverine vegetation; in the Ituri forest, *Colobus guereza* appears to be most frequent in swamp and riverine forest, while *C. angolensis* and *Procolobus badius* are more typical of dry-land forest. In Asia, four species may occur in the same general locality in some parts of Borneo, but this is usually a consequence of an inland forest community with two *Presbytis* species meeting a riverine community containing silvered langurs and proboscis monkeys (Rodman, 1978; Davies & Payne, 1982). In contrast to African forests, however, Asian forests have cercopithecine diversity that is typically low; there is only a single Asian cercopithecine genus (*Macaca*), which is most commonly represented by one or two species, and only very rarely by as many as four (e.g. in parts of Thailand (Eudey, 1980)). This low species-diversity may in part be the result of the presence of gibbons (*Hylobates* spp.), which occupy niches somewhat similar to those of cercopithecines, and in part be a consequence of the low productivity of the dipterocarp-dominated forests of Sundaland. Dipterocarps are rarely used as food by colobines or macaques (Caldecott, 1986).

We suggest that the typically low number of sympatric colobines in non-riverine forests is largely a consequence of the key importance of their food-processing adaptations, rather than of their foraging techniques. In any given ecosystem, only a limited number of different colobine gut configurations and body sizes may be both workable and sufficiently different in function to provide niche separation. This may be particularly crucial during periods of food scarcity, when similar-sized colobines of similar morphology would be likely to converge on the same set of resources. Some evidence of such competitive exclusion is provided by *Presbytis hosei* and *P. rubicunda* in Borneo; where one is relatively abundant, the other is very rare (Chapter 10).

When several colobines are sympatric, these species are very often of different sizes or have different morphology; each species eats a different selection of foods, but unlike cercopithecines they do not search for them or gather them in distinctly different ways. For instance, in the three-species West African community, the olive colobus is about half the size of sympatric red or black-and-white colobus; the red colobus has a distinctly four-chambered stomach, whereas in the black-and-white colobus it is difficult to distinguish between the first two chambers (presaccus and saccus gastricus) (Kuhn, 1964). The single large chamber in the black-and-white colobus may be an adaptation to a relatively uniform, high-fibre diet, while separate chambers in the other species could be adaptations to a more varied intake and the separation of more and less digestible components (D. Chivers, personal communication). In Asian forests, *Trachypithecus* species have larger stomachs than sympatric *Presbytis* (discussed in Chapter 7). The proboscis monkey is sympatric in parts of Borneo with members of these two genera, but it is considerably larger than either of them. Where similarly-sized colobines co-occur it is possible that niche separation is at least in part a consequence of different abilities to handle particular plant defence strategies, as demonstrated in lemur communities by Ganzhorn (1989); this is an area that requires further research.

Distribution patterns

The composition of colobine communities has also been affected by long-term historical processes that have interacted with more proximate ecological conditions to produce the presence or absence of species in particular areas: patterns of geographical distribution. Throughout much of colobine history, Old World forests have experienced cycles of contraction and expansion as climate has changed, but the overall pattern has been one of forest reduction since the Miocene (Delson, Chapter 2; Bonnefille, 1985). During the Pleistocene there were particularly marked cyclical swings in tropical climates, from cool, dry conditions during glacial maxima to warm, moist conditions in interglacials (Hamilton, 1988). At times of forest retraction, colobines will have found themselves at a competitive disadvantage with cercopithecines in more open habitats. In consequence, colobine populations must have been frequently fragmented, and often driven to extinction.

Present geographical patterns of colobine species-diversity are probably in part a consequence of these historical patterns of climate change. Rates of colobine species extinction are likely to have been greatest where moist forests growing in warm conditions have been most greatly reduced by climate change. Europe, where no colobines live today, has at times lost all semblance of moist forest, and most of the continent currently experiences cold winters. Africa suffered more than Asia from forest shrinkage in the Pleistocene, and there are presently fewer colobine species in Africa than in Asia. Borneo has a richer colobine fauna (six species) than any area in the world of similar size, perhaps in part because the island has had long-persistent tropical moist forest. Borneo also appears to be an area where there has been relatively recent faunal mixing and where a radiation of *Presbytis* species has occurred.

The effects of long-term climate change are also reflected in the distribution patterns of two colobine species-groups at opposite geographical ends of the monkeys' present distribution: the red colobus monkeys of West Africa and the snub-nosed monkeys of Vietnam and southern China. These two groups contain what are probably the two most threatened of all colobine forms: Bouvier's red colobus (*Procolobus badius bouvieri*) of the Congo Republic, and the Tonkin snub-nosed monkey (*Pygathrix (Rhinopithecus) avunculus*) of northern Vietnam. In West Africa, there is a gap of almost 1000 km between the population of red colobus (*P. badius preussi*) on the Nigeria–Cameroon border and populations of *P. b. oustaleti* and *P. b. bouvieri* in Congo. This fragmented distribution (and the absence of red colobus monkeys from much of western equatorial Africa) is unlikely to have resulted primarily from a recently acting ecological factor, such as hunting by humans. Locally, hunting has probably caused some range shrinkage, for there are museum specimens of *P. b. preussi* from localities in western Cameroon (e.g. the Yabassi district) where the monkey no longer exists. Yet there are no museum specimens or other historical records of red colobus from eastern Cameroon, mainland Equatorial Guinea or Gabon, and these regions have a less dense human population than other parts of Africa where red colobus still occur. Most significantly, the taxonomic distinctiveness of *P. b. preussi, bouvieri* and *oustaleti* suggests a long-standing separation of populations, not a recent fragmentation caused by hunting.

Similarly, red colobus monkeys in East Africa have a much more limited distribution than do members of the black-and-white colobus and blue monkey (*Cercopithecus albogularis/mitis*) groups, with which they are often sympatric. Different forms of red colobus survive in forest fragments along the western Rift Valley, in the Uzungwa Mountains, on Zanzibar and on the Tana River. Each of these forms is regarded as taxonomically distinct, again suggesting a long history of separation.

Paralleling the situation of Africa's red colobus are Asia's snub-nosed monkeys. The four *Rhinopithecus* forms in southern China and northern Vietnam have a disjunct distribution. While there has been debate as to whether the three Chinese forms should be regarded as separate species or as subspecies of *P. (R.) roxellana*, they are clearly different from one another, and the different forms occur in geographically distinct areas. In turn, the Chinese forms are clearly distinct from *P. (R.) avunculus* in northern Vietnam. Therefore, although each of the snub-nosed species has suffered a reduction in its range in recent times as a result of habitat destruction and hunting by humans, the different forms must have been geographically restricted and isolated for a considerable time, probably at least since the later Pleistocene (Jablonski, 1993).

The present distributions of the red colobus and snub-nosed monkeys are therefore the result not only of recent human impacts, discussed below, but also of the long-term interaction between historical changes in their environments and their own adaptive features (which appear to include relatively limited ecological tolerances and limited powers of dispersal following isolation events).

Threats to colobine survival

While colobines can be very successful forest-canopy dwellers, their adaptive strategy has made them vulnerable to changes in their habitat. Cercopithecines are much more successful in drier habitats and/or habitats with sparse tree cover. More than any other factor, it is surely their dependence on forest that has brought so many of the colobines to their precarious situation today. In both Africa and Asia, the least threatened colobine species are those most willing to travel on the ground and best able to use dry or gallery forest: *Colobus guereza* in Africa and *Semnopithecus entellus* in Asia. Many other colobines are among the world's most threatened primate species or subspecies; the threats to their survival stem from reductions in forest cover caused by the expansion of human populations and economic development, and from hunting by predatory humans.

In the Action Plans for African and Asian Primate Conservation produced by the Primate Specialist Group of the World Conservation Union (IUCN), 13 of 37 listed colobine species (35%) are rated as endangered or highly endangered from extinction, compared with 15 of 57 cercopithecines (26%) (Table 12.1). Seven colobines, compared with four non-colobines, are rated as highly endangered (meaning that there are estimated to be fewer than 10 000 individuals remaining, with no large section of the population really secure). Fifteen further colobine species are rated as vulnerable or highly vulnerable. Even more striking are the number of threatened colobine subspecies; of 62 primate subspecies listed as requiring special attention in Africa and Asia, half are colobines. Six of the threatened subspecies are forms of red colobus, including the Tana River red colobus (*P. b. rufomitratus*).

All the threatened colobines are forest-dependent forms, and most occur in a limited geographical area where their habitat is being reduced and fragmented. The greatest long-term threat to the survival of colobines (and of other forest primates) is the complete loss of their habitat through subsistence farming and commercial agricultural development (Marsh *et al.*, 1987; Mittermeier & Cheney, 1987). Selective logging of forest usually results in a population decline, related to the extent of canopy damage and other disturbance (Johns & Skorupa, 1987; and see Chapter 10), but logging alone appears to cause extinction rarely; after some years, populations typically recover (e.g. A. D. Johns, 1985, 1986). However, different colobines are not equally sensitive to logging and other forms of habitat disturbance. As noted in Chapter 10, a careful study of the effect of selective logging on the primates of Kibale Forest by Skorupa (1986, 1988) found that heavy logging damage greatly reduced the population density of red colobus (*P. badius*), while the density of black-and-white colobus (*C. guereza*) was actually greater in logged forest than in unlogged forest. These different responses apparently resulted from a major loss of red colobus food trees, and a low degree of dietary flexibility in these monkeys, compared with a probable increase of *C. guereza* foods immediately after logging.

Habitat damage and loss are not the only factors threatening many colobine populations with extinction, and in some cases they are not the primary factors. As was also noted in Chapter 10, colobine monkeys tend to be very vulnerable to hunting by humans. In many of the forested regions of Africa and South-east Asia where colobines occur, monkey meat has long been a common item of human diet. Compared with sympatric cercopithecines, colobines tend to be both easy for hunters to detect (their leaping movements in the canopy are noisy, and adult males commonly produce loud calls) and easy to shoot (they are often large-bodied, and they frequently use the upper part of the canopy, where they are more visible than in low thickets). Their relatively large size means that they can produce a high return of meat per round of ammunition, and people who eat monkey meat usually say that colobine meat tastes good. As a result of hunting, black-and-white colobus have been virtually eliminated from forest near villages in north-eastern Gabon, while three smaller-bodied *Cercopithecus* still occur at moderate densities (Lahm, 1993).

Even when colobine habitats are not destroyed, a combination of selective logging and hunting can cause very great population declines or even extinctions (Struhsaker, 1975). For instance, while Tiwai Island in Sierra Leone has high densities of both red and black-and-white colobus, the nearby Gola West Forest Reserve has very low densities, almost certainly the result of commercial hunting by Liberians (Davies, 1987*a*). This hunting has been facilitated by a network of roads within the reserve created during a selective logging operation. Similarly, the opening up of mature forests in northern Congo to commercial logging, with the creation of logging camps, logging roads and forest inventory trails, has recently led to a massive increase in monkey hunting in those forests; the hunting is apparently devastating populations of forest monkeys, including black-and-white colobus, and causing a much greater impact on monkey populations than the logging itself (Wilkie *et al.*, 1992).

Paradoxically, then, the adaptive features that often made colobines very successful animals in the past now threaten their future survival. With a few exceptions (such as the guereza, the Zanzibar red colobus and the Hanuman langur) colobines have not survived where their natural habitat has been very greatly altered. In addition, most colobines are difficult to maintain in captivity (Collins & Roberts, 1978), perhaps in part because of the peculiarities of their digestive systems; therefore captive breeding probably has even less potential as a conservation tool for colobine monkeys than it does for other

Table 12.1. *Primate species listed as endangered (E) or highly endangered (H) in the IUCN/SSC Action Plans for African and Asian Primate Conservation*

Species (as named in Action Plans)	Family or subfamily	Threat status	Range	Name in this book if different from Action Plan
Africa				
Macaca sylvanus	Cercopithecinae	E	North Africa	
Mandrillus leucophaeus	Cercopithecinae	E	W. Equatorial Africa	
Cercopithecus preussi	Cercopithecinae	E	W. Cameroon–E. Nigeria	
Cercopithecus erythrogaster	Cercopithecinae	E	S.W. Nigeria, Benin	
Cercopithecus sclateri	Cercopithecinae	H	S.E. Nigeria	
Procolobus pennanti	Colobinae	E	W. Equatorial Africa	*P. badius pennantii*, *preussi* and *bouvieri*
Procolobus kirkii	Colobinae	H	Zanzibar	*P. badius kirkii*
Procolobus gordonorum	Colobinae	H	Tanzania	*P. badius gordonorum*
Asia				
Macaca silenus	Cercopithecinae	H	S. India	
Macaca maura	Cercopithecinae	E	Sulawesi	
Macaca ochreata	Cercopithecinae	E	Sulawesi	
Macaca brunnescens	Cercopithecinae	E	Sulawesi	
Macaca tonkeana	Cercopithecinae	E	Sulawesi	
Macaca hecki	Cercopithecinae	E	Sulawesi	

Macaca pagensis	Cercopithecinae	H	Mentawai Islands	
Macaca thibetana	Cercopithecinae	E	S. China	
Macaca cyclopis	Cercopithecinae	E	Taiwan	
Macaca fuscata	Cercopithecinae	E	Japan	
Presbytis comata	Colobinae	H	Java	
Presbytis potenziani	Colobinae	E	Mentawai Islands	
Trachypithecus francoisi	Colobinae	E	N. Vietnam, Laos, S. China	
Simias concolor	Colobinae	H	Mentawai Islands	
Pygathrix nemaeus	Colobinae	E	N. Vietnam	*P. nemaeus nemaeus*
Pygathrix nigripes	Colobinae	E	S. Vietnam	*P. nemaeus nigripes*
Rhinopithecus avunculus	Colobinae	H	Vietnam	*Pygathrix (R.) avunculus*
Rhinopithecus roxellana	Colobinae	E	S. China	*P. (R.) roxellana*
Rhinopithecus bieti	Colobinae	H	S. China	*P. (R.) bieti*
Rhinopithecus brelichi	Colobinae	H	S. China	*P. (R.) brelichi*
Hylobates concolor	Hylobatidae	E	Indo-China	
Hylobates hoolock	Hylobatidae	E	Burma, E. India, E. Bangladesh	
Hylobates klossii	Hylobatidae	E	Mentawai Islands	
Hylobates moloch	Hylobatidae	H	Java	
Hylobates pileatus	Hylobatidae	E	W. Cambodia, S. Thailand	

Source: From Oates, 1986; Eudey, 1987.

groups of primates. For these reasons, conservation of the extant diversity of colobine monkeys will depend on the effective management of a network of protected forests, such as those listed in the Action Plans for African and Asian Primate Conservation (Oates, 1986; Eudey, 1987). In these protected forests, strenuous efforts will have to be made to prevent hunting. The protected areas must also include relatively extensive areas in which logging is prohibited, especially where the targets of conservation are taxa intolerant of habitat disturbance (such as proboscis monkeys and some populations of red colobus).

Such conservation efforts will be most effective if they have the support of local human communities, and if these communities can be encouraged to adopt lifestyles and economic activities that pose less of a threat to the integrity of the forest and its wildlife populations. We caution, however, against the currently-popular management approach that places more emphasis on the agricultural and economic development of communities adjacent to a forest of concern than on the protection of the forest itself. A likely medium-term outcome of such programmes is a larger and more prosperous human population which has both greater resource needs and a greater ability to impinge destructively on the forest. The international agencies that are promoting the conservation and development approach will rarely be in a position to control the potentially disastrous long-term consequences of having larger and richer human populations living in immediate proximity to nominally-protected forests. Community development schemes should therefore be pursued with great circumspection, and only after durable checks and balances have been put in place that regulate the use of the forest and its resources.

References

Aerts, J. V., DeBrabander, D. L., Cottyn, B. G., Buysse, F. X., Carlier, L. A. & Moermans, R. J. (1978). Some remarks on the analytical procedures of van Soest for the prediction of forage digestibility. *Animal Feed Science and Technology*, **3**, 309–22.
Agoramoorthy, G. & Mohnot, S. M. (1988). Infanticide and juvenilicide in hanuman langurs (*Presbytis entellus*) around Jodhpur, India. *Human Evolution*, **3**, 279–96.
Agoramoorthy, G., Mohnot, S. M., Sommer, V. & Srivastava, A. (1988). Abortions in free ranging hanuman langurs (*Presbytis entellus*) – a male induced strategy. *Human Evolution*, **3**, 297–308.
Aldrich-Blake, F. P. G. (1970). Problems of social structure in forest monkeys. In *Social Behaviour in Birds and Mammals*, ed. J. H. Crook, pp. 79–101. London: Academic Press.
Alexander, R. D. (1974). The evolution of social behavior. *Annual Review of Ecology and Systematics*, **5**, 325–83.
Alexander, R. M. (1991). Optimization of gut structure and diet for higher vertebrates. *Philosophical Transactions of the Royal Society London*, **B333**, 249–55.
Ali, R., Johnson, J. M. & Moore, J. (1985). Female emigration in *Presbytis johnii*: a life-history strategy. *Journal of the Bombay Natural History Society*, **82**, 249–52.
Allen, J. A. (1925). Primates collected by the American Museum Congo expedition. *Bulletin of the American Museum of Natural History*, **47**, 283–499.
Allison, M. (1978). The role of ruminal microbes in the metabolism of toxic constituents from plants. In *Effects of Poisonous Plants on Livestock*, ed. R. F. Keeler, K. R. van Kampen & L. F. James, pp. 101–18. New York: Academic Press.
Altmann, J. (1990). Primate males go where the females are. *Animal Behaviour*, **39**, 193–5.
Amer, M. M. & Court, W. E. (1980). Leaf alkaloids of *Rauvolfia vomitoria. Phytochemistry*, **19**, 1833–6.
An, Z. & Ho, C. K. (1989). New magnetostratigraphic dates of Lantian *Homo erectus. Quaternary Research*, **32**, 213–21.
Anadu, P. A., Elamah, P. O. & Oates, J. F. (1988). The bushmeat trade in southwestern Nigeria: a case study. *Human Ecology*, **16**, 199–208.
Andelman, S. J. (1986). Ecological and social determinants of cercopithecine

mating systems. In *Ecological Aspects of Social Evolution*, ed. D. I. Rubenstein & R. W. Wrangham, pp. 201–16. Princeton: Princeton University Press.

Anderson, C. M. (1986). Predation and primate evolution. *Primates*, **27**, 15–39.

Anderson, J. A. R. (1980). *A Checklist of the Trees of Sarawak*. Kuching: Forest Department of Sarawak.

Andrews, P. (1981). Species diversity and diet in monkeys and apes during the Miocene. In *Aspects of Human Evolution*, ed. C. B. Stringer, pp. 25–61. London: Taylor and Francis.

Andrews, P. J. (1989). Palaeoecology of Laetoli. *Journal of Human Evolution*, **18**, 173–81.

Andrews, P. J. & Aiello, L. (1984). An evolutionary model for feeding and positional behaviour. In *Food Acquisition and Processing in Primates*, ed. D. J. Chivers, B. A. Wood & A. Bilsborough, pp. 429–66. New York: Plenum Press.

Andrews, P. J., Harrison, T., Martin, L., Delson, E. & Bernor, R. L. (1995). Systematics and biochronology and biogeography of European Neogene catarrhine primates. In *Evolution of Neogene Continental Biotopes in Central Europe and the Eastern Mediterranean*, ed. R. L. Bernor, V. Fahlbusch & S. Rietschel. New York: Columbia University Press, in press.

Andrews, P. J., Lord, J. M. & Evans, E. M. N. (1981). Patterns of ecological diversity of fossil and modern mammalian faunas. *Biological Journal of the Linnean Society*, **11**, 177–205.

Anon. (1980). *Saving Siberut: a Conservation Master Plan*. Bogor: World Wildlife Fund Indonesia.

Arambourg, C. (1959). Vertébrés Continentaux du Miocène Supérieur de l'Afrique du Nord. *Publications Service Carte Géologique Algérie, (Nouvelle Série), Paléontologie*, **4**, 5–159.

Aronson, J. L. & Taieb, M. (1986). Geology and paleogeography of the Hadar hominid site, Ethiopia. In *Hominid Sites: their Geologic Settings*, ed. G. Rapp Jr & C. F. Vondra, pp. 165–95. Boulder: Westview.

Ashby, M. F. (1992). *Materials Selection in Mechanical Design*. Oxford: Pergamon Press.

Asibey, E. O. A. (1974). Wildlife as a source of protein in Africa south of the Sahara. *Biological Conservation*, **6**, 32–9.

Atkins, A. G. & Mai, Y.-W. (1979). On the guillotining of materials. *Journal of Materials Science*, **14**, 2747–54.

Austin, P. J., Suchar, L. A., Robbins, C. T. & Hagerman, A. E. (1989). Tannin-binding proteins in saliva of deer and their absence in saliva of sheep and cattle. *Journal of Chemical Ecology*, **15**, 1335–47.

Ayer, A. A. (1948). *The Anatomy of Semnopithecus entellus*. Madras: The Indian Publishing House.

Badam, G. L. (1979). *Pleistocene Fauna of India*. Pune (India): Deccan College.

Bailey, R. W. (1973). Structural carbohydrates. In *Chemistry and Biochemistry of Herbage*, vol. 1, ed. G. W. Butler & R. W. Bailey, pp. 157–211. London: Academic Press.

Baranga, D. (1983). Changes in chemical composition of food plants in the diet of colobus monkeys. *Ecology*, **64**, 668–73.

Barry, J. C. (1987). The history and chronology of Siwalik cercopithecids. *Human Evolution*, **2**, 47–58.

Barry, J. C. & Flynn, L. J. (1990). Key biostratigraphic events in the Siwalik sequence. In *European Neogene Mammal Chronology*, ed. E. H. Lindsay, V. Fahlbusch & F. Steininger, pp. 557–71. New York: Plenum.

Bate-Smith, E. C. (1962). The phenolic constituents of plants and their taxonomic significance. *Journal of the Linnaean Society (Botany)*, **58**, 95–173.

Bauchop, T. (1978). Digestion of leaves in vertebrate arboreal folivores. In *The Ecology of Arboreal Folivores*, ed. G. G. Montgomery, pp. 193–204. Washington DC: Smithsonian Institution Press.

Bauchop, T. & Martucci, R. W. (1968). Ruminant-like digestion of the langur monkey. *Science*, **161**, 698–700.

Becker, P. & Martin, J. S. (1982). Protein-precipitating capacity of tannins in *Shorea* (Dipterocarpaceae) seedling leaves. *Journal of Chemical Ecology*, **8**, 1353–67.

Beever, D. E., Coelho da Silva, J. F., Prescott, J. H. D. & Armstrong, D. G. (1972). The effect in sheep of physical form and stage of growth on sites of digestion of a dried grass. 1. Sites of digestion of organic matter, energy and carbohydrate. *British Journal of Nutrition*, **28**, 347–56.

Beintema, J. J., Scheffer, A. J., van Dijk, H., Welling, G. W. & Zwiers, H. (1973). Pancreatic ribonuclease distribution and comparisons in mammals. *Nature New Biology*, **241**, 76–8.

Benefit, B. R. (1987). The molar morphology, natural history, and phylogenetic position of the Middle Miocene monkey *Victoriapithecus*. PhD dissertation, New York University.

Benefit, B. R. (1993). The permanent dentition and phylogenetic position of *Victoriapithecus* from Maboko Island, Kenya. *Journal of Human Evolution*, **25**, 83–172.

Benefit, B. R. & McCrossin, M. L. (1991). The ancestral facial morphology of Old World higher primates. *Proceedings of the National Academy of Sciences, USA*, **88**, 5267–71.

Benefit, B. R. & McCrossin, M. L. (1993). The facial anatomy of *Victoriapithecus* and its relevance to the ancestral cranial morphology of Old World monkeys and apes. *American Journal of Physical Anthropology*, **92**, 329–70.

Benefit, B. & Pickford, M. (1986). Miocene fossil cercopithecoids from Kenya. *American Journal of Physical Anthropology*, **69**, 441–64.

Bennett, E. L. (1983). The banded langur: ecology of a colobine in a West Malaysian rain-forest. PhD thesis, University of Cambridge, England.

Bennett, E. L. (1986*a*). *Proboscis Monkeys in Sarawak: their Ecology, Status, Conservation and Management*. Kuala Lumpur, Malaysia: World Wildlife Fund.

Bennett, E. L. (1986*b*). Environmental correlates of ranging behaviour in the banded langur, *Presbytis melalophos*. *Folia Primatologica*, **47**, 26–38.

Bennett, E. L. (1987). Big noses of Borneo. *Animal Kingdom*, **90**, 9–15.

Bennett, E. L. (1988). Cyrano of the swamps. *BBC Wildlife*, **6**, 71–5.

Bennett, E. L. (1992). *A wildlife survey of Sarawak*. Report to Wildlife Conservation International, New York.

Bennett, E. L. & Bennett, J. M. (1988). A survey of primates in Ulu Sebuyan. Report to Sarawak Forest Department, Kuching, Sarawak, Malaysia.

Bennett, E. L. & Caldecott, J. O. (1989). Primates of peninsular Malaysia. In *Ecosystems of the World, 14B: Tropical Rain Forest Ecosystems*, ed. H. Lieth & M. J. A. Werger, pp. 355–63. Amsterdam: Elsevier.

Bennett, E. L. & Sebastian, A. C. (1988). Social organization and ecology of proboscis monkeys (*Nasalis larvatus*) in mixed coastal forest in Sarawak. *International Journal of Primatology*, **9**, 233–55.

Berenbaum, M. R. & Neal, J. J. (1987). Interactions among allelochemicals and insect resistance in crop plants. In *Allelochemicals: Role in Agriculture and Forestry*, ed. G. R. Waller, pp. 416–30. Washington, DC: American Chemical Society.

Berggren, W. A. & Prothero, D. R. (1992). Eocene–Oligocene climatic and biotic evolution: an overview. In *Eocene–Oligocene Climatic and Biotic Evolution*, ed. D. R. Prothero & W. A. Berggren, pp. 1–28. Princeton: Princeton University Press.

Bernays, E. A., Cooper-Driver, G. & Bilgener, M. (1989). Herbivores and plant tannins. *Advances in Ecological Research*, **19**, 263–302.

Bernhard-Reversat, F., Huttel, C. & Lemée, G. (1978). Structure and functioning of evergreen rain forest ecosystems of the Ivory Coast. In *Tropical Forest Ecosystems*, pp. 557–74. Paris: Unesco.

Bernor, R. L. (1983). Geochronology and zoogeographic relationships of Miocene Hominoidea. In *New Interpretations of Ape and Human Ancestry*, ed. R. L. Ciochon & R. S. Corruccini, pp. 21–64. New York: Plenum.

Bernor, R. L. (1985). A zoogeographic theater and biochronologic play: the time/biofacies phenomena of Eurasian and African Miocene mammal provinces. *Paléobiologie Continentale*, **14**, 121–42.

Bernor, R. L., Kovar-Eder, J., Lipscomb, D., Rôgl, F., Sen, S. & Tobien, H. (1988). Systematic, stratigraphic and paleoenvironmental contexts of first-appearing Hipparion in the Vienna Basin, Austria. *Journal of Vertebrate Paleontology*, **8**, 427–52.

Bernstein, I. S. (1968). The lutong of Kuala Selangor. *Behaviour*, **32**, 1–16.

Bessedik, M., Aguilar, J.-P., Cappetta, H. & Michaux, J. (1984). Le climat du Néogène dans le sud de la France (Provence, Languedoc, Roussillon), d'après l'analyse des faunes (rongeurs, sélaciens), et des flores polliniques. *Paléobiologie Continentale*, **14**, 181–90.

Bestland, E. A. & Retallack, G. J. (1993). Volcanically influenced calcareous paleosols from the Miocene Kiahera Formation, Rusinga Island, Kenya. *Journal of the Geological Society, London*, **150**, 293–310.

Birchette, M. (1981). Postcranial remains of *Cercopithecoides*. *American Journal of Physical Anthropology*, **54**, 201.

Birchette, M. (1982). The postcranial skeleton of *Paracolobus chemeroni*. PhD dissertation, Harvard University, Cambridge, MA.

Bishop, N. H. (1979). Himalayan langurs: temperate colobines. *Journal of Human Evolution*, **8**, 251–81.

Bleisch, W., Cheng, A.-S., Ren, X.-D. & Xie, J.-H. (1993). Preliminary results from a field study of wild Guizhou snubnosed monkeys (*Rhinopithecus brelichi*). *Folia Primatologica*, **60**, 172–82.

Bodmer, R. E. (1989). Frugivory in Amazonian Artiodactyla: evidence for the evolution of the ruminant stomach. *Journal of Zoology, London*, **219**, 457–67.

Bodmer, R. E. (1991). Strategies of seed dispersal and seed predation in Amazonian ungulates. *Biotropica*, **23**, 255–61.

Bodmer, R. E., Mather, R. J. & Chivers, D. J. (1991). Rain forests of central Borneo – threatened by modern development. *Oryx*, **25**, 21–6.

Boesch, C. & Boesch, H. (1989). Hunting behavior of wild chimpanzees in the Tai National Park. *American Journal of Physical Anthropology*, **78**, 547–73.

Boggess, J. (1976). The social behavior of the Himalayan langur (*Presbytis entellus*) in eastern Nepal. PhD dissertation, University of California, Berkeley.

Boggess, J. E. (1979). Troop male membership changes and infant killing in langurs (*Presbytis entellus*). *Folia Primatologica*, **32**, 65–107.

Boggess, J. (1980). Intermale relations and troop male membership changes in langurs (*Presbytis entellus*) in Nepal. *International Journal of Primatology*, **1**, 233–73.

Boggess, J. (1984). Infant killing and male reproductive strategies in langurs (*Presbytis entellus*). In *Infanticide: Comparative and Evolutionary Perspectives*, ed. G. Hausfater & S. B. Hrdy, pp. 283–310. New York: Aldine.

Bonnefille, R. (1984). Cenozoic vegetation and environments of early hominids in East Africa. In *The Evolution of the East Asian Environment*, vol. II, ed. R. O. Whyte, pp. 579–612. Hong Kong: University of Hong Kong Centre of Asian Studies.

Bonnefille, R. (1985). Evolution of the continental vegetation. The paleobotanical record from East Africa. *South African Journal of Science*, **81**, 267–70.

Bonnefille, R. & Riollet, G. (1987). Palynological spectra from the upper Laetoli Beds. In *The Pliocene Site of Laetoli, Northern Tanzania*, ed. M. D. Leakey & J. M. Harris, pp. 52–61. Oxford: Oxford University Press.

Bonnefille, R. & Vincens, A. (1985). Apport de la palynologie à l'environnement des Hominidés d'Afrique orientale. In *L'Environnement des Hominidés au Plio-Pléistocène*, ed. Y. Coppens, pp. 237–278. Paris: Masson.

Bonnefille, R., Vincens, A. & Buchet, G. (1987). Palynology, stratigraphy and paleoenvironment of a Pliocene hominid site (2.9–3.3 MY) at Hadar, Ethiopia. *Palaeogeography, Palaeoclimatology, Palaeoecology*, **60**, 249–81.

Booth, A. H. (1956). The distribution of primates in the Gold Coast. *Journal of the West African Science Association*, **2**, 122–33.

Booth, A. H. (1957). Observations on the natural history of the olive colobus monkey, *Procolobus verus* (van Beneden). *Proceedings of the Zoological Society of London*, **129**, 421–30.

Booth, A. H. (1958). The zoogeography of West African primates: a review. *Bulletin de l'Institut français d'Afrique noire, série A.*, **20**, 587–622.

Borissoglebskaya, M. B. (1981). New species of monkey (Mammalia, Primates) from the Pliocene of North Mongolia. *Fossil Vertebrates of Mongolia, Papers of the Joint Soviet–Mongolian Paleontological Expeditions*, **15**, 95–108.

Borries, C. (1986). Grandmaternal behaviour in free-ranging hanuman langurs (*Presbytis entellus*). *Primate Report*, **14**, 207–8.

Bourlière, F. (1983). Tropical rain forest ecosystems: structure and function. In *Ecosystems of the World*, vol. 14A, ed. F. B. Golley. Amsterdam: Elsevier.

Bourlière, F. (1985). Primate communities: their structure and role in tropical ecosystems. *International Journal of Primatology*, **6**, 1–26

Bramblett, C. A. (1969). Non-metric skeletal age changes in the Darajani baboon. *American Journal of Physical Anthropology*, **30**, 161–72.

Brandon-Jones, D. (1977). The evolution of recent Asian Colobinae. In *Recent Advances in Primatology*, vol. 3, ed. D. J. Chivers & K. A. Joysey, pp. 323–5. London: Academic Press.

Brandon-Jones, D. (1984). Colobus and leaf monkeys. In *Encyclopaedia of Mammals*, ed. I. D. Macdonald, pp. 398–408. London: George Allen and Unwin.

Brattsten, L. B. (1979). Biochemical defense mechanisms in herbivores against allelochemicals. In *Herbivores: their Interaction with Secondary Plant Compounds*, ed. G. A. Rosenthal & D. J. Janzen, pp. 199–270. New York: Academic

Brown, F. H. & Feibel, C. S. (1989). 'Robust' hominids and Plio-Pleistocene paleogeography of the Turkana Basin, Kenya and Ethiopia. In *Evolutionary*

History of the 'Robust' Australopithecines, ed. F. E. Grine, pp. 325–41. New York: Aldine de Gruyter.
Brown, F. H. & Feibel, C. S. (1991). Stratigraphy, depositional environments and palaeogeography of the Koobi Fora Formation. In *Koobi Fora Research Project, vol. 3, The Fossil Ungulates: Geology, Fossil Artiodactyls and Palaeoenvironments*, ed. J. M. Harris, pp. 1–30. Oxford: Clarendon Press.
Bruce, H. M. (1960). A block to pregnancy in the house mouse caused by the proximity of strange males. *Journal of Reproduction and Fertility*, **1**, 96–103.
Brunet, M., Heintz, E. & Battail, B. (1984). Molayan (Afghanistan) and the Khaur Siwaliks of Pakistan: an example of biogeographic isolation of Late Miocene mammalian faunas. *Geologie en Mijnbouw*, **63**, 31–8.
Burton, G. J. (1981). The relationship between body weight and gonadal weights of dusky leaf monkeys (*Presbytis obscurus*). *International Journal of Primatology*, **2**, 351–69.
Busse, C. D. (1977). Chimpanzee predation as a possible factor in the evolution of red colobus monkey social organization. *Evolution*, **31**, 907–11.
Cadman, A. & Rayner, R. J. (1989). Climatic change and the appearance of *Australopithecus africanus* in the Makapansgat sediments. *Journal of Human Evolution*, **18**, 107–13.
Caldecott, J. O. (1986). An ecological and behavioural study of the pig-tailed macaque. *Contributions to Primatology*, **21**, 1–259.
Calvert, J. J. (1985). Food selection by western gorilla (*G. g. gorilla*) in relation to food chemistry. *Oecologia (Berlin)*, **65**, 236–46.
Cant, J. G. H. (1980). What limits primates? *Primates*, **21**, 538–44.
Caton, J. M. (1990). Structure and function in the catarrhine stomach, with particular reference to the family Colobidae. MA thesis, Australian National University, Canberra.
Caton, J. M. (1991). The Old World monkeys, I: the leaf-eating monkeys of the family Colobidae. *Australian Primatology*, **6**, 12–17.
Cerling, T. E. (1992). Development of grasslands and savannas in East Africa during the Neogene. *Palaeogeography, Palaeoclimatology, Palaeoecology*, **97**, 241–7.
Cerling, T. E., Bowman, J. R. & O'Neil, J. R. (1988). An isotopic study of a fluvial-lacustrine sequence: The Plio-Pleistocene Koobi Fora sequence, East Africa. *Palaeogeography, Palaeoclimatology, Palaeoecology* **63**, 335–356.
Cerling, T. E., Kappelman, J., Quade, J., Ambrose, S. H., Sikes, N. E. & Andrews, P. (1992). Reply to comments on the paleoenvironment of *Kenyapithecus* at Fort Ternan. *Journal of Human Evolution*, **23**, 371–7.
Cerling, T. E., Wang, Y. & Quade, J. (1993). Expansion of C4 ecosystems as an indicator of global ecological change in the late Miocene. *Nature*, **361**, 344–5.
Chai, P. P. K. (1982). Ecological studies of mangrove forest in Sarawak. PhD thesis, Universiti Kebangsaan Malaysia.
Chalmers, N. (1979). *Social Behaviour in Primates*. London: Edward Arnold.
Champion, H. G. & Seth, S. K. (1968). *A Revised Survey of the Forest Types of India*. Delhi: Manager of Publications.
Chapman, M. & Hausfater, G. (1979). The reproductive consequences of infanticide in langurs: a mathematical model. *Behavioral Ecology and Sociobiology*, **5**, 227–40.
Chasen, F. N. (1940). A handlist of Malaysian mammals. *Bulletin of the Raffles Museum*, **15**, 1–209.
Cheney, D. L. (1987). Interactions and relationships between groups. In *Primate Societies*, ed. B. B. Smuts, D. L. Cheney, R. M. Seyfarth, R. W. Wrangham & T. T. Struhsaker, pp. 267–81. Chicago: University of Chicago Press.

Cheney, D. L. & Wrangham, R. W. (1987). Predation. In *Primate Societies*, ed. B. B. Smuts, D. L. Cheney, R. M. Seyfarth, R. W. Wrangham & T. T. Struhsaker, pp. 227–39. Chicago: University of Chicago Press.

Chism, J. B. & Rowell, T. E. (1986). Mating and residence patterns of male patas monkeys. *Ethology*, **72**, 31–9.

Chivers, D. J. (1974). The siamang in Malaya: a field study of a primate in a tropical rain forest. *Contributions to Primatology*, **4**, 1–335. Basel: S. Karger.

Chivers, D. J. (ed.) (1980). *Malayan Forest Primates*. New York: Plenum Press.

Chivers, D. J. (1992). Diet and guts. In *Cambridge Encyclopedia of Human Evolution*, ed. S. Jones, R. Martin & D. Pilbeam, pp. 60–2. Cambridge: Cambridge University Press.

Chivers, D. J. & Burton, K. M. (1988). Some observations on the primates of Kalimantan Tengah, Indonesia. *Primate Conservation*, **9**, 138–46.

Chivers, D. J. and Davies, A. G. (1979). Abundance of primates in the Krau Game Reserve, Peninsular Malaysia. In *The Abundance of Animals in the Malesian Rainforests*, Proceedings of the 6th Aberdeen–Hull Symposium on Malesian Ecology, ed. A. G. Marshall, pp. 9–33. Department of Geography, University of Hull.

Chivers, D. J. & Hladik, C. M. (1980). Morphology of the gastrointestinal tract in primates: comparisons with other mammals in relation to diet. *Journal of Morphology*, **166**, 337–86.

Choo, G. M., Waterman, P. G., McKey, D. B. & Gartlan, J. S. (1981). A simple enzyme assay for dry matter digestibility and its value in studying food selection by generalist herbivores. *Oecologia (Berlin)*, **49**, 170–8.

Choong, M. F., Lucas, P. W., Ong, J. Y. S., Tan, H. T. W. & Turner, I. M. (1992). Leaf fracture toughness and sclerophylly: their correlations and ecological implications. *New Phytologist*, **121**, 597–610.

Choudhury, A. (1988). Phayre's leaf monkey (*Trachypithecus phayrei*) in Cachar. *Journal of the Bombay Natural History Society*, **85**, 485–92.

Choudhury, A. (1992). Golden langur – distribution confusion. *Oryx*, **26**, 172–3.

Ciochon, R. L., Olsen, J. W. & James, J. (1990). *Other Origins*. New York: Bantam Books.

Clutton-Brock, T. H. (1972). Feeding and ranging behaviour of the red colobus monkey. PhD thesis, University of Cambridge, England.

Clutton-Brock, T. H. (1974*a*). Activity patterns of red colobus (*Colobus badius tephrosceles*). *Folia Primatologica*, **21**, 161–87.

Clutton-Brock, T. H. (1974*b*). Primate social organisation and ecology. *Nature*, **250**, 539–42.

Clutton-Brock, T. H. (1975*a*). Feeding behaviour of red colobus and black and white colobus in East Africa. *Folia Primatologica*, **23**, 165–207.

Clutton-Brock, T. H. (1975*b*). Ranging behaviour of red colobus (*Colobus badius tephrosceles*) in the Gombe National Park. *Animal Behaviour*, **23**, 706–22.

Clutton-Brock, T. H. & Harvey, P. H. (1976). Evolutionary rules and primate societies. In *Growing Points in Ethology*, ed. P. P. G. Bateson & R. A. Hinde, pp. 195–237. Cambridge: Cambridge University Press.

Clutton-Brock, T. H. & Harvey, P. H. (1977). Primate ecology and social organisation. *Journal of Zoology, London*, 183, 1–39.

Clutton-Brock, T. H. & Harvey, P. H. (1978). Mammals, resources and reproductive strategies. *Nature*, **273**, 191–5.

Colbert, E. H. & Hooijer, D. A. (1953). Pleistocene mammals from the limestone fissures of Szechuan, China. *Bulletin of the American Museum of Natural History*, **103**, 1–134.

Coley, P. D. (1983). Herbivory and defensive characteristics of tree species in a lowland tropical forest. *Ecological Monoraphs*, **53**, 209–33.
Coley, P. D., Bryant, J. P. & Chapin III, F. S. (1985). Resource availability and plant antiherbivore defense. *Science,* **230**, 895–89.
Collias, N. & Southwick, C. (1952). A field study of population density and social organization in howling monkeys. *Proceedings of the American Philosophical Society*, **96**, 143–56.
Collins, L. & Roberts, M. (1978). Arboreal folivores in captivity – maintenance of a delicate minority. In *The Ecology of Arboreal Folivores*, ed. G. G. Montgomery, ed. pp. 5–12. Washington, DC: Smithsonian Institution Press.
Collins, N. M., Sayer, J. A. & Whitmore, T. C. (1991). *The Conservation Atlas of Tropical Forests – Asia and the Pacific*. London: Macmillan.
Collinson, M. E. & Hooker, J. J. (1991). Fossil evidence of interaction between plants and plant-eating mammals. *Philosophical Transactions of the Royal Society London*, **B333**, 197–208.
Colyer, F. (1936). *Variations and Diseases of the Teeth of Animals*. London: John Bale Sons and Danielsson.
Colyn, M. M. (1987). Les primates des forêts ombrophiles de la Cuvette du Zaire: interprétations zoogéographiques des modèles de distribution. *Revue de Zoologie africaine*, **101**, 183–96.
Colyn, M. M. (1991). L'importance zoogéographique du bassin du fleuve Zaire pour la spéciation: le cas des primates simiens. *Annales de le Musée Royal de l'Afrique Centrale, Tervuren, Sciences Zoologiques*, **264**, 1–250.
Colyn, M. M. & Verheyen, W. N. (1987). *Colobus rufomitratus parmentieri*: une nouvelle sous-espece du Zaire (Primates, Cercopithecidae). *Revue de Zoologie Africaine*, **101**, 125–32.
Coppens, Y. (ed.) (1985). *L'Environnement des Hominidés au Plio–Pléistocène*. Paris: Masson.
Coppens, Y. (1989). Hominid evolution and the evolution of the environment. *Ossa*, **9**, 157–63.
Cords, M. (1987). Forest guenons and patas monkeys: male–male competition in one-male groups. In *Primate Societies*, ed. B. B. Smuts, D. L. Cheney, R. M. Seyfarth, R. W. Wrangham & T. T. Struhsaker, pp. 98–111. Chicago: University of Chicago Press.
Cork, S. J. & Foley, W. J. (1991). Digestive and metabolic strategies of arboreal mammalian folivores in relation to chemical defenses in temperate and tropical forests. In *Plant Defenses Against Mammalian Herbivory*, Chapter 8, ed. R. T. Palo & C. T. Robbins, pp. 133–66. Boca Raton: CRC Press.
Corlett, R. T. & Lucas, P. W. (1990). Alternative seed-handling strategies in primates: seed-spitting by long-tailed macaques (*Macaca fascicularis*). *Oecologia*, **82**, 166–71.
Corner, E. J. H. (1976). *The Seeds of Dicotyledons*, vol. 1. Cambridge: Cambridge University Press.
Cotterell, B. & Mai, Y.-W. (1989). On mixed-mode plane stress ductile fracture. In *Advances in Fracture Research*, ed. K. Salama, K. Ravi-Chandar, D. M. R. Taplin & P. R. Rama Rao, pp. 2269–78. Oxford: Pergamon Press.
Cowlishaw, G. & Dunbar, R. I. M. (1991). Dominance rank and mating success in male primates. *Animal Behaviour*, **41**, 1045–56.
Crockett, C. M. & Eisenberg, J. F. (1987). Howlers – variation in group size and demography. In *Primate Societies*, ed. B. B. Smuts, D. L. Cheney, R. M. Seyfarth, R. W. Wrangham & T. T. Struhsaker, pp. 54–68. Chicago: University of Chicago Press.

Crompton, A. W. & Sita-Lumsden, A. G. (1970). Functional significance of therian molar pattern. *Nature*, **222**, 678–9.

Culvenor, C. C. J., Jago, M. V., Peterson, J. E., Smith, L. W., Payne, A. L., Campbell, D. G., Edgar, J. A. & Frahn, J. L. (1984). Toxicity of *Echium plantagineum* (Paterson's Curse). I. Marginal toxic effects in Merino wethers from long-term feeding. *Australian Journal of Agricultural Research*, **35**, 293–304.

Curtin, R. A. (1975). The socio-ecology of the common langur, *Presbytis entellus* in the Nepal Himalaya. PhD thesis, University of California, Berkeley, California.

Curtin, R. A. (1977). Langur social behavior and infant mortality. *Kroeber Anthropological Society Papers*, **50**, 27–36.

Curtin, R. A. & Dolhinow, P. (1978*a*). Primate social behavior in a changing world. *American Scientist*, **66**, 468–75.

Curtin, R. A. & Dolhinow, P. (1978*b*). Letter to the editor. *American Scientist*, **66**, 668–9.

Curtin, S. H. (1980). Dusky and banded leaf monkeys. In *Malayan Forest Primates*, ed. D. J. Chivers, pp. 107–45. New York: Plenum Press.

Curtin, S. H. & Chivers, D. J. (1978). Leaf-eating primates of Peninsular Malaysia: the siamang and the dusky leaf-monkey. In *The Ecology of Arboreal Folivores*, ed. G. G. Montgomery, 441–64. Washington, DC: Smithsonian Institution Press.

Curtin, S. H. & Olson, D. K. (1984). Ranging patterns of black-and-white colobus and diana monkeys in the Bia National Park, Ghana. Paper presented at the *53rd Annual Meeting of the American Association of Physical Anthropologists*, Philadelphia, Pennsylvania.

Dandelot, P. (1971). Order Primates, suborder Anthropoidea. In *The Mammals of Africa: an Identification Manual*, ed. J. Meester & H. W. Setzer, pp. 1–43. Washington, DC: Smithsonian Institution Press.

Darwin, C. (1871). *The Descent of Man and Selection in Relation to Sex*. London: John Murray.

Dasilva, G. L. (1989). The ecology of the western black and white colobus (*Colobus polykomos polykomos* Zimmerman 1780) on a riverine island in southeastern Sierra Leone. DPhil thesis, University of Oxford, England.

Dasilva, G. L. (1992). The western black-and-white colobus as a low-energy strategist: activity budgets, energy expenditure and energy intake. *Journal of Animal Ecology*, **61**, 79–91.

Dasilva, G. L. (1993). Postural changes and behavioural thermoregulation in *Colobus polykomos*: the effect of climate and diet. *African Journal of Ecology*, **31**, 226–41.

Dasilva, G. L. (1994). Diet of *Colobus polykomos* on Tiwai Island: selection of food in relation to its seasonal abundance and nutritional quality. *International Journal of Primatology*, **15**, 1–26.

David, G. F. X. & Ramaswami, L. S. (1969). Studies on menstrual cycles and other related phenomena in the langur (*Presbytis entellus entellus*). *Folia Primatologica*, **11**, 300–16.

Davies, A. G. (1984). An ecological study of the red leaf monkey (*Presbytis rubicunda*) in the dipterocarp forest of northern Borneo. PhD thesis, University of Cambridge, England.

Davies, A. G. (1987*a*). *The Gola Forest Reserves, Sierra Leone: Wildlife Conservation and Forest Management*. Gland: IUCN.

Davies, A. G. (1987*b*). Adult male replacement and group formation in *Presbytis rubicunda. Folia Primatologica*, **49**, 111–14.

Davies, A. G. (1991). Seed-eating by red leaf monkeys (*Presbytis rubicunda*) in dipterocarp forest of northern Borneo. *International Journal of Primatology*, **12**, 119–44.

Davies, A. G. & Baillie, I. C. (1988). Soil-eating by the red leaf monkey (*Presbytis rubicunda*) in Sabah, North Borneo. *Biotropica*, **20**, 252–8.

Davies, A. G., Bennett, E. L. & Waterman, P. G. (1988). Food selection by two South-East Asian colobine monkeys (*Presbytis rubicunda* and *P. melalophos*) in relation to plant chemistry. *Biological Journal of the Linnaean Society*, **34**, 33–56.

Davies, A. G., Caldecott, J. O. & Chivers, D. J. (1984). Natural foods as a guide to nutrition of Old World Primates. In *Standards in Laboratory Animal Management*, ed. J. Ramfrey, pp. 225–44. Potters Bar, England: Universities Federation for Animal Welfare.

Davies, A. G. & Payne, J. B. (1982). A faunal survey of Sabah, Malaysia. Kuala Lumpur: World Wildlife Fund.

Davis, D. D. (1962). Mammals of North Borneo. *Bulletin of the National Museum of Singapore*, **31**, 1–129.

de Bonis, L., Bouvrain, G., Geraads, D. & Koufos, G. (1990). New remains of *Mesopithecus* (Primates, Cercopithecoidea) from the Late Miocene of Macedonia (Greece), with the description of a new species. *Journal of Vertebrate Paleontology*, **10**, 473–83.

de Bonis, L., Bouvrain, G., Geraads, D. & Koufos, G. (1992). Diversity and paleoecology of Greek late Miocene mammalian faunas. *Palaeogeography, Palaeoclimatology, Palaeoecology*, **91**, 99–121.

Decker, B. S. (1989). Effects of habitat disturbance on the behavioral ecology and demographics of the Tana red colobus (*Colobus badius rufomitratus*). PhD thesis, Emory University, Atlanta, GA.

Delson, E. (1973). Fossil colobine monkeys of the circum-Mediterranean region and the evolutionary history of the Cercopithecidae (Primates, Mammalia). PhD dissertation, Columbia University.

Delson, E. (1975*a*). Evolutionary history of the Cercopithecidae. In *Approaches to Primate Paleobiology; Contributions to Primatology*, Vol. 5, ed. F. S. Szalay, pp. 167–217. Basel: S. Karger.

Delson, E. (1975*b*). Paleoecology and zoogeography of the Old World monkeys. In *Primate Functional Morphology and Evolution*, ed. R. Tuttle, pp. 37–64. The Hague: Mouton.

Delson, E. (1977). Catarrhine phylogeny and classification: principles, methods and comments. *Journal of Human Evolution*, **6**, 433–59.

Delson, E. (1980). Fossil macaques, phyletic relationships and a scenario of deployment. In *The Macaques: Studies in Ecology, Behavior and Evolution*, ed. D. E. Lindburg, pp. 10–30. New York: Van Nostrand.

Delson, E. (1984). Cercopithecid biochronology of the African Plio-Pleistocene: correlation among eastern and southern hominid-bearing localities. *Courier Forschungs-Institut Senckenberg*, **69**, 199–218.

Delson, E. (1988). Catarrhini. In *Encyclopedia of Human Evolution and Prehistory*, ed. I. Tattersall, E. Delson & J. A. Van Couvering, pp. 111–16. New York: Garland.

Delson, E. (1989). Chronology of South African australopith site units. In *Evolutionary History of the 'Robust' Australopithecines*, ed. F. E. Grine, pp. 317–24. New York: Aldine-de Gruyter.

Delson, E. (1992). Evolution of Old World monkeys. In *Cambridge Encyclopedia of Human Evolution*, ed. J. S. Jones, R. D. Martin, D. Pilbeam & S. Bunney, pp. 217–22. Cambridge: Cambridge University Press.

Delson, E. & Dean, D. (1993). Are *P. baringensis* R. Leakey, 1969, and *P. quadratirostris* Iwamoto, 1982, species of *Papio* or *Theropithecus*? In *Theropithecus: Rise and Fall of a Primate Genus*, ed. N. Jablonski, pp. 125–56. Cambridge: Cambridge University Press.

Delson, E., Groves, C. P., Grubb, P., Miu, C. A. & Wang, S. (1982). Family Cercopithecidae. In *Mammal Species of the World*, ed. J. H. Honacki, K. E. Kinman & J. W. Koeppl, pp. 230–42. Lawrence, KS: Association of Systematics Collections.

Demarcq, G., Ballesio, R., Rage, J.-C., Guérin, C., Mein, P. & Méon, H. (1983). Données paléoclimatiques du Néogène de la Vallée du Rhône (France). *Palaeogeography, Palaeoclimatology, Palaeoecology*, **42**, 247–72.

Demment, M. W. & Van Soest, P. J. (1985). A nutritional explanation for body-size patterns of ruminant and non-ruminant herbivores. *American Naturalist*, **125**, 641–72.

Depéret, C. (1890). Les animaux pliocènes du Roussillon. *Memoires Société Geologique France (Paléontologie)*, **3**, 1–126.

De Vos, J. (1989). The environment of *Homo erectus* from Trinil H. K. In *Hominidae*, ed. G. Giacobini, pp. 225–9. Milan: Jaca Book.

Diamond, A. W. & Hamilton, A. H. (1980). The distribution of forest passerine birds and Quaternary climatic change in tropical Africa. *Journal of Zoology, London*, **191**, 379–402.

Dittus, W. P. J. (1977). The ecology of a semi-evergreen forest community in Sri Lanka. *Biotropica*, **9**, 268–86.

Dittus, W. P. J. (1985). The influence of cyclones on the dry evergreen forest of Sri Lanka. *Biotropica*, **17**, 1–14.

Dobson, D. E., Prager, E. M. & Wilson, A. C. (1984). Stomach lysozymes of ruminants. 1. Distribution and catalytic properties. *Journal of Biological Chemistry*, **259**, 11607–16.

Dolhinow, P. & DeMay, M. G. (1982). Adoption – the importance of infant choice. *Journal of Human Evolution*, **11**, 391–420.

Dolhinow, P., McKenna, J. J. & Vonder Haar Laws, J. (1979). Rank and reproduction among female langur monkeys: ageing and improvement (they're not just getting older, they're getting better). *Aggressive Behavior*, **5**, 19–30.

Drawert, F., Kuhn, H. J. & Rapp, A. (1962). Reactions-Gaschromatographic, III. Gaschromatographische Bestimmung der niederfluchtigen Fettsauren im Magen von Schlankaffen (Colobinae). *Zeitschrift für physiologische Chemie*, **329**, 84–9.

Dubost, G. (1984). Comparison of the diets of frugivorous forest ruminants of Gabon. *Journal of Mammalogy*, **65**, 298–316.

Dunbar, R. I. M. (1984). *Reproductive Decisions; an Economic Analysis of Gelada Baboon Social Strategies*. Princeton: Princeton University Press.

Dunbar, R. I. M. (1987). Habitat quality, population dynamics, and group composition in colobus monkeys (*Colobus guereza*). *International Journal of Primatology*, **8**, 299–330.

Dunbar, R. I. M. (1988). *Primate Social Systems*. Beckenham: Croom Helm.

Dunbar, R. I. M. & Dunbar, E. P. (1974). Ecology and population dynamics of *Colobus guereza* in Ethiopia. *Folia Primatologica*, **21**, 188–208.

Dunbar, R. I. M. & Dunbar, E. P. (1976). Contrasts in social structure among black-and-white colobus monkey groups. *Animal Behaviour*, **24**, 84–92.

Eisentraut, M. (1973). Die Wirbeltierfauna von Fernando Poo und Westkamerun. *Bonner Zoologische Monographien*, **3**, 1–428.

Ellerman, J. R. & Morrison-Scott, T. C. S. (1951). *Checklist of Palaearctic and Indian Mammals*. London: British Museum (Natural History).

Emel, L. M. & Swindler, D. R. (1992). Underbite and the scaling of facial dimensions in colobine monkeys. *Folia Primatologica*, **58**, 177–189.

Emlen, S. & Oring, L. (1977). Ecology, sexual selection and the evolution of mating systems. *Science* **197**, 215–223.

Emmons, L. H. (1980). Ecology and resource partitioning among nine species of African rain forest squirrels. *Ecological Monographs*, **50**, 31–54.

Emmons, L. H., Gautier-Hion, A. & Dubost, G. (1983). Community structure of the frugivorous–folivorous forest mammals of Gabon. *Journal of Zoology, London*, **199**, 209–22.

Engelhardt, W. V. & Sallmann, H. P. (1972). Resorption und Sekretion im Panzen des Guanakos (*Lama guanacoe*). *Zentralblatt für Veterinarmedizin A*, **19**, 117–132.

Eudey, A. A. (1980). Pleistocene glacial phenomena and the evolution of Asian macaques. In *The Macaques: Studies in Ecology, Behavior and Evolution*, ed. D. G. Lindburg, pp. 52–83. New York: Van Nostrand Reinhold.

Eudey, A. A. (1987). *Action Plan for Asian Primate Conservation: 1987–91*. Stony Brook, NY: IUCN/SSC Primate Specialist Group.

Evans, S. V., Fellows, L. E. & Bell, E. A. (1985). Distribution and systematic significance of basic non-protein amino acids and amines in the Tephrosieae. *Biochemical Systematics and Ecology*, **13**, 271–302.

Faithful, N. Y. (1985). The *in vitro* digestibility of feedstuffs: a century of ferment. *Journal of the Science of Food and Agriculture*, **35**, 818–26.

Farid Ahsan, M. (1984). Study of primates in Bangladesh: determination of population status and distribution of non-human primates in Bangladesh with emphasis on rhesus monkey. MPhil thesis, University of Dhaka, Bangladesh.

Fay, J. M. (1985). Range extensions for four *Cercopithecus* species in the Central African Republic. *Primate Conservation*, **6**, 63–68.

Feeny, P. (1976). Plant apparency and chemical defense. *Recent Advances in Phytochemistry*, **10**, 1–40.

Feibel, C. S., Brown, F. H. & McDougall, I. (1989). Stratigraphic context of fossil hominids from the Omo Group deposits: northern Turkana Basin, Kenya and Ethiopia. *American Journal of Physical Anthropology*, **78**, 595–622.

Feibel, C. S., Harris, J. M. & Brown, F. H. (1991). Palaeoenvironmental context for the Late Neogene of the Turkana Basin. In *Koobi Fora Research Project, vol. 3. The Fossil Ungulates: Geology, Fossil Artiodactyls and Palaeoenvironments*, ed. J. M. Harris, pp. 321–70. Oxford: Clarendon Press.

Fisher, R. A. (1958). *The Genetical Theory of Natural Selection*, 2nd ed. New York: Dover.

Fleagle, J. G. (1976). Locomotor behavior and skeletal anatomy of sympatric Malaysian leaf-monkeys (*Presbytis obscura* and *Presbytis melalophos*). *Yearbook of Physical Anthropology*, **20**, 440–53.

Fleagle, J. G. (1977). Locomotor behavior and muscular anatomy of sympatric Malaysian leaf-monkeys (*Presbytis obscura* and *Presbytis melalophos*). *American Journal of Physical Anthropology*, **46**, 297–308.

Fleagle, J. G. (1980). Locomotion and posture. In *Malayan Forest Primates*, ed. D. J. Chivers, pp. 191–207. New York: Plenum Press.

Fleagle, J. G. (1988). *Primate Adaptation and Evolution*. San Diego: Academic Press.

Fooden, J. (1971). Report of primates collected in western Thailand, *Fieldiana: Zoology*, **59**, 1–62.

Fooden, J. (1976). Primates observed in peninsular Thailand June–July, 1973, with notes on the distribution of continental southest Asian leaf-monkeys (*Presbytis*). *Primates*, **17**, 95–118.

Foster, R. B. (1982). The seasonal rhythm of fruitfall on Barro Colorado Island. In *The Ecology of a Tropical Forest*, ed. E. G. Leigh, A. S. Rand & D. M. Windsor, pp. 151–72. Washington, DC: Smithsonian Institution Press.

Fox, J. E. D. (1973). Kabili–Sepilok Forest Reserve. *Sabah Forest Record No. 9*. Kuching, Sarawak: Borneo Literature Bureau.

Freeland, W. J. (1991). Plant secondary metabolites: biochemical coevolution with herbivores. In *Plant Defenses Against Mammalian Herbivory*, Chapter 4, ed. R. T. Palo & C. T. Robbins, pp. 161–81. Boca Raton: CRC Press.

Freeland, W. J., Calcott, P. H. & Anderson, L. R. (1985*a*). Tannins and saponins: interaction in herbivore diets. *Biochemical Systematics and Ecology*, **13**, 189–93.

Freeland, W. J., Calcott, P. H. & Geiss, C. P. (1985*b*). Allelochemicals, minerals and herbivore population size. *Biochemical Systematics and Ecology*, **13**, 195–206.

Freeland, W. J. & Janzen, D. H. (1974). Strategies in herbivory by mammals: the role of plant secondary compounds. *American Naturalist*, **108**, 269–89.

Freeman, P. W. (1988). Frugivorous and animalivorous bats (Microchiroptera): dental and cranial adaptations. *Biological Journal of the Linnaean Society*, **33**, 387–408.

Freeman, P. W. (1992). Canine teeth of bats (Microchiroptera): size, shape and role in crack propagation. *Biological Journal of the Linnaean Society*, **45**, 97–115.

Freese, C. H., Heltne, P. G., Castro, N. R. & Whitesides, G. H. (1982). Patterns and determinants of monkey densities in Peru and Bolivia, with notes on distributions. *International Journal of Primatology*, **3**, 53–90.

Galat, G. (1977). Enquête sur les pygmées de Lobaye: rapports trophiques avec les mammifères et pression de prédation. Adiopodoumé: ORSTOM.

Galat, G. & Galat-Luong, A. (1985). La communauté de primates diurnes de la forêt de Taï, Cote d'Ivoire. *Revue d'Écologie (Terre Vie)*, **40**, 3–32.

Galat-Luong, A. (1983). Socio-écologie de trois colobes sympatriques, *Colobus badius, C. polykomos* et *C. verus* du Parc National de Taï, Côte d'Ivoire. PhD thesis, Université Pierre et Marie Curie, Paris, France.

Galat-Luong, A. & Galat, G. (1978). Abondance relative et associations plurispécifiques de primates diurnes du Parc National de Taï, Côte d'Ivoire. Adiopodoumé: ORSTOM.

Galat-Luong, A. & Galat, G. (1979). Quelques observations sur l'écologie de *Colobus badius oustaleti* en Empire Centrafricain. *Mammalia*, **43**, 309–12.

Ganzhorn, J. U. (1985). Habitat use and feeding behaviour in *Lemur catta* and *Lemur fulvus*. PhD thesis, Universität Tubingen, Germany.

Ganzhorn, J. U. (1989). Niche separation of seven lemur species in the eastern rainforest of Madagascar. *Oecologia*, **79**, 279–86.

Gartlan, J. S., McKey, D. B., Waterman, P. G., Mbi, C. N. & Struhsaker, T. T. (1980). A comparative study of the phytochemistry of two African rainforests. *Biochemical Systematics and Ecology*, **8**, 401–22.

Gatinot, B. L. (1975). Écologie d'un colobe bai (*Colobus badius temmincki*, Kuhn 1820) dans un milieu marginal au Sénégal. PhD thesis, Université de Paris VI, France.

Gatinot, B. L. (1976). Les milieux fréquentés par le colobe bai d'Afrique de l'Ouest (*Colobus badius temmincki*, Kuhn 1820) en Sénégambie. *Mammalia*, **40**, 1–12.

Gatinot, B. L. (1977). Le régime alimentaire du colobe bai au Sénégal. *Mammalia*, **41**, 373–402.

Gaudry, A. (1862). *Animaux Fossiles et Géologie de l'Attique*. Paris: F. Savy.

Gautier-Hion, A. (1980). Seasonal variations of diet related to species and sex in a community of *Cercopithecus* monkeys. *Journal of Animal Ecology*, **49**, 237–70.

Gautier-Hion, A., Duplantier, J.-M., Quris, R., Feer, F., Sourd, C., Decoux, J.-P., Dubost, G., Emmons, L., Erard, C., Hecketsweiler, P., Moungazi, A., Roussilhon, C., & Thiollay, J.-M. (1985). Fruit character as a basis of fruit choice and seed dispersal in a tropical forest vertebrate community. *Oecologia*, **65**, 324–37.

Gee, E. P. (1961). The distribution and feeding habitats of the golden langur, *Presbytis geei* (Khajuria 1956). *Journal of the Bombay Natural History Society*, **58**, 1–12.

Gee, E. P. (1964). *The Wildlife of India*. London: Collins.

Geerling, C. & Bokdam, J. (1973). Fauna of the Comoe National Park, Ivory Coast. *Biological Conservation*, **5**, 251–7.

Geraads, D. (1982). Paléobiogéographie de l'Afrique du Nord depuis le Miocène terminal, d'après les grands mammifères. *Geobios, Mémoires speciaux*, **6**, 473–81.

Geraads, D. (1987). Dating the northern African cercopithecid fossil record. *Human Evolution*, **2**, 19–27.

Ghiglieri, M. P. (1984). *The Chimpanzees of Kibale Forest*. New York: Columbia University Press.

Ghosh, A. K. & Biswas, S. (1975). A note on the ecology of the golden langur (*Presbytis geei* Khajuria). *Journal of the Bombay Natural History Society*, **72**, 524–8.

Gibson, L. J. & Ashby, M. F. (1988). *Cellular Solids*. Oxford: Pergamon Press.

Gingerich, P. D. (1986). *Plesiadapis* and the delineation of the order Primates. In *Major Topics in Primate and Human Evolution*, ed. B. Wood, L. Martin & P. Andrews, pp. 32–46. Cambridge: Cambridge University Press.

Gittins, S. P. (1980). A survey of the primates of Bangladesh. Report, University of Cambridge.

Gittins, S. P. & Raemaekers, J. J. (1980). Siamang, lar and agile gibbons. In *Malayan Forest Primates*, ed. D. J. Chivers, pp. 63–105. New York: Plenum Press.

Glander, K. E. (1982). The impact of plant secondary compounds on primate feeding behavior. *Yearbook of Physical Anthropology*, **25**, 1–18.

Gochfeld, M. (1974). Douc langurs. *Nature*, **247**, 167.

Golley, F. B., Yankto, J. & Jordan, C. (1980*a*). Biogeochemistry of tropical forests: 2. The frequency distribution and mean concentrations of selected elements near San Carlos de Rio Negro, Venezuela. *Tropical Ecology*, **21**, 71–80.

Golley, F. B., Yantko, J., Richardson, T. & Klinge, H. (1980*b*). Biogeochemistry of tropical forests. 1. The frequency distribution and mean concentrations of selected elements in a forest near Manaus, Brazil. *Tropical Ecology*, **21**, 59–70.

Goltenboth, R. (1976). Nonhuman primates (apes, monkeys and prosimians). In *The Handbook of Zoo Medicine*, ed. H.-G. Kloss & E. M. Lang, pp. 46–85. New York: Van Nostrand Reinhold.

Gomes, C. M. R., Gottlieb, O. R., Marini Bettolo, B. B., Delle Monache, F. &

Polhill, R. (1981). Systematic significance of flavonoids in *Derris* and *Lonchocarpus*. *Biochemical Systematics and Ecology*, **9**, 129–47.
Goodall, A. G. (1977). Feeding and ranging behaviour of a mountain gorilla group (*Gorilla gorilla beringei*) in the Tshibinda–Kahuzi region (Zaire). In *Primate Ecology: Studies of Feeding and Ranging Behaviour in Lemurs, Monkeys and Apes*, ed. T. H. Clutton-Brock, pp. 449–79. London: Academic Press.
Gordon, K. D. (1982). A study of microwear on chimpanzee molars: implications for dental microwear analysis. *American Journal of Physical Anthropology*, **59**, 195–215.
Gordon, K. D. (1984). Hominoid dental microwear: complications in the use of microwear analysis to detect diet. *Journal of Dental Research*, **63**, 1043–6.
Green, K. M. (1978). Primates of Bangladesh: a preliminary survey of population and habitat. *Biological Conservation*, **13**, 141–60.
Green, K. M. (1981). Preliminary observations on the ecology and behaviour of the capped langur, *Presbytis pileatus*, in the Madhupur Forest of Bangladesh. *International Journal of Primatology*, **2**, 131–51.
Groves, C. P. (1970). The forgotten leaf-eaters, and the evolution of the Colobinae. In *Old World Monkeys: Evolution, Systemetics, and Behavior*, ed. J. R. Napier & P. N. Napier, pp. 555–87, New York: Academic Press.
Groves, C. P. (1973). Notes on the ecology and behaviour of the Angola colobus (*Colobus angolensis* P. L. Sclater 1860) in N. E. Tanzania. *Folia Primatologica*, **20**, 12–26.
Groves, C. P. (1989). *A Theory of Human and Primate Evolution*. Oxford: Clarendon Press.
Groves, C. P., Angst, R. & Westwood, C. (1993). The status of *Colobus polykomos dollmani* Schwarz. *International Journal of Primatology*, **14**, 573–86.
Grubb, P. (1978). Patterns of speciation in African mammals. *Bulletin of the Carnegie Museum of Natural History*, **6**, 152–67.
Gu, Y. & Hu, C. (1991). A fossil cranium of *Rhinopithecus* found in Xinan, Henan Province. *Vertebrata PalAsiatica*, **29**, 55–8.
Gu, Y. & Jablonski, N. (1989). A reassessment of *Megamacaca lantianensis* of Gongwangling, Shaanxi Province. *Acta Anthropologica Sinica*, **8**, 343–6.
Gurmaya, K. J. (1986). Ecology and behavior of *Presbytis thomasi* in Northern Sumatra. *Primates*, **27**(2), 151–72.
Hagerman, A. E. & Butler, L. G. (1981). The specificity of proanthocyanidin–protein interactions. *Journal of Biological Chemistry*, **256**, 4494–7.
Hagerman, A. E. & Butler, L. G. (1991). Tannins and lignin. In *Herbivores, their Interactions with Secondary Plant Metabolites*, vol. 1, ed. G. A. Rosenthal & M. R. Berenbaum, pp. 355–88. New York: Academic Press.
Hamilton, A. C. (1988). Guenon evolution and forest history. In *A Primate Radiation: Evolutionary Biology of the African Guenons*, ed. A. Gautier-Hion, F. Bourlière, J.-P. Gautier & J. Kingdon, pp. 13–34. Cambridge: Cambridge University Press.
Happel, R. (1988). Seed-eating by West African cercopithecines, with reference to the possible evolution of bilophodont molars. *American Journal of Physical Anthropology*, **75**, 303–27.
Happel, R. & Cheek, T. (1986). Evolutionary biology and ecology of *Rhinopithecus*. In *Current Perspectives in Primate Social Dynamics*, ed. D. M. Taub & F. A. King, pp. 305–24. New York: Van Nostrand Reinhold.
Haq, B. U., Hardenbol, J. & Vail, P. R. (1987). Chronology of fluctuating sea levels since the Triassic. *Science*, **234**, 1156–67.

Harborne, J. B. (1984). *Phytochemical Methods*, 2nd edn. London: Chapman & Hall.
Harborne, J. B. (1988). *Introduction to Ecological Biochemistry*, 3rd edn. London: Academic Press.
Harborne, J. B. & Turner, B. L. (1983). *Plant Chemosystematics* London: Academic Press.
Harding, R. S. O. (1984). Primates of the Kilimi area, northwest Sierra Leone. *Folia Primatologica*, **42**, 96–114.
Hardy, G. M. (1990). Comparative socio-ecology of dusky langurs at Kuala Lompat, West Malaysia. PhD thesis, University of Cambridge, England.
Harkin, J. M. (1973). Lignin. In *Chemistry and Biochemistry of Herbage*, vol. 1, ed. G. W. Butler & R. W. Bailey, pp. 323–73. London: Academic Press.
Harley, D. (1985). Birth spacing in langur monkeys (*Presbytis entellus*). *International Journal of Primatology*, **6**, 227–42.
Harley, D. (1988). Patterns of reproduction and mortality in two captive colonies of hanuman langur monkeys (*Presbytis entellus*). *American Journal of Primatology*, **15**, 103–14.
Harris, E. E. & Harrison, T. (1991). Undescribed fossil cercopithecids from the Plio-Pleistocene of Kanam East in Western Kenya (abstract). *American Journal of Physical Anthropology*, Supplement **12**, 88.
Harris, J. M. (1985). Age and paleoecology of the Upper Laetoli Beds, Laetolil, Tanzania. In *Ancestors: the Hard Evidence*, ed. E. Delson, pp. 76–81. New York: Alan R. Liss.
Harris, J. M., Brown, F. H. & Leakey, M. G. (1988). Stratigraphy and paleontology of Pliocene and Pleistocene localities west of Lake Turkana. *Contributions to Science (Natural History Museum of Los Angeles County)*, **399**, 1–128.
Harrison, M. J. S. (1986). Feeding ecology of black colobus, *Colobus satanas*, in central Gabon. In *Primate Ecology and Conservation*, ed. J. G. Else & P. C. Lee, pp. 31–7. Cambridge: Cambridge University Press.
Harrison, M. J. S. & Hladik, C. M. (1986). Un primate granivore: le colobe noir dans la forêt du Gabon; potentialité d'évolution du comportement alimentaire. *Revue d'Ecologie (Terre Vie)*, **41**, 281–98.
Harrison, T. (1989). New postcranial remains of *Victoriapithecus* from the middle Miocene of Kenya. *Journal of Human Evolution*, **18**, 3–54.
Harrison, T. (1992). A reassessment of the taxonomic and phylogenetic affinities of the fossil catarrhines from Fort Ternan, Kenya. *Primates*, **33**, 501–22.
Hart, J. A., Hart, T. B. & Thomas, S. (1986). The Ituri Forest of Zaire: primate diversity and prospects for conservation. *Primate Conservation*, **7**, 42–4.
Hart, T. B., Hart, J. A. & Murphy, P. G. (1989). Monodominant and species-rich forests of the humid tropics: causes for their co-occurrence. *American Naturalist*, **133**, 613–33.
Hasan, C. M., Healey, T. M. & Waterman, P. G. (1982). 7β-Acetoxy-trachyloban-18-oic acid from the stem bark of *Xylopia aethiopica*. *Phytochemistry*, **21**, 177–9.
Hasegawa, Y. (1993). Japan's oldest! 2,500,000 year old monkey fossil discovery. *Kagaku Asahi (Monthly Journal of Science)*, **53**(6), 136–9.
Haslam, E. (1985). *Metabolites and Metabolism*. Oxford: Clarendon Press.
Haslam, E. (1989). *Plant Polyphenols*. Cambridge: Cambridge University Press.
Hauser, M. D. (1990). Do chimpanzee copulatory calls incite male–male competition? *Animal Behaviour*, **39**, 596–7.

Hausfater, G. (1984). Infanticide in langurs: strategies, counterstrategies and parameter values. In *Infanticide: Comparative and Evolutionary Perspectives*, ed. G. Hausfater & S. B. Hrdy, pp. 257–81. New York: Aldine.

Hausfater, G., Aref, S. & Cairns, S. J. (1982). Infanticide as an alternative male reproductive strategy in langurs: a mathematical model. *Journal of Theoretical Biology*, **94**, 391–412.

Hausfater, G. & Hrdy, S. B. (eds.) (1984). *Infanticide: Comparative and Evolutionary Perspectives.* New York: Aldine.

Heintz, E., Brunet, M. & Battail, B. (1981). Cercopithecid primates from the late Miocene of Molayan, Afghanistan, with remarks on *Mesopithecus. International Journal of Primatology*, **2**, 273–84.

Hiiemae, K. M. & Crompton, A. W. (1985). Mastication, food transport, and swallowing. In *Functional Vertebrate Morphology*, ed. M. Hildebrand, D. M. Bramble, K. F. Liem & D. B. Wake, pp. 262–90. Cambridge, MA: Harvard University Press.

Hill, A. (1988). Causes of perceived faunal change in the later Neogene of East Africa. *Journal of Human Evolution*, **16**, 583–96.

Hill, A., Behrensmeyer, A. K., Brown, B., Deino, A., Rose, M., Saunders, J., Ward, S. & Winkler, A. (1991). Kipsaramon: a Lower Miocene hominoid site in the Tugen Hills, Baringo District, Kenya. *Journal of Human Evolution*, **19**, 67–77.

Hill, A., Drake, R., Tauxe, L., Monaghan, M., Barry, J. C., Behrensmeyer, A. K., Curtis, G., Jacobs, B. F., Jacobs, L., Johnson, N. & Pilbeam, D. (1985). Neogene palaeontology and geochronology of the Baringo Basin, Kenya. *Journal of Human Evolution*, **14**, 759–73.

Hill, W. C. O. (1934). A monograph on the purple-faced leaf-monkeys (*Pithecus vetulus*). *Ceylon Journal of Science (B)*, **9**, 23–88.

Hill, W. C. O. (1952). On the external and visceral anatomy of the olive colobus monkey (*Procolobus verus*). *Proceedings of the Zoological Society of London*, **122**, 127–86.

Hill, W. C. O. (1958). Pharynx, oesophagus, stomach, small and large intestine: form and position. *Primatologia*, **3**, 139–207.

Hill, W. C. O. & Booth, A. H. (1957). Voice and larynx in African and Asiatic Colobidae. *Journal of the Bombay Natural History Society*, **54**, 309–21.

Hladik, C. M. (1975). Ecology, diet and social patterning in Old and New World primates. In *Sociology and Psychology of Primates*, ed. R. H. Tuttle, pp. 3–35. The Hague: Mouton.

Hladik, C. M. (1977*a*). Chimpanzees of Gabon and chimpanzees of Gombe: some comparative data on the diet. In *Primate Ecology: Studies of Feeding and Ranging Behaviour in Lemurs, Monkeys and Apes*, ed. T. H. Clutton-Brock, pp. 481–501. London: Academic Press.

Hladik, C. M. (1977*b*). A comparative study of the feeding strategies of two sympatric species of leaf monkeys: *Presbytis senex* and *Presbytis entellus*. In *Primate Ecology: Studies of Feeding and Ranging Behaviour in Lemurs, Monkeys and Apes*, ed. T. H. Clutton-Brock, pp. 323–53. London: Academic Press.

Hladik, C. M. (1978). Adaptive strategies of primates in relation to leaf-eating. In *The Ecology of Arboreal Folivores*, ed. G. G. Montgomery, pp. 373–95. Washington, DC: Smithsonian Institution Press.

Hladik, C. M. & Gueguen, L. (1974). Géophagie et nutrition minérale chez les primates sauvages. *Comptes rendus de l'Académie des Sciences Paris, Série D*, **279**, 1393–6.

Hladik, C. M. & Hladik, A. (1972). Disponibilités alimentaires et domaines vitaux des primates à Céylan. *Terre et Vie*, **26**, 149–215.
Hobson, P. N. (ed.) (1988). *The Rumen Microbial Ecosystem*. London: Elsevier Applied Science.
Hofmann, R. R. (1973). *The Ruminant Stomach*. Nairobi: East African Literature Bureau.
Hofmann, R. R. (1989). Evolutionary steps of ecophysiological adaptation and diversification of ruminants: a comparative view of their digestive system. *Oecologia*, **78**, 443–57.
Hogberg, P. (1986). Soil nutrient availability, root symbioses and tree species composition in tropical Africa: a review. *Journal of Tropical Ecology*, **2**, 359–72.
Hohmann, G. (1989). Group fission in Nilgiri langurs (*Presbytis johnii*). *International Journal of Primatology*, **10**, 441–54.
Hooijer, D. A. (1962). Quaternary langurs and macaques from the Malay archipelago. *Zool Verhandelingen*, **55**, 1–64.
Hooker, J. J. (1989). British mammals in the Tertiary period. *Biological Journal of the Linnaean Society*, **38**, 9–21.
Hoppe, P. P. (1977*a*). Comparison of voluntary food and water consumption and digestion in Kirk's dik-dik and suni. *East African Wildlife Journal*, **15**, 41–8.
Hoppe, P. P. (1977*b*). Rumen fermentation and body weight in African ruminants. In *Proceedings of 13th Congress of Game Biologists, Atlanta. USA*, ed. T. J. Peterle, pp. 141–50. Washington, DC: Wildlife Society.
Hoppe-Dominik, B. (1984). Étude du spectre des proies de la panthère, *Panthera pardus*, dans le Parc National de Taï en Côte d'Ivoire. *Mammalia*, **48**, 477–87.
Hörnicke, H. & Björnhag, G. (1980). Coprophagy and related strategies for digesta utilization. In *Digestive Physiology and Metabolism in Ruminants*, ed. Y. Ruckebusch & P. Thivend, pp. 707–30. Lancaster: MTP Press.
Horwich, R. H. (1972). Home range and food habits of the Nilgiri langur, *Presbytis johnii*. *Journal of the Bombay Natural History Society*, **69**, 255–67.
Howe, H. F. & Smallwood, J. (1982). Ecology of seed dispersal. *Annual Review of Ecology and Systematics*, **13**, 201–18.
Hrdy, S. B. (1974). Male-male competition and infanticide among the langurs (*Presbytis entellus*) of Abu, Rajasthan. *Folia Primatologica*, **22**, 19–58.
Hrdy, S. B. (1976). The care and exploitation of non-human primate infants by conspecifics other than the mother. In *Advances in the Study of Behaviour*, vol. 6, ed. J. S. Rosenblatt, R. A. Hinde, E. Shaw & C. Beer, pp. 101–58. New York: Academic Press.
Hrdy, S. B. (1977). *The Langurs of Abu: Female and Male Strategies of Reproduction*. Cambridge, MA: Harvard University Press.
Hrdy, S. B. (1979). Infanticide among animals: a review, classification and examination of the implications for the reproductive strategies of females. *Ethology and Sociobiology*, **1**, 13–40.
Hrdy, S. B. & Hausfater, G. (1984). Comparative and evolutionary perspectives on infanticide: an overview and introduction. In *Infanticide: Comparative and Evolutionary Perspectives*, ed. G. Hausfater & S. B. Hrdy, pp. xiii–xxxv. New York: Aldine.
Hrdy, S. B. & Hrdy, D. B. (1976). Hierarchical relations among female hanuman langurs (Primates: Colobinae, *Presbytis entellus*). *Science*, **193**, 913–15.
Hrdy, S. B. & Whitten, P. L. (1987). Patterning of sexual activity. In *Primate Societies*, ed. B. B. Smuts, D. L. Cheney, R. M. Seyfarth, R. W. Wrangham & T. T. Struhsaker, pp. 370–84. Chicago: University of Chicago Press.

Hu, C. & Qi, T. (1978). Gongwangling Pleistocene mammalian fauna of Lantian, Shaanxi. *Palaeontologica Sinica, No. 155, Ser. C*, no. 21, 1–64.

Hu, J. C. (1981). *Giant Pandas, Takin and Golden Monkey*. Sichuan: Sichuan People's Press.

Hull, D. B. (1979). A craniometric study of the black and white *Colobus* Illiger 1811 (Primates: Cercopithecoidea). *American Journal of Physical Anthropology*, **51**, 163–82.

Huxley, J. S. (1934). A natural experiment on the territorial instinct. *British Birds*, **27**, 270–7.

Hylander, W. L. (1975). Incisor size and diet in anthropoids with special reference to Cercopithecidae. *Science*, **189**, 1095–8.

Hylander, W. L., Johnson, K. R. & Crompton, A. W. (1987). Loading patterns and jaw movements during mastication in *Macaca fascicularis*: a bone-strain, electromyographic, and cineradiographic analysis. *American Journal of Physical Anthropology*, **72**, 287–314.

Isbell, L. A. (1984). Daily ranging behavior of red colobus (*Colobus badius tephrosceles*) in Kibale Forest, Uganda. *Folia Primatologica*, **41**, 34–48.

Islam, M. A. & Husain, K. Z. (1982). A preliminary study on the ecology of the capped langur. *Folia Primatologica*, **39**, 145–59.

Jablonski, N. G. (1993). Quarternary environments and the evolution of primates in East Asia, with notes on two new specimens of fossil Cercopithecidae from China. *Folia Primatologica*, **60**, 118–32.

Jablonski, N. G. & Gu, Y. (1988). A reassessment of *Megamacaca lantianensis* from the Pleistocene of Shaanxi province. *China American Journal of Physical Anthropology*, **75**, 225.

Jablonski, N. G. & Gu, Y. (1991). A reassessment of *Megamacaca lantianensis*, a large monkey from the Pleistocene of north-central China. *Journal of Human Evolution*, **20**, 51–66.

Jablonski, N. G. & Pan, Y. (1988). The evolution and palaeobiogeography of monkeys in China. In *The Palaeoenvironment of East Asia from the mid-Tertiary*, vol. II, ed. P. Whyte, J. S. Aigner, N. G. Jablonski, G. Taylor, D. Walker, P. Wang & C.-L. So, pp. 849–67. Hong Kong: University of Hong Kong Centre of Asian Studies.

Jablonski, N. G. & Pan, R.-L. (1991). Sexual dimorphism in *Rhinopithecus bieti* and other species of *Rhinopithecus*. In *Primatology Today*, ed. A. Ehara, T. Kimura, O. Takenaka & M. Iwamoto. Amsterdam: Elsevier.

Jablonski, N. G. & Peng, Y.-Z. (1993). The phylogenetic relationships and classification of the doucs and snub-nosed langurs of China and Vietnam. *Folia Primatologica*, **60**, 36–55.

Janis, C. (1976). The evolutionary strategy of the Equidae and the origins of rumen and caecal digestion. *Evolution*, **30**, 757–74.

Janis, C. M. (1984). Prediction of primate diets from molar wear patterns. In *Food Acquisition and Processing in Primates*, ed. D. J. Chivers, B. A. Wood & A. Bilsborough, pp. 331–40. New York: Plenum Press.

Janzen, D. H. (1974). Tropical blackwater rivers, animals, and mast fruiting by the Dipterocarpaceae. *Biotropica*, **6**, 69–103.

Janzen, D. H. (1975). *The Ecology of Plants in the Tropics*. London: Edward Arnold.

Janzen, D. H. (1983). Physiological ecology of fruits and their seeds. In *Physiological Plant Ecology*, vol. 3., ed. O. L. Lange, P. S. Nobel, C. B. Osmond & H. Ziegler, pp. 625–55. Berlin: Springer Verlag.

Janzen, D. H. & Waterman, P. G. (1984). A seasonal census of phenolics, fibre and alkaloids in foliage of forest trees in Costa Rica: some factors influencing their distribution and relation to host selection by Sphingidae and Saturniidae. *Biological Journal of the Linnaean Society*, **21**, 439–54.

Jarman, P. J. (1974). The social organisation of antelope in relation to their ecology. *Behaviour*, **48**, 215–61.

Jay, P. C. (1965). The common langur of North India. In *Primate Behavior: Field Studies of Monkeys and Apes*, ed. I. DeVore, pp. 197–249. New York: Holt, Rinehart & Winston.

Johns, A. D. (1983). Ecological effects of selective logging in a West Malaysian rain forest. PhD thesis, University of Cambridge, England.

Johns, A. D. (1985). Selective logging and wildlife conservation in tropical rainforest: problems and recommendations. *Biological Conservation*, **31**, 355–75.

Johns, A. D. (1986). Effects of selective logging on the behavioral ecology of West Malaysian primates. *Ecology*, **67**, 684–94.

Johns, A. D. & Skorupa, J. P. (1987). Responses of rain-forest primates to habitat disturbance: a review. *International Journal of Primatology*, **8**, 157–91.

Johns, T. (1986). Detoxification function of geophagy and domestication of the potato. *Journal of Chemical Ecology*, **12**, 635–46.

Jones, C. (1970). Stomach contents and gastro-intestinal relationships of monkeys collected in Rio Muni, West Africa. *Mammalia*, **34**, 107–17.

Jones, W. T. & Mangan, J. L. (1977). Complexes of the condensed tannins of sainfoin (*Onobrychis viciifolia* Scop.) with fraction-1 leaf protein and with submaxillary mucoprotein, and their reversal by polyethylene glycol and pH. *Journal of the Science of Food and Agriculture*, **28**, 126–36.

Kahlke, H. D. (1973). A review of the Pleistocene history of the Orang-Utan (*Pongo* Lacépède, 1799). *Asian Perspectives*, **15**, 5–14.

Kalmykov, N. P. & Maschenko, E. N. (1992). The most northern representative of early Pliocene Cercopithecidae from Asia. *Paleontologichesky zhurnal (Moscow)*, No. 2, 136–8.

Kavanagh, M. (1972). Food-sharing behaviour within a group of douc monkeys (*Pygathrix nemaeus nemaeus*). *Nature*, **239**, 406–7.

Kawabe, M. & Mano, T. (1972). Ecology and behaviour of the wild proboscis monkey, *Nasalis larvatus* (Wurmb) in Sabah, Malaysia. *Primates*, **13**, 213–27.

Kay, R. F. (1975). The functional adaptations of primate molar teeth. *American Journal of Physical Anthropology*, **43**, 195–216.

Kay, R. F. (1977*a*). The evolution of molar occlusion in the Cercopithecidae and early catarrhines. *American Journal of Physical Anthropology*, **46**, 327–52.

Kay, R. F. (1977*b*). Diets of early Miocene African hominoids. *Nature*, **268**, 628–30.

Kay, R. F. (1978). Molar structure and diet in extant Cercopithecidae. In *Development, Function and Evolution of Teeth*, ed. P. M. Butler & K. A. Joysey, pp. 309–39. New York: Academic Press.

Kay, R. F. & Hylander, W. L. (1978). The dental structure of mammalian folivores with special reference to primates and phalangeroids (Marsupialia). In *The Ecology of Arboreal Folivores*, ed. G. G. Montgomery, pp. 173–92. Washington, DC: Smithsonian Institution Press.

Kay, R. N. B. (1966). The influence of saliva on digestion in ruminants. *World Review of Nutrition and Dietetics*, **6**, 292–325.

Kay, R. N. B. (1985). Comparative studies of food propulsion in ruminants. In *Physiological and Pharmacological Aspects of the Reticulo-rumen*, ed. L. A. A. Ooms, A. D. Degryse & A. S. J. A. M. van Miert, pp. 155–70. Dordrecht: Martinus Nijhoff.

Kay, R. N. B. (1987). Weights of salivary glands in some ruminant animals. *Journal of Zoology, London*, **211**, 431–6.

Kay, R. N. B., Engelhardt, W. V. & White, R. G. (1980). The digestive physiology of wild ruminants. In *Digestive Physiology and Metabolism in Ruminants*, ed. Y. Ruckebusch & P. Thivend, pp. 743–61. Lancaster: MTP Press.

Kay, R. N. B., Hoppe, P. P. & Maloiy, G. M. O. (1976). Fermentative digestion of food in the colobus monkey, *Colobus polykomos*. *Experientia*, **32**, 485–6.

Keeler, R. F., van Kampen, K. A. & James, L. F. (eds.) (1978). *Effects of Poisonous Plants on Livestock*. New York: Academic Press.

Kern, J. A. (1964). Observations on the habitats of the proboscis monkey, *Nasalis larvatus* (Wurmb), made in the Brunei Bay area, Borneo. *Zoologica (New York)*, **49**, 183–92.

Khajuria, H. (1977). Ecological observations on the golden langur, *Presbytis geei* Khajuria, with remarks on its conservation. In *Use of Non-Human Primates in Bio-medical Research*, ed. M. R. N. Prasad & T. C. Anand Kumar, pp. 52–61. New Delhi: Indian National Science Academy.

Kinnaird, M. F. (1992). Phenology of flowering and fruiting of an East African riverine forest ecosystem. *Biotropica*, **24**, 187–94.

Kleiber, N. (1961). *The Fire of Life*. New York: John Wiley & Sons.

Kolattukudy, P. E. (1980). Biopolyester membranes of plants: cutin and suberin. *Science*, **208**, 990–1000.

Kool, K. M. (1989). Behavioural ecology of the silver leaf monkey, *Trachypithecus auratus sondaicus*, in the Pangandaran Nature Reserve, West Java, Indonesia. PhD thesis, University of New South Wales, Sydney, Australia.

Kool, K. M. (1992). Food selection by the silver leaf monkey, *Trachypithecus auratus sondaicus*, in relation to plant chemistry. *Oecologia*, **90**, 527–33.

Koufos, G. D., Syrides, G. E. & Koliadimou, K. K. (1991). A Pliocene primate from Macedonia (Greece). *Journal of Human Evolution*, **21**, 283–94.

Kovar-Eder, J. (1987). Pannonian (Upper Miocene) vegetational character and climatic inferences in the central Paratethys area. *Annales Naturhistorisches Museum Wien*, **88A**, 117–29.

Kubitzki, K. & Gottlieb, O. R. (1984). Phytochemical aspects of angiosperm origin and evolution. *Acta Botanica Neerlandica*, **33**, 457–68.

Kuhn, H.-J. (1964). Zur Kenntnis von Bau und Funktion des Magens der Schlankaffen (Colobinae). *Folia Primatologica*, **2**, 193–221.

Kuhn, H.-J. (1967). Zur Systematik der Cercopithecidae. In *Neue Ergebnisse der Primatologie*, ed. D. Starck, R. Schneider & H.-J. Kuhn, pp. 25–46. Stuttgart: Gustav Fischer.

Kunkin, K. J. (1986). Ecology and behaviour of *Presbytis thomasi* in North Sumatra. *Primates*, **27**, 151–72.

Kursar, T. A. & Coley, P. D. (1991). Nitrogen content and expansion rate of young leaves of rain forest species: implications for herbivory. *Biotropica*, **23**, 141–50.

Lahm, S. A. (1993). Ecology and economics of human/wildlife interaction in northeastern Gabon. PhD thesis, New York University, New York.

Lambert, F. (1990). Some notes on fig-eating by arboreal mammals in Malaysia. *Primates*, **31**, 453–8.

Lancaster, J. (1971). Play-mothering: the relations between juvenile females and young infants among free-ranging vervet monkeys (*Cercopithecus aethiops*). *Folia Primatologica*, **15**, 161–82.

Langenheim, J. H., Stubblebine, W. H., Lincoln, D. E. & Foster, C. E. (1978). Implications of variation in resin content among organs, tissues and popula-

tions of the tropical legume *Hymenaea. Biochemical Systematics and Ecology*, **6**, 299–313.
Langer, P. (1986). Large mammalian herbivores in tropical forests with either hindgut- or forestomach-fermentation. *Zeitschrift für Saugetierkunde*, **51**, 173–87.
Langer, P. (1987). Evolutionary patterns of Perissodactyla and Artiodactyla (Mammalia) with different types of digestion. *Zeitschrift für Zoologisches Systematik und Evolutionsforschung*, **25**, 212–36.
Laws, J. W. & Vonder Haar Laws, J. (1984). Social interactions among adult male langurs (*Presbytis entellus*) at Rajaji Wildlife Sanctuary. *International Journal of Primatology*, **5**, 31–50.
Leakey, L. S. B. (1963). East African fossil Hominoidea and the classification within this super-family. In *Classification and Human Evolution*, ed. S. L. Washburn, pp. 32–49. Chicago: Aldine.
Leakey, M. D. (1987). Animal prints and trails. In *The Pliocene Site of Laetoli, northern Tanzania*, ed. M. D. Leakey & J. M. Harris, pp. 451–89. Oxford: Oxford University Press.
Leakey, M. D. & Harris, J. M. (eds.) (1987). *The Pliocene Site of Laetoli, northern Tanzania*. Oxford: Oxford University Press.
Leakey, M. D. & Hay, R. L. (1979). Pliocene footprints in the Laetoli Beds at Laetolil, northern Tanzania. *Nature*, **278**, 317–23.
Leakey, M. G. (1982). Extinct large colobines from the Plio-Pleistocene of Africa. *American Journal of Physical Anthropology*, **58**, 153–72.
Leakey, M. G. (1987). Colobinae (Mammalia, Primates) from the Omo Valley, Ethiopia. In *Cahiers de paléontologie. Travaux de paléontologie est-africaine. Les faunes Plio-Pléistocènes de la vallée de l'Omo (Éthiopie). Vol. 3, Cercopithecidae de la Formation de Shungura*, pp. 147–69. Paris: Éditions du C. N. R. S.
Leakey, M. G. & Delson, E. (1987). Fossil Cercopithecidae from the Laetolil Beds, Tanzania. In *The Pliocene Site of Laetoli, northern Tanzania*, ed. M. D. Leakey & J. M. Harris, pp. 91–107. Oxford: Oxford University Press.
Lebreton, P. (1982). Tannins ou alcaloides: deux tactiques phytochimiques de dissuasion des herbivores. *Revue Ecologie* (*Terre Vie*), **36**, 539–72.
Lee, P. C. (1983). Caretaking of infants and mother-infant relationships. In *Primate Social Relationships*, ed. R. A. Hinde, pp. 145–51. Oxford: Blackwell Scientific Publications.
Leigh, C. (1926). Weights and measurements of the Nilgiri langur (*Pithecus johnii*). *Journal of the Bombay Natural History Society*, **31**, 223.
Lekagul, B. & McNeely, J. A. (1977). *Mammals of Thailand*. Bangkok: Association for the Conservation of Wildlife.
Leland, L., Struhsaker, T. T. & Butynski, T. M. (1984). Infanticide by adult males in three primate species of the Kibale Forest, Uganda: A test of hypotheses. In *Infanticide: Comparative and Evolutionary Perspectives*, ed. G. Hausfater & S. B. Hrdy, pp. 151–72. New York: Aldine.
Leskes, A. & Acheson, N. H. (1971). Social organization of a free-ranging troop of black and white colobus monkeys (*Colobus abyssinicus*). In *Proceedings of the Third International Congress of Primatology, Zurich 1970, Volume 3, Behavior*, ed. H. Kummer, pp. 22–31. Basel: S. Karger.
Leutenegger, W. (1971). Metric variability of the anterior dentition of African colobines. *American Journal of Physical Anthropology*, **35**, 91–100.
Li, Z. X., Ma, S. L., Hue, C. H. & Wang, Y. X. (1982). The distribution and

habits of the Yunnan golden monkey, *Rhinopithecus bieti*. *Journal of Human Evolution*, **11**, 633–8.

Lippold, L. K. (1977). The douc langur: a time for conservation. In *Primate Conservation*, ed. H. S. H. Prince Rainier & G. H. Bourne, pp. 513–38. New York: Academic Press.

Long, Y., Kirkpatrick, C. R., Zhongtai & Xiaolin (1994). Report on the distribution, population, and ecology of the Yunnan snub-nosed monkey (*Rhinopithecus bieti*). *Primates*, **35**, 241–50.

Lucas, P. W. (1989). A new theory relating seed processing by primates to their relative tooth sizes. In *The Growing Scope of Human Biology, Proceedings of the Australasian Society for Human Biology 2*, ed. L. H. Schmitt, L. Freedman & N. W. Bruce, pp. 37–49. Perth, Western Australia: Centre for Human Biology, University of Western Australia.

Lucas, P. W. (1991). Fundamental physical properties of fruits and seeds in primate diets. In *Primatology Today*, ed. A. Ehara *et al.*, pp. 125–8. Amsterdam: Elsevier.

Lucas, P. W. (1994). Categorization of food items for oral processing. In *The Digestive System of Mammals*, ed. D. J. Chivers & P. Langer, pp. 197–218. Cambridge: Cambridge University Press.

Lucas, P. W., Choong, M. F., Tan, H. T. W., Turner, I. M. & Berrick, A. J. (1991). The fracture toughness of the leaf of the dicotyledon *Calophyllum inophyllum* L. (Guttiferae). *Philosophical Transactions of the Royal Society, London*, **B334**, 95–106.

Lucas, P. W., Corlett, R. T. & Luke, D. A. (1986*a*). Sexual dimorphism of tooth size in anthropoids. In *Sexual Dimorphism in Living and Fossil Primates*, ed. M. Pickford & B. Chiarelli, pp. 23–39. Florence: Il Sedicesimo.

Lucas, P. W. Corlett, R. T. & Luke, D. A. (1986*b*). Postcanine tooth size and diet in anthropoids. *Zeitschrift für Morphologie und Anthropologie*, **76**, 253–76.

Lucas, P. W. & Pereira, B. (1991). Thickness effect in cutting systems. *Journal of Materials Science Letters*, **10**, 235–6.

Lydekker, R. (1884). Rodents and new ruminants from the Siwaliks, and synopsis of Mammalia. *Memoirs of the Geological Survey of India, Palaeontologica India, Series X*, **3**, 105–34.

Macdonald, D. W. (1982). Notes on the size and composition of groups of proboscis monkey, *Nasalis larvatus*. *Folia Primatologica*, **37**, 95–8.

Machado, A. de B. (1969). Mamíferos de Angola ainda não citados ou pouco conhecidos. *Publicações Culturais da Companhia de Diamantes de Angola*, **46**, 93–232.

MacKinnon, J. & MacKinnon, K. (1987). Conservation status of the primates of the Indo-Chinese subregion. *Primate Conservation*, **8**, 187–95.

MacKinnon, J. R. & MacKinnon, K. S. (1980). Niche differentiation in a primate community. In *Malayan Forest Primates*, ed. D. J. Chivers, pp. 167–90, New York: Plenum Press.

MacKinnon, K. S. (1986). Survey to determine the status and conservation needs of the golden snub-nosed monkey *Pygathrix roxellana* in Sichuan Province, China. Gland, Switzerland: International Union for the Conservation of Nature and Natural Resources (IUCN).

MacLarnon, A. M., Chivers, D. J. & Martin, R. D. (1986). Gastro-intestinal allometry in primates and other mammals including new species. In *Primate Ecology and Conservation*, ed. J. G. Else & P. C. Lee, pp. 75–85. Cambridge: Cambridge University Press.

Maier, W. (1977). Die Evolution der bilophodonten Molaren der Cercopithecoidea. *Zeitschrift für Morphologie und Anthropologie*, **68**, 5–56.

Maisels, F., Gautier-Hion, A. & Gautier, J.-P. (1994) Diets of two sympatric colobines in Zaire: more evidence on seed-eating in forests on poor soils. *International Journal of Primatology*, in press.

Makwana, S. C. (1979). Infanticide and social change in two groups of the hanuman langur, *Presbytis entellus*, at Jodhpur. *Primates*, **20**, 293–300.

Marler, P. (1969). *Colobus guereza*: territoriality and group composition. *Science*, **163**, 93–5.

Marler, P. (1970). Vocalizations of East African monkeys, I: red colobus. *Folia Primatologica*, **13**, 81–91.

Marler, P. (1972). Vocalizations of East African monkeys, II: black and white colobus. *Behaviour*, **42**, 175–97.

Marsh, C. W. (1976). A management plan for the Tana River Game Reserve, Kenya. Report to the Kenya Game Department, Nairobi.

Marsh, C. W. (1978*a*). Ecology and social organization of the Tana River red colobus, *Colobus badius rufomitratus*. PhD thesis, University of Bristol, England.

Marsh, C. W. (1978*b*). Tree phenology in a gallery forest on the Tana River, Kenya. *East African Agricultural and Forestry Journal*, **43**, 305–16.

Marsh, C. W. (1978*c*). Comparative activity budgets of red colobus. In *Recent Advances in Primatology, vol. 1*, ed. D. J. Chivers & J. Herbert, pp. 249–51. London: Academic Press.

Marsh, C. W. (1979*a*). Female transference and mate choice among Tana River red colobus. *Nature*, **281**, 568–9.

Marsh, C. W. (1979*b*). Comparative aspects of social organization in the Tana River red colobus, *Colobus badius rufomitratus. Zeitschrift für Tierpsychologie*, **51**, 337–62.

Marsh, C. W. (1981*a*). Ranging behaviour and its relation to diet selection in Tana River red colobus (*Colobus badius rufomitratus*). *Journal of Zoology, London*, **195**, 473–92.

Marsh, C. W. (1981*b*). Diet choice among red colobus (*Colobus badius rufomitratus*) on the Tana River, Kenya. *Folia Primatologica*, **35**, 147–78.

Marsh, C. W. (1981*c*). Time budget of Tana River red colobus. *Folia Primatologica*, **35**, 30–50.

Marsh, C. W. (1986). A resurvey of Tana primates and their forest habitat. *Primate Conservation*, **7**, 72–81.

Marsh, C. W., Johns, A. D. & Ayres, J. M. (1987). Effects of habitat disturbance on rain forest primates. In *Primate Conservation in the Tropical Rain Forest*, ed. C. W. Marsh & R. A. Mittermeier, pp. 83–107. New York: Alan R. Liss.

Marsh, C. W. & Wilson, W. W. (1981). *A Survey of Primates in Peninsular Malaysian Forests*. Kuala Lumpur: Universiti Kebangsaan, Malaysia.

Martin, C. & Asibey, E. O. A. (1979). Effect of timber exploitation on primate population and distribution in the Bia rain forest area of Ghana. Paper presented at the *Seventh Congress of the International Primatological Society*, Bangalore, India.

Martin, J. S. & Martin, M. M. (1982). Tannin assays in ecological studies: lack of correlation between total phenolics, proanthocyanidins and protein precipitating constituents in mature foliage of six oak species. *Oecologia (Berlin)*, **54**, 205–11.

Martin, J. S. & Martin, M. M. (1983). Tannin assays in ecological studies. *Journal of Chemical Ecology*, **9**, 285–94.

Martin, J. S., Martin, M. M. & Bernays, E. A. (1987). Failure of tannic acid to inhibit digestion or reduce digestibility of plant protein in gut fluids of insect herbivores: implications for theories of plant defense. *Journal of Chemical Ecology*, **13**, 605–21.

Martin, M. M., Rockholm, D. C. & Martin, J. S. (1985). Effects of surfactants, pH and certain cations on the precipitation of proteins by tannins. *Journal of Chemical Ecology*, **11**, 485–94.

Martin, R. D., Chivers, D. J., MacLarnon, A. M. & Hladik, C. M. (1985). Gastrointestinal allometry in primates and other mammals. In *Size and Scaling in Primate Biology*, ed. W. L. Jungers, pp. 61–89. New York: Plenum Press.

Maschenko, E. N. (1991). Tooth system and taxonomic status of early Pliocene cercopithecid monkey *Dolichopithecus hipsulophus* (Primates, Cercopithecoidea). *Biol. Moscow Ov. Ist. Prir. Geol.* **66**, 61–74.

Mathur, R. & Manohar, B. R. (1991). Departure of juvenile male *Presbytis entellus* from the natal group. *International Journal of Primatology*, **12**, 39–43.

Matthew, W. D. & Granger, W. (1923). New fossil mammals from the Pliocene of Szechuan, China. *Bulletin of the American Museum of Natural History*, **48**, 563–98.

McCann, C. (1928). Notes on the common Indian langur (*Pithecus entellus*). *Journal of the Bombay Natural History Society*, **33**, 192–4.

McCann, C. (1933). Observations on some of the Indian langurs. *Journal of the Bombay Natural History Society*, **36**, 618–28.

McKenna, J. J. (1978). Biosocial factors of grooming behaviour among the common Indian langur monkey (*Presbytis entellus*). *American Journal of Physical Anthropology*, **48**, 503–10.

McKenna, J. J. (1979). The evolution of allomothering behavior among colobine monkeys: function and opportunism in evolution. *American Anthropologist*, **81**, 818–40.

McKey, D. B. (1978*a*). Plant chemical defenses and the feeding and ranging behavior of colobus monkeys in African rain forests. PhD thesis, University of Michigan, Ann Arbor, MI.

McKey, D. B. (1978*b*). Soils vegetation, and seed-eating by black colobus monkeys. In *The Ecology of Arboreal Folivores*, ed. G. G. Montgomery, pp. 423–38. Washington, DC: Smithsonian Institution Press.

McKey, D. B., Gartlan, J. S., Waterman, P. G. & Choo, G. M. (1981). Food selection by black colobus monkeys (*Colobus satanas*) in relation to food chemistry. *Biological Journal of the Linnaean Society*, **16**, 115–46.

McKey, D. B. & Waterman, P. G. (1982). Ranging behavior of a group of black colobus (*Colobus satanas*) in the Douala-Edea Reserve, Cameroon. *Folia Primatologica*, **39**, 264–304.

McKey, D. B., Waterman, P. G., Mbi, C. N., Gartlan, J. S. & Struhsaker, T. T. (1978). Phenolic content of vegetation in two African rain forests: ecological implications. *Science*, **202**, 61–4.

McNaughton, S. J., Tarrants, J. L., McNaughton, M. M. & Davis, R. H. (1985). Silica as a defense against herbivory and a growth promoter in African grasses. *Ecology*, **66**, 528–35.

Medley, K. E. (1990). Forest ecology and conservation in the Tana River National Primate Reserve, Kenya. PhD thesis, Michigan State University, East Lansing, MI.

Medway, Lord (1970). The monkeys of Sundaland: ecology and systematics of the cercopithecids of a humid equatorial environment. In *Old World Monkeys*, ed. J. R. Napier and P. H. Napier, pp. 513–53. New York: Academic Press.

Medway, Lord (1972). Phenology of a tropical rain forest in Malaya. *Biological Journal of the Linnaean Society*, **4**, 117–46.

Medway, Lord (1977). Mammals of Borneo. Field Keys and Annotated Checklist. *Monographs of the Malaysian Branch of the Royal Asiatic Society*, no. 7. Kuala Lumpur: Royal Asiatic Society.

Mehansho, H., Butler, L. G. & Carlson, D. M. (1987). Dietary tannins and salivary proline-rich proteins: interactions, induction and defense mechanisms. *Annual Review of Nutrition*, **7**, 423–40.

Mehansho, H., Clements, S., Shearer, B. T., Smith, S. & Carlson, D. M. (1985). Induction of proline-rich glycoprotein synthesis in mouse salivary glands by isoproterenol and by tannins. *Journal of Biological Chemistry*, **260**, 4418–23.

Meikle, E. (1987). Fossil Cercopithecidae from the Sahabi Formation. In *Neogene Paleontology and Geology of Sahabi*, ed. N. T. Boaz, A. El-Arnauti, A. W. Gaziry, J. de Heinzelin & D. D. Boaz, pp. 119–27. New York: Alan R. Liss.

Melnick, D. J. & Pearl, M. C. (1987). Cercopithecines in multi-male groups: genetic diversity and population structure. In *Primate Societies*, ed. B. B. Smuts, D. L. Cheney, R. M. Seyfarth, R. W., Wrangham & T. T. Struhsaker, pp. 121–34. Chicago: University of Chicago Press.

Menzies, J. I. (1970). An eastward extension to the known range of the olive colobus monkey (*Colobus verus*, Van Beneden). *Journal of the West African Science Association*, **15**, 83–4.

Millburn, P. (1978). Biotransformations of xenobiotics by animals. In *Biochemical Aspects of Plant and Animal Coevolution*, ed. J. B. Harborne, pp. 35–73. London: Academic Press.

Milligan, L. P., Grovum, W. L. & Dobson, A. (eds.) (1986) *Control of Digestion and Metabolism in Ruminants*. Englewood Cliffs: Prentice-Hall.

Milton, K. (1979). Factors influencing leaf choice by howler monkeys: a test of some hypotheses of food selection by generalist herbivores. *American Naturalist*, **114**, 362–78.

Milton, K. (1981). Food choice and digestive strategies by two sympatric primate species. *The American Naturalist*, **117**, 496–505.

Milton, K. (1982). Dietary quality and demographic regulation in a howler monkey population. In *The Ecology of a Tropical Forest*, ed. E. G. Leigh, A. S. Rand & D. M. Windsor, pp. 273–89. Washington, DC: Smithsonian Institution Press.

Milton, K. (1984). The role of food-processing factors in primate food choice. In *Adaptations for Foraging in Nonhuman Primates*, ed. P. S. Rodman & J. G. H. Cant, pp. 249–79. New York: Columbia University Press.

Milton, K. & McBee, R. H. (1983). Rates of fermentative digestion in the howler monkey, *Alouatta palliata* (Primates: Ceboidea). *Comparative Biochemistry Physiology*, **74A**, 29–31.

Mitani, M. (1990). A note on the present situation of the primate fauna found from south-eastern Cameroon to northern Congo. *Primates*, **31**, 625–34.

Mitani, M. (1992). Preliminary results of the studies on wild western lowland gorillas and other sympatric diurnal primates in the Ndoki Forest, Northern Congo. In *Topics in Primatology, Vol. 2: Behavior, Ecology and Conservation*, ed. N. Itoigawa, Y. Sugiyama, G. P. Sackett & R. K. R. Thompson, pp. 215–24. Tokyo: University of Tokyo Press.

Mittermeier, R. A. (1973). Group activity and population dynamics of the howler monkey on Barro Colorado Island. *Primates*, **14**, 1–19.

Mittermeier, R. A. & Cheney, D. L. (1987). Conservation of primates and their habitats. In *Primate Societies*, ed. B. B. Smuts, D. L. Cheney, R. M. Seyfarth,

R. W. Wrangham & T. T. Struhsaker, pp. 477–90. Chicago: University of Chicago Press.

Mittermeier, R. A. & Fleagle, J. G. (1976). The locomotor and postural repertoires of *Ateles geoffroyi* and *Colobus guereza* and a re-evaluation of the locomotor category semibrachiation. *American Journal of Physical Anthropology*, **45**, 235–56.

Mohnot, S. M. (1971). Some aspects of social changes and infant-killing in the hanuman langur, *Presbytis entellus* (Primates: Cercopithecidae) in Western India. *Mammalia*, **35**, 175–98.

Mohnot, S. M. (1978). Peripheralisation of weaned male juveniles in *Presbytis entellus*. In *Recent Advances in Primatology 1*, ed. D. J. Chivers & J. Herbert, pp. 87–91. London: Academic Press.

Mohnot, S. M. (1980). Intergroup infant kidnapping in hanuman langur. *Folia Primatologica*, **34**, 259–77.

Mohnot, S. M. (1984). Some observations on all-male bands of hanuman langurs (*Presbytis entellus*). In *Current Primate Researches*, ed. M. L. Roonwal, S. M. Mohnot & N. S. Rathore, pp. 343–56. Jodhpur: Department of Zoology, University of Jodhpur, India.

Mohnot, S. M., Gadgil, M. and Makwana, S. C. (1981). On the dynamics of the hanuman langur populations of Jodphur (Rajasthan, India). *Primates*, **22**, 182–91.

Moir, R. J. (1968). Ruminant digestion and evolution. In *Handbook of Physiology, Section 6: Alimentary Canal*, vol. 5, ed. C. F. Code, pp. 2673–94. Washington, DC: American Physiological Society.

Mole, S., Butler, L. G. & Iason, G. (1990). Defense against dietary tannin in herbivores: a survey for proline-rich proteins in mammals. *Biochemical Systematics and Ecology*, **18**, 287–93.

Mole, S., Ross, J. A. M. & Waterman, P. G. (1988). Light-induced variation in phenolic levels in foliage of rain-forest plants. *Journal of Chemical Ecology*, **14**, 1–21.

Mole, S. & Waterman, P. G. (1985). Stimulatory effects of tannins and cholic acid on tryptic hydrolysis of proteins: ecological implications. *Journal of Chemical Ecology*, **11**, 1323–32.

Mole, S. & Waterman, P. G. (1987*a*). Tannic acid and proteolytic enzymes: enzyme inhibition or substrate deprivation? *Phytochemistry*, **26**, 99–102.

Mole, S. & Waterman, P. G. (1987*b*). Tannins as antifeedants to mammalian herbivores – still an open question? In *Allelochemicals: Role in Agriculture and Forestry*, ed. G. R. Waller, pp. 572–87. Washington DC: American Chemical Society.

Mole, S. & Waterman, P. G. (1987*c*). A critical analysis of techniques for measuring tannins in ecological studies. I. Techniques for chemically defining tannins. *Oecologia (Berlin)*, **72**, 137–47.

Mole, S. & Waterman, P. G. (1987*d*). A critical analysis of techniques for measuring tannins in ecological studies. II. Techniques for biochemically defining tannins. *Oecologia (Berlin)*, **72**, 148–56.

Mole, S. & Waterman, P. G. (1988). Light-induced variation in phenolic levels in foliage of rain-forest plants. II. Potential significance to herbivores. *Journal of Chemical Ecology*, **14**, 23–34.

Moore, J. (1984). Female transfer in primates. *International Journal of Primatology*, **5**, 537–89.

Moore, J. (1985*a*). Demography and sociality in primates. PhD thesis, Harvard University, Cambridge, MA.

Moore, J. (1985*b*). Insectivory by grey langurs. *Journal of the Bombay Natural History Society*, **82**, 38–44.
Moore, J. (1986). Paternity in all-male groups of langurs (*Presbytis entellus*). *Primate Report*, **14**, 30–1.
Morbeck, M. E. (1977). Positional behavior, selective use of habitat substrate and associated non-positional behavior in free-ranging *Colobus guereza* (Rüppell, 1835). *Primates*, **18**, 35–58.
Moreau, R. E. (1966). *The Bird Faunas of Africa and its Islands*. London: Academic Press.
Moreno-Black, G. S. & Bent, E. F. (1982). Secondary compounds in the diet of *Colobus angolensis*. *African Journal of Ecology*, **20**, 29–36.
Moreno-Black, G. S. & Maples, W. R. (1977). Differential habitat utilization of four Cercopithecidae in a Kenyan forest. *Folia Primatologica*, **27**, 85–107.
Mould, F. L., Saadullah, M., Haque, M., Davis, D., Dolberg, F. & Orskov, E. R. (1982). Investigation of some of the physiological factors influencing intake and digestion of rice straw by native cattle in Bangladesh. *Tropical Animal Production*, **7**, 174–81.
Mturi, F. A. (1993). Ecology of the Zanzibar red colobus monkey, *Colobus badius kirkii* (Gray, 1868), in comparison with other red colobines. In *Biogeography and Ecology of the Rain Forests of Eastern Africa*, ed. J. C. Lovett & S. K. Wasser, pp. 243–66. Cambridge: Cambridge University Press.
Mukherjee, R. P. (1978). Further observations on the golden langur (*Presbytis geei* Khajuria, 1956), with a note on capped langur (*Presbytis pileata* Blyth, 1843) of Assam. *Primates*, **19**, 737–47.
Mukherjee, R. P. & Saha, S. S. (1974). The golden langurs (*Presbytis geei* Khajuria, 1956) of Assam. *Primates*, **15**, 327–40.
Müller, E. F., Kamau, J. M. Z. & Maloiy, G. M. O. (1983). A comparative study of basal metabolism and thermoregulation in a folivorous (*Colobus guereza*) and an omnivorous (*Cercopithecus mitis*) primate species. *Comparative Biochemistry and Physiology*, **74A**, 319–22.
Murray, P. (1975). The role of cheek pouches in cercopithecine monkey adaptive strategy. In *Primate Functional Morphology and Evolution*, ed. R. H. Tuttle, pp. 151–94. The Hague: Mouton.
Nagy, K. A. & Milton, K. (1979). Energy metabolism and food consumption by wild howler monkeys (*Alouatta palliata*). *Ecology*, **60**, 475–80.
Napier, J. R. (1970). Paleoecology and catarrhine evolution. In *Old World Monkeys: Ecology, Systematics, and Behavior*, ed. J. R. Napier & P. H. Napier, pp. 53–95. New York: Academic Press.
Napier, J. R. & Napier, P. H. (1967). *A Handbook of Living Primates*. London: Academic Press.
Napier, P. H. (1985). *Catalogue of Primates in the British Museum (Natural History) and Elsewhere in the British Isles. Part III: Family Cercopithecidae, subfamily Colobinae*. London: British Museum (Natural History).
Nesbit-Evans, E. M., Van Couvering, J. A. H. & Andrews, P. (1981). Paleoecology of Miocene sites in western Kenya. *Journal of Human Evolution*, **10**, 99–116.
Neville, A. C. & Levy, S. (1985). The helicoidal concept of plant cell wall ultrastructure and morphogenesis. In *Biochemistry of Plant Cell Walls*, ed. C. T. Brett & J. C. Hillman, pp. 99–124. Cambridge: Cambridge University Press.
Newton, P. N. (1984). The ecology and social organisation of Hanuman langurs (*Presbytis entellus*, Dufresne 1797) in Kanha Tiger Reserve, Central Indian Highlands. DPhil thesis, University of Oxford, England.
Newton, P. N. (1985). The behavioral ecology of forest Hanuman langurs. *Tigerpaper*, **12**, 3–7.

Newton, P. N. (1986). Infanticide in an undisturbed forest population of Hanuman langurs, (*Presbytis entellus*). *Animal Behaviour*, **34**, 785–9.
Newton, P. N. (1987). The social organisation of forest Hanuman langurs (*Presbytis entellus*). *International Journal of Primatology*, **8**, 199–232.
Newton, P. N. (1988*a*). The variable social organization of Hanuman langurs (*Presbytis entellus*), infanticide, and the monopolization of females. *International Journal of Primatology*, **9**, 59–77.
Newton, P. N. (1988*b*). The structure and phenology of a moist deciduous forest in the central Indian highlands. *Vegetatio*, **75**, 3–16.
Newton, P. N. (1992). Feeding and ranging patterns of forest Hanuman langurs (*Presbytis entellus*). *International Journal of Primatology*, **13**, 245–85.
Newton, P. N. (1994). Social change and stability among forest Hanuman langurs. *Primates*, **35**, in press.
Nordin, M. (1978). Voluntary food intake and digestion by the lesser mousedeer. *Journal of Wildlife Management*, **42**, 185–7.
Northcote, D. H. (1985). Control of cell wall formation during growth. In *Biochemistry of Plant Cell Walls*, ed. C. T. Brett & J. C. Hillman, pp. 177–97. Cambridge: Cambridge University Press.
Oates, J. F. (1974). The ecology and behaviour of the black-and-white colobus monkey (*Colobus guereza* Rüppell) in East Africa. PhD thesis, University of London, England.
Oates, J. F. (1977*a*). The guereza and its food. In *Primate Ecology: Studies of Feeding and Ranging Behaviour in Lemurs, Monkeys and Apes*, ed. T. H. Clutton-Brock, pp. 275–321. London: Academic Press.
Oates, J. F. (1977*b*). The social life of a black-and-white colobus monkey, *Colobus guereza. Zeitschrift für Tierpsychologie*, **45**, 1–60.
Oates, J. F. (1977*c*). The guereza and man. In *Primate Conservation*, ed. H.S.H. Prince Rainier III & G. H. Bourne, pp. 419–67. London: Academic Press.
Oates, J. F. (1978). Water-plant and soil consumption by guereza monkeys (*Colobus guereza*): a relationship with minerals and toxins in the diet? *Biotropica*, **10**, 241–53.
Oates, J. F. (1979). Comments on the geographical distribution and status of the South Indian black leaf-monkey (*Presbytis johnii*). *Mammalia*, **43**, 485–93.
Oates, J. F. (1981). Mapping the distribution of West African rain-forest monkeys: issues, methods and preliminary results. *Annals of the New York Academy of Sciences*, **376**, 53–64.
Oates, J. F. (1986). *Action Plan for African Primate Conservation: 1986–90*. Stony Brook, NY: IUCN/SSC Primate Specialist Group.
Oates, J. F. (1987). Food distribution and foraging behaviour. In *Primate Societies*, ed. B. B. Smuts, D. L. Cheney, R. M. Seyfarth, R. W. Wrangham & T. T. Struhsaker, pp. 197–209. Chicago: University of Chicago Press.
Oates, J. F. (1988*a*). The distribution of *Cercopithecus* monkeys in West African forests. In *A Primate Radiation: Evolutionary Biology of the African Guenons*, ed. A. Gautier-Hion, F. Bourlière, J.-P. Gautier & J. Kingdon, pp. 79–103. Cambridge: Cambridge University Press.
Oates, J. F. (1988*b*). The diet of the olive colobus monkey, *Procolobus verus*, in Sierra Leone. *International Journal of Primatology*, **9**, 457–78.
Oates, J. F., Swain, T. & Zantovska, J. (1977). Secondary compounds and food selection by colobus monkeys. *Biochemical Systematics and Ecology*, **5**, 317–21.
Oates, J. F. & Trocco, T. F. (1983). Taxonomy and phylogeny of black-and-white colobus monkeys; inferences from an analysis of loud call variation. *Folia Primatologica*, **40**, 83–113.

Oates, J. F., Waterman, P. G. & Choo, G. M. (1980). Food selection by the south Indian leaf-monkey, *Presbytis johnii*, in relation to leaf chemistry. *Oecologia (Berlin)*, **45**, 45–56.
Oates, J. F. & Whitesides, G. H. (1990). Association between olive colobus (*Procolobus verus*), Diana guenons (*Cercopithecus diana*) and other forest monkeys in Sierra Leone. *American Journal of Primatology*, **21**, 129–46.
Oates, J. F., Whitesides, G. H., Davies, A. G., Waterman, P. G., Green, S. M., Dasilva, G. L. & Mole, S. (1990). Determinants of variation in tropical forest primate biomass: new evidence from West Africa. *Ecology*, **71**, 328–43.
Oboussier, H. & von Maydell, G. A. (1959). Zur Kenntnis des Indischen Goldlangurs. *Zeitschrift für Morphologie und Ökologie der Tiere (Berlin)*, **48**, 102–14.
Ochiago, W. O. (1991). The demography of the Tana River red colobus, *Colobus badius rufomitratus*. MSc thesis, University of Nairobi, Nairobi, Kenya.
Oh, H. K., Jones, M. B. & Longhurst, W. M. (1968). Comparison of rumen microbial inhibition resulting from various essential oils isolated from relatively unpalatable plant species. *Applied Microbiology*, **16**, 39–44.
Ohwaki, K., Hungate, R. E., Lotter, L., Hofmann, R. R. & Maloiy, G. M. O. (1974). Stomach fermentation in East African colobus monkeys in their natural state. *Applied Microbiology*, **27**, 713–23.
Olson, D. K. (1980). Male interactions and troop split among black-and-white colobus monkeys (*Colobus polykomos vellerosus*). Paper presented at the *Eighth Congress of the International Primatological Society*, Florence, Italy.
Olson, D. K. (1986). Determining range size for arboreal monkeys: methods, assumptions, and accuracy. In *Current Perspectives in Primate Social Dynamics*, ed. D. M. Taub & F. A. King, pp. 212–27. New York: Van Nostrand Reinhold.
Olson, D. K. & Curtin, S. (1984). The role of economic timber species in the ecology of black-and-white colobus and diana monkeys in Bia National Park, Ghana. Paper presented at the *Tenth Congress of the International Primatological Society*, Nairobi, Kenya.
Oppenheimer, J. R. (1977). *Presbytis entellus*, the Hanuman langur. In *Primate Conservation*, ed. H. S. H. Prince Rainier & G. H. Bourne, pp. 469–512. New York: Academic Press
Owen-Smith, N. (1989). Megafaunal extinctions: the conservation message from 11,000 years BP. *Conservation Biology*, **3**, 405–12.
Owen-Smith, N. & Novellie, P. (1982). What should a clever ungulate eat? *American Naturalist*, **119**, 151–78.
Oxnard, C. E. (1966). Vitamin B12 nutrition in some primates in captivity. *Folia Primatologica*, **4**, 424–31.
Pan, Y. & Jablonski, N. (1987). The age and geographical distribution of fossil cercopithecids in China. *Human Evolution*, **2**, 59–69.
Parra, R. (1978). Comparison of foregut and hindgut fermentation in herbivores. In *The Ecology of Arboreal Folivores*, ed. G. G. Montgomery, pp. 205–29. Washington, DC: Smithsonian Institution Press.
Parthasarathy, M. D. & Rahman, H. (1974). Infant killing and dominance assertion among the hanuman langur. Paper presented at the *Fifth Congress of the International Primatological Society*, Nagoya, Japan.
Payne, J. B. (1980). Competitors. In *Malayan Forest Primates*, ed. D. J. Chivers, pp. 261–77. New York: Plenum Press.
Payne, J., Francis, C. M. & Phillipps, K. (1985). *A Field Guide to the Mammals of Borneo*. Kuala Lumpur: The Sabah Society and the World Wildlife Fund.
Pickford, M. (1983). Sequence and environments of the Lower and Middle

Miocene hominoids of western Kenya. In *New Interpretations of Ape and Human Ancestry*, ed. R. L. Ciochon & R. S. Corruccini, pp. 421–39. New York: Plenum.

Pickford, M. (1987). The chronology of the Cercopithecoidea of East Africa. *Human Evolution*, **2**, 1–17.

Pocock, R. I. (1928). The langurs, or leaf monkeys, of British India. *Journal of the Bombay Natural History Society*, **32**, 472–504.

Pocock, R. I. (1935). The monkeys of the genera *Pithecus* (or *Presbytis*) and *Pygathrix* found to the East of the Bay of Bengal. *Proceedings of the Zoological Society of London (1934)*, pp. 895–961.

Pocock, R. I. (1936). The external characters of a female red colobus monkey (*Procolobus badius waldroni*). *Proceedings of the Zoological Society of London (1935)*, pp. 939–44.

Pocock, R. I. (1939). *The Fauna of British India, Including Ceylon and Burma: Mammals. 1. Primates and Carnivores (in part), Families Felidae and Viverridae*, 2nd ed. London: Taylor & Francis.

Poirier, F. E. (1968). The Nilgiri langur (*Presbytis johnii*) mother–infant dyad. *Primates*, **9**, 45–68.

Poirier, F. E. (1969*a*). The Nilgiri Langur (*Presbytis johnii*) troop: its composition, structure, function and change. *Folia Primatologica*, **10**, 20–47.

Poirier, F. E. (1969*b*). Nilgiri langur (*Presbytis johnii*) territorial behavior. *Proceedings of the Second International Congress of Primatology*, **1**, 31–5.

Poirier, F. E. (1970*a*). Dominance structure of the Nilgiri langur (*Presbytis johnii*) of South India. *Folia Primatologica*, **12**, 161–86.

Poirier, F. E. (1970*b*). The Nilgiri langur (*Presbytis johnii*) of South India. In *Primate Behaviour: Developments in Field and Laboratory Research*, ed. L. A. Rosenblum, pp. 251–383. New York: Academic Press.

Pollock, J. I. (1977). The ecology and sociology of feeding in *Indri indri*. In *Primate Ecology: Studies of Feeding and Ranging Behaviour in Lemurs, Monkeys and Apes*, ed. T. H. Clutton-Brock, pp. 37–69. London: Academic Press.

Pope, G. G. (1988). Current issues in Far Eastern paleoanthropology. In *The Palaeoenvironment of East Asia from the mid-Tertiary*, vol. II, ed. P. Whyte, J. S. Aigner, N. G. Jablonski, G. Taylor, D. Walker, P. Wang & C.-L. So, pp. 1097–123. Hong Kong: University of Hong Kong Centre of Asian Studies.

Prentice, M. L. & Denton, G. H. (1989). The deep-sea oxygen isotope record, the global ice-sheet system and hominid evolution. In *Evolutionary History of the 'Robust' Australopithecines*, ed. F. E. Grine, pp. 383–403. New York: Aldine de Gruyter.

Preston, C. M. & Sayer, B. G. (1992). What's in a nutshell: an investigation of structure by carbon-13 cross-polarization magic-angle spinning nuclear magnetic resonance spectroscopy. *Journal of Agricultural and Food Chemistry*, **40**, 206–10.

Pusey, A. E. & Packer, C. (1987). Dispersion and philopatry. in *Primate Societies*, ed. B. B. Smuts, D. L. Cheney, R. M. Seyfarth, R. W. Wrangham & T. T. Struhsaker, pp. 250–66. Chicago: University of Chicago Press.

Quade, J., Cerling, T. E. & Bowman, J. R. (1989). Development of Asian monsoon revealed by marked ecological shift during the latest Miocene in northern Pakistan. *Nature*, **342**, 163–6.

Quiatt, D. (1979). Aunts and mothers: adaptive implications of allomaternal behavior of non-human primates. *American Anthropologist*, **81**, 310–19.

Raemaekers, J. J. (1978). Changes through the day in the food choice of wild gibbons. *Folia Primatologica*, **30**, 194–205.

Raemaekers, J. J., Aldrich-Blake, F. P. G. & Payne, J. B. (1980). The forest. In *Malayan Forest Primates*, ed. D. J. Chivers, pp. 29–61. New York: Plenum Press.

Rahm, U. H. (1970). Ecology, zoogeography and systematics of some African forest monkeys. In *Old World Monkeys: Evolution, Systematics and Behavior*, ed. J. R. & P. M. Napier, pp. 589–626. New York: Academic Press.

Rahm, U. H. & Christiaensen, A. R. (1960). Notes sur *Colobus polycomos cordieri* (Rahm) du Congo Belge. *Revue de Zoologie et de Botanique Africaines*, **61**, 215–20.

Rajanathan, R. & Bennett, E. L. (1990). Notes on the social behaviour of wild proboscis monkeys (*Nasalis larvatus*). *Malayan Nature Journal*, **44**, 35–44.

Rajpurohit, L. S. (1991). Resident male replacement, formation of a new male band and paternal behaviour in *Presbytis entellus*. *Folia Primatologica*, **57**, 159–64.

Rajpurohit, L. S., Mohnot, S. M., Agoramoorthy, G. & Srivastava, A. (1986). Observations on ousted alpha males of bisexual groups of hanuman langurs (*Presbytis entellus*). *Primate Report*, **14**, 209.

Rajpurohit, L. S. & Sommer, V. (1991). Sex differences in mortality among langurs (*Presbytis entellus*) of Jodhpur, Rajasthan. *Folia Primatologica*, **56**, 17–27.

Ratajszczak, R. (1988). Notes on the current status and conservation of primates in Vietnam. *Primate Conservation*, **9**, 134–6.

Reena, M. & Ram, M. B. (1992). Rate of takeover in groups of hanuman langurs (*Presbytis entellus*) at Jaipur. *Folia Primatologica*, **58**, 61–71.

Reichenbach, H. G. L. (1862). Die vollständigste Naturgeschichte der Affen. In *Die vollständigste Naturschichte In- und Auslandes*. Dresden: Central-Atlas für Zoologischer Garten.

Retallack, G. J. (1992). Middle Miocene fossil plants from Fort Ternan (Kenya) and evolution of African grasslands. *Paleobiology*, **18**, 383–400.

Rhoades, D. F. & Cates, R. G. (1976). Toward a general theory of plant antiherbivore chemistry. *Recent Advances in Phytochemistry*, **10**, 168–213.

Rice, E. L. & Pancholy, S. A. (1974). Inhibition of nitrification by climax vegetation. II. Additional evidence and possible role of tannins. *American Journal of Botany*, **60**, 691–702.

Richards, P. W. (1952). *The Tropical Rain Forest*. Cambridge: Cambridge University Press.

Ridley, M. (1986). The number of males in a primate troop. *Animal Behaviour*, **34**, 1848–58.

Rijksen, H. D. (1978). *A Field Study on Sumatran Orang-utans* (Pongo pygmaeus abelii *Lesson 1827): Ecology, Behaviour and Conservation*. Wageningen: H. Veenman & Zonen BV.

Ripley, S. (1967). Intertroop encounters among Ceylon gray langurs (*Presbytis entellus*). In *Social Communication among Primates*, ed. S. A. Altmann, pp. 237–53. Chicago: University of Chicago Press.

Ripley, S. (1970). Leaves and leaf-monkeys: the social organization of foraging in gray langurs. In *Old World Monkeys: Evolution, Systematics and Behavior*, ed. J. R. Napier & P. H. Napier, pp. 481–509. New York: Academic Press.

Ripley, S. (1984). Environmental grain, niche diversification and feeding behaviour in primates. In *Food Acquisition and Processing in Primates*, ed. D. J. Chivers, B. A. Wood & A. Bilsborough, pp. 33–72. New York: Plenum Press.

Robbins, C. T., Mole, S., Hagerman, A. E. & Hanley, T. A. (1987). Role of tan-

nins in defending plants against ruminants: reduction in dry matter digestion? *Ecology*, **68**, 1606–15.
Rodgers, W. A. (1981). The distribution and conservation status of colobus monkeys in Tanzania. *Primates*, **22**, 33–45.
Rodman, P. S. (1978). Diets, densities and distribution of Bornean primates. In *The Ecology of Arboreal Folivores*, ed. G. G. Montgomery, pp. 465–78. Washington, DC: Smithsonian Institution Press.
Rodriguez, E., Aregullin, M., Nishida, T., Uehara, S., Wrangham, R., Abramowski, Z., Finlayson, A. & Towers, G. H. N. (1985). Thiarubin A, a bioactive constituent of *Aspilia* (Asteraceae) consumed by wild chimpanzees. *Experientia*, **41**, 419–420.
Rogers, M. E., Maisels, F., Williamson, E. A., Fernandez, M. & Tutin, C. E. G. (1990). Gorilla diet in the Lopé Reserve, Gabon. *Oecologia (Berlin)*, **84**, 326–39.
Roonwal, M. L. (1981). Intraspecific variation in size, proportion of body parts and weight in the Hanuman langur, *Presbytis entellus* (Primates), in South Asia, with remarks on subspeciation. *Records of the Zoological Survey of India*, **79**, 125–158.
Roonwal, M. L. & Mohnot, S. M. (1977). *Primates of South Asia: Ecology, Sociobiology and Behavior*. Cambridge, MA: Harvard University Press.
Rose, M. D. (1977). Interspecific play between free ranging guerezas (*Colobus guereza*) and vervet monkeys (*Cercopithecus aethiops*). *Primates*, **18**, 957–64.
Rose, M. D. (1978). Feeding and associated positional behavior of black and white colobus monkeys (*Colobus guereza*). In *The Ecology of Arboreal Folivores*, ed. G. G. Montgomery, pp. 253–62. Washington, DC: Smithsonian Institution Press.
Rosenthal, G. A. & Berenbaum, M. R. (eds.) (1991). *Herbivores, their Interaction with Secondary Plant Metabolites*, 2nd edn, vol. 1. New York: Academic Press.
Rosenthal, G. A. & Janzen, D. H. (1979). *Herbivores: their Interaction with Secondary Plant Metabolites*. New York: Academic Press.
Rosenthal, G. A. & Janzen, D. H. (1985). Ammonia utilization by the bruchid beetle *Caryedes brasiliensis* (Bruchidae). *Journal of Chemical Ecology*, **11**, 539–44.
Ross, C. (1993). Takeover and infanticide in south Indian Hanuman langurs (*Presbytis entellus*). *American Journal of Primatology*, **30**, 75–82.
Rucks, M. (1978). Monkey miscellany means safety in numbers. *Wildlife*, **20**, 268-70.
Rudran, R. (1973*a*). Adult male replacement in one-male troop of purple-faced langurs (*Presbytis senex senex*) and its effect on population structure. *Folia Primatologia*, **19**, 166–92.
Rudran, R. (1973*b*). The reproductive cycles of two subspecies of purple-faced langurs (*Presbytis senex*) with relation to environmental factors. *Folia Primatologia*, **19**, 41–60.
Ruhiyat, Y. (1983). Socio-ecological study of *Presbytis aygula* in West Java. *Primates*, **24**, 344–59.
Sabater-Pi, J. (1973). Contribution to the ecology of *Colobus polykomos satanas* (Waterhouse, 1838) of Rio Muni (Republic of Equatorial Guinea). *Folia Primatologica*, **19**, 193–207.
Sabater-Pi, J. & Jones, C. (1967). Notes on the distribution and ecology of the

higher primates of Rio Muni, West Africa. *Tulane Studies in Zoology*, **14**, 101–9.
Salter, R. E., MacKenzie, N. A., Nightingale, N., Aken, K. M. & Chai, P. P. K. (1985). Habitat use, ranging behaviour, and food habitats of the proboscis monkey, *Nasalis larvatus* (van Wurmb) in Sarawak. *Primates*, **26**(4), 436–51.
Sayer, J. A. & Green, A. A. (1984). The distribution and status of large mammals in Benin. *Mammal Review*, **14**, 37–50.
Schaller, G. B. (1985). China's golden treasure. *International Wildlife*, **5**, 29–31.
Schaller, G. B., Hu, J. C., Pan, W. S. & Zhu, J. (1985). *The Giant Pandas of Wolong*. Chicago: University of Chicago Press.
Schenkel, R. & Schenkel-Hulliger, L. (1967). On the sociology of free-ranging colobus (*Colobus guereza caudatus* Thomas 1885). In *Neue Ergebnisse der Primatologie*, ed. D. Starck, R. Schneider & H.-J. Kuhn, pp. 185–94. Stuttgart: Gustav Fischer.
Schubert, G. (1982). Infanticide by usurper hanuman langur males: a sociobiological myth. *Social Science Information*, **21**, 199–244.
Schultz, J. C., Baldwin, I. T. & Nothnagle, P. J. (1981). Hemoglobin as a binding substrate in the quantitative analysis of plant tannins. *Journal of Agriculture and Food Chemistry*, **29**, 823–6.
Schwarz, E. (1929). On the local races and distribution of the black and white colobus monkeys. *Proceedings of the Zoological Society of London (1929)*, pp. 585–98.
Scollay, P. A. & DeBold, P. (1980). Allomothering in a captive colony of hanuman langurs (*Presbytis entellus*). *Ethology and Sociobiology*, **1**, 291–9.
Shipman, P. & Harris, J. M. (1989). Habitat preference and paleoecology of *Australopithecus boisei* in eastern Africa. In *Evolutionary History of the 'Robust' Australopithecines*, ed. F. E. Grine, pp. 343–81. New York: Aldine de Gruyter.
Silkiluwasha, F. (1981). The distribution and conservation status of the Zanzibar red colobus. *African Journal of Ecology*, **19**, 187–94.
Sim, B. J., Lucas, P. W., Pereira, B. P. & Oates, C. G. (1993). Mechanical and sensory assessment of the texture of refrigerator-stored spring roll pastry. *Journal of Texture Studies*, **24**, 27–44.
Simberloff, D. & Connor, E. F. (1981). Missing species combinations. *American Naturalist*, **118**, 215–39.
Simons, E. L. (1970). The deployment and history of Old World monkeys (Cercopithecidae, Primates). In *Old World Monkeys*, ed. J. R. Napier & P. H. Napier, pp. 97–137. New York: Academic Press.
Sirianni, J. E. (1979). Craniofacial morphology of the underbite trait in *Presbytis*. *Journal of Dental Research*, **58**, 1655.
Skelton, R. R. (1990). Beneath the surface: the promise and problems of the Laetoli site. *American Journal of Primatology*, **20**, 57–62.
Skorupa, J. P. (1986). Responses of rainforest primates to selective logging in Kibale Forest, Uganda: a summary report. In *Primates: the Road to Self-sustaining Populations*, ed. K. Benirschke, pp. 57–70. New York: Springer-Verlag.
Skorupa, J. P. (1988). The effects of selective timber harvesting on rain-forest primates in Kibale Forest, Uganda. PhD thesis, University of Michigan, Ann Arbor, MI.
Skorupa, J. P. (1989). Crowned eagles *Stephanoaetus coronatus* in rainforest: observations on breeding chronology and diet at a nest in Uganda. *Ibis*, **131**, 294–8.
Smith, H. W. (1965). Observations on the flora of the alimentary tract of animals

and factors affecting its composition. *Journal of Pathology and Bacteriology*, **89**, 95–122.

Smuts, B. B. (1983). Special relationships between adult male and female olive baboons: selective advantages. In *Primate Social Relationships: an Integrated Approach*, ed. R. A. Hinde, pp. 262–6. Oxford: Blackwell Scientific Publications.

Solounias, N. & Dawson-Saunders, B. (1988). Dietary adaptations and paleoecology of the Late Miocene ruminants from Pikermi and Samos in Greece. *Palaeogeography, Palaeoclimatology, Palaeoecology*, **65**, 149–72.

Solounias, N. & Moelleken, S. M. C. (1994). Dietary adaptation, cranial restoration, and evolutionary considerations of a Miocene gazelle. *Journal of Evolutionary Biology*, in press.

Sommer, V. (1987). Infanticide among free-ranging langurs (*Presbytis entellus*) at Jodhpur (Rajasthan/India): recent observations and a reconsideration of hypotheses. *Primates*, **28**, 163–97.

Sommer, V. (1988). Male competition and coalitions in langurs (*Presbytis entellus*) at Jodhpur, Rajasthan, India. *Human Evolution*, **3**, 261–78.

Sommer, V. (1989*a*). Sexual harassment in langur monkeys (*Presbytis entellus*), competition for ova, sperm or nurture? *Ethology*, **80**, 205–17.

Sommer, V. (1989*b*). Infant mistreatment in langur monkeys – sociobiology tackled from the wrong end? In *The Sociobiology of Sexual and Reproductive Strategies*, ed. A. E. Rasa, C. Vogel & E. Voland, pp. 110–27. London: Chapman & Hall.

Sommer, V. & Mohnot, S. M. (1985). New observations of infanticide among hanuman langurs (*Presbytis entellus*) near Jodhpur (Rajasthan/India). *Behavioral Ecology and Sociobiology*, **16**, 245–8.

Sommer, V. & Rajpurohit, L. S. (1989). Male reproductive success in harem troops of Hanuman langurs (*Presbytis entellus*). *International Journal of Primatology*, **10**, 293–317.

Sommer, V., Srivastava, A. & Borries, C. (1992). Cycles, sexuality and conception in free-ranging langurs (*Presbytis entellus*). *American Journal of Primatology*, **28**, 1–27.

Southwood, T. R. E., May, R. M., Hassel, M. P. & Conway, G. K. (1974). Ecological strategies and population parameters. *American Naturalist*, **118**, 215–39.

Spalinger, D. E., Robbins, C. T. & Hanley, T. A. (1986). The assessment of handling time in ruminants: the effect of plant chemical and physical structure on the rate of breakdown of plant particles in the rumen of the mule deer and elk. *Canadian Journal of Zoology*, **64**, 312–31.

Srivastava, A., Borries, C. & Sommer, V. (1991). Homosexual mounting in free-ranging female Hanuman langurs (*Presbytis entellus*). *Archives of Sexual Behaviour*, **20**, 487–512.

Stanford, C. B. (1989). Predation on capped langurs (*Presbytis pileata*) by cooperatively hunting jackals (*Canis aureus*). *American Journal of Primatology*, **19**, 53–6.

Stanford, C. B. (1990). The capped langur in Bangladesh: behavioral ecology and reproductive tactics. *Contributions to Primatology*, **26**, 1–179.

Stanford, C. B. (1991). The diet of the capped langur (*Presbytis pileata*) in a moist deciduous forest in Bangladesh. *International Journal of Primatology*, **12**, 199–216.

Starin, E. D. (1981). Monkey moves. *Natural History*, **90**, 36–43.

Starin, E. D. (1991). Socioecology of the red colobus monkey in the Gambia with

particular reference to female–male differences and transfer patterns. PhD thesis, City University of New York, New York, NY.
Stevens, C. E., Argenzio, R. A. & Clemens, E. T. (1980). Microbial digestion: rumen versus large intestine. In *Digestive Physiology and Metabolism in Ruminants*, ed. Y. Ruckebusch & P. Thivend, pp. 685–706. Lancaster: MTP Press.
Strait, S. G. (1991). Dietary reconstruction in small-bodied fossil primates. PhD thesis, State University of New York, Stony Brook, NY.
Strasser, E. (1988). Pedal evidence for the origin and diversification of cercopithecid clades. *Journal of Human Evolution*, **17**, 225–45.
Strasser, E. (1989). Form, function, and allometry of the cercopithecid foot. PhD thesis, City University of New York, New York, NY.
Strasser, E. (1992). Hindlimb proportions, allometry, and biomechanics in Old World Monkeys (Primates, Cercopithecidae). *American Journal of Physical Anthropology*, **87**, 187–213.
Strasser, E. (1994). The relative size of the hallux and pedal digit formulas in Cercopithecidae. *Journal of Human Evolution*, in press.
Strasser, E. & Delson, E. (1987). Cladistic analysis of cercopithecid relationships. *Journal of Human Evolution*, **16**, 81–99.
Struhsaker, T. T. (1974). Correlates of ranging behavior in a group of red colobus monkeys (*Colobus badius tephrosceles*). *American Zoologist*, **14**, 177–84.
Struhsaker, T. T. (1975). *The Red Colobus Monkey*. Chicago: University of Chicago Press.
Struhsaker, T. T. (1978). Interrelations of red colobus monkeys and rain-forest trees in the Kibale Forest, Uganda. In *The Ecology of Arboreal Folivores*, ed. G. G. Montgomery, pp. 397–422. Washington, DC: Smithsonian Institution Press.
Struhsaker, T. T. (1981*a*). Vocalizations, phylogeny and palaeogeography of red colobus monkeys (*Colobus badius*). *African Journal of Ecology*, **19**, 265–83.
Struhsaker, T. T. (1981*b*). Polyspecific associations among tropical rain-forest primates. *Zeitshrift für Tierpsychologie*, **57**, 268–304.
Struhsaker, T. T. & Leakey, M. (1990). Prey selectivity by crowned hawk-eagles on monkeys in the Kibale Forest, Uganda. *Behavioral Ecology and Sociobiology*, **26**, 435–43.
Struhsaker, T. T. & Leland, L. (1979). Socioecology of five sympatric monkey species in the Kibale Forest, Uganda. In *Advances in the Study of Behavior*, vol. 9, ed. J. Rosenblatt, R. A. Hinde, C. Beer & M. C. Busnel, pp. 158–228. New York: Academic Press.
Struhsaker, T. T. & Leland, L. (1985). Infanticide in a patrilineal society of red colobus monkeys. *Zeitschrift für Tierpsychologie*, **69**, 89–132.
Struhsaker, T. T. & Leland, L. (1987). Colobines: infanticide by adult males. In *Primate Societies*, ed. B. B. Smuts, D. L. Cheney, R. M. Seyfarth, R. W. Wrangham & T. T. Struhsaker, pp. 83–97. Chicago: University of Chicago Press.
Struhsaker, T. T. & Oates, J. F. (1975). Comparison of the behavior and ecology of red colobus and black-and-white colobus monkeys in Uganda: a summary. In *Socioecology and Psychology of Primates*, ed. R. H. Tuttle, pp. 103–23. The Hague: Mouton.
Sugiyama, Y. (1964). Group composition, population density and some sociological observations of Hanuman langurs (*Presbytis entellus*). *Primates*, **5**, 7–37.
Sugiyama, Y. (1965). On the social change of hanuman langurs (*Presbytis entellus*) in their natural conditions. *Primates*, **6**, 381–417.

Sugiyama, Y. (1966). An artificial social change in a hanuman langur troop (*Presbytis entellus*). *Primates*, **7**, 41–72.

Sugiyama, Y. (1967). Social organization of hanuman langurs. In *Social Communication Among Primates*, ed. S. Altmann, pp. 221–36. Chicago: University of Chicago Press.

Sugiyama, Y. (1976). Characteristics of the ecology of the Himalayan langurs. *Journal of Human Evolution*, **5**, 249–77.

Supriatna, J., Manullang, B. O. & Soekara, E. (1986). Group composition, home range and diet of the maroon leaf monkey (*Presbytis rubicunda*) at Tanjung Puting Reserve, Central Kalimantan, Indonesia. *Primates*, **27**, 185–90.

Suzuki, A. (1979). The variation and adaptation of social groups of chimpanzees and black and white colobus monkeys. In *Primate Ecology and Human Origins*, ed. I. S. Bernstein & E. O. Smith, pp. 153–73. New York: Garland STPM Press.

Swain, T. (1978). Plant–animal co-evolution; a synoptic view of the Paleozoic and Mesozoic. In *Biochemical Aspects of Plant and Animal Co-evolution*, ed. J. B. Harborne, pp. 3–19. London: Academic Press.

Swain, T. (1979). Tannins and lignins. In *Herbivores: their Interactions with Secondary Plant Metabolites*, ed. G. A. Rosenthal & D. H. Janzen, pp. 637–82. New York: Academic Press.

Swindler, D. R. (1976). *Dentition of Living Primates*. New York: Academic Press.

Swindler, D. R. (1979). The incidence of underbite occlusion in leaf-eating monkeys. *Ossa*, **6**, 261–72.

Swindler, D. R., McCoy, H. A. & Hornbeck, P. V. (1967). The dentition of the baboon (*Papio anubis*). In *The Baboon in Medical Research*, vol. II, ed. H. Vagtborg, pp. 133–50. Austin: University of Texas Press.

Swindler, D. R. & Orlosky, F. J. (1974). Metric and morphological variability in the dentition of colobine monkeys. *Journal of Human Evolution*, **3**, 135–60.

Szalay, F. S. & Delson, E. (1979). *Evolutionary History of the Primates*. New York: Academic Press.

Takahata, Y., Hasegawa, T. & Nishida, T. (1984). Chimpanzee predation in the Mahale Mountains from August 1979 to May 1982. *International Journal of Primatology*, **5**, 213–33.

Tan, B. J. (1985). The status of primates in China. *Primate Conservation*, **5**, 63–81.

Tan, B. & Poirier, F. E. (1988). Status report on some Chinese primates. *Primate Conservation*, **9**, 129–31.

Tanaka, J. (1965). Social structure of Nilgiri langurs. *Primates*, **6**, 107–22.

Teaford, M. F. (1983*a*). The morphology and wear of the lingual notch in macaques and langurs. *American Journal of Physical Anthropology*, **60**, 7–14.

Teaford, M. F. (1983*b*). Functional morphology of the underbite in two species of langurs. *Journal of Dental Research*, **62**, 183.

Teaford, M. F. (1988). A review of dental microwear and diet in modern mammals. *Scanning Microscopy*, **2**, 1149–66.

Teaford, M. F. (1993). Dental microwear and diet in extant and extinct *Theropithecus*: preliminary analyses. In *Theropithecus: the Life and Death of a Primate Genus*, ed. N. G. Jablonski, pp. 331–49. Cambridge: Cambridge University Press.

Teaford, M. F. & Glander, K. E. (1991). Dental microwear in live, wild-trapped *Alouatta* from Costa Rica. *American Journal of Physical Anthropology*, **85**, 313–19.

Teaford, M. F. & Leakey, M. G. (1992). Dental microwear and diet in Plio-

Alouatta from Costa Rica. *American Journal of Physical Anthropology*, **85**, 313–19.

Teaford, M. F. & Leakey, M. G. (1992). Dental microwear and diet in Plio-Pleistocene cercopithecoids from Kenya. *American Journal of Physical Anthropology*, **Suppl. 14**, 160–11.

Teaford, M. F. & Runestad, J. A. (1992). Dental microwear and diet in Venezuelan primates. *American Journal of Physical Anthropology*, **88**, 347–64.

Temerin, L. A. & Cant, J. G. H. (1983). The evolutionary divergence of Old World monkeys and apes. *American Naturalist*, **122**, 335–51.

Tenaza, R. R. (1989). Female sexual swellings in the Asian colobine *Simias concolor*. *American Journal of Primatology*, **17**, 81–6.

Terborgh, J. (1983). *Five New World Primates*. Princeton: Princeton University Press.

Terborgh, J. (1986). The social systems of New World primates: an adaptionist's view. In *Primate Ecology and Conservation*, ed. J. Else & P. Lee, pp. 199–211. Cambridge: Cambridge University Press.

Terborgh, J. & Janson, C. H. (1986). The socioecology of primate groups. *Annual Review of Ecology and Systematics*, **17**, 111–35.

Thapar, V. (1986). Tiger – Portrait of a predator. *Sanctuary/Asia*, **6** (4), 344–53, 370–83.

Thomas, H. (1985). The Early and Middle Miocene land connection of the Afro-Arabian plate and Asia: a major event for hominoid dispersal? In *Ancestors: The Hard Evidence*, ed. E. Delson, pp. 42–50. New York: Alan R. Liss.

Thomas, S. C. (1991). Population densities and patterns of habitat use among anthropoid primates of the Ituri Forest, Zaire. *Biotropica*, **23**, 68–83.

Thorington, R. W. & Groves, C. P. (1970). An annotated classification of the Cercopithecoidea. In *Old World Monkeys: Ecology, Systematics, and Behavior*, ed. J. R. Napier & P. N. Napier, pp. 629–47. New York: Academic Press.

Tiercelin, J. J. (1986). The Pliocene Hadar Formation, Afar depression of Ethiopia. In *Sedimentation in the African Rifts*, ed. L. E. Frostick, Geological Society of London Special Publication 25; pp. 221–40. Oxford: Blackwell Scientific.

Tilson, R. L. (1976). Infant coloration and taxonomic affinity of the Mentawai Islands leaf monkey, *Presbytis potenziani*. *Journal of Mammalogy*, **57**, 766–9.

Tilson, R. (1977). Social organization of Simakobu monkeys (*Nasalis concolor*) in Siberut Island, Indonesia. *Journal of Mammalogy*, **58**, 202–12.

Tilson, R. & Tenaza, R. (1976). Monogamy and duetting in an Old World monkey. *Nature*, **262**, 320–1.

Tsingalia, H. M. & Rowell, T. E. (1984). The behaviour of adult male blue monkeys. *Zeitschrift für Tierpsychologie*, **64**, 253–68.

Ullrich, W. (1961). Zur Biologie und Soziologie der Colobusaffen (*Colobus guereza caudatus* Thomas 1885). *Der Zoologischer Garten*, **25**, 305–68.

Ungar, P. S. (1992). Incisor microwear and feeding behavior of four Sumatran anthropoids. PhD thesis, State University of New York, Stony Brook, NY.

Van Couvering, J. A. & Kukla, G. (1988*a*). Glaciation. In *Encyclopedia of Human Evolution and Prehistory*, ed. I. Tattersall, E. Delson, & J. A. Van Couvering, pp. 226–33. New York: Garland Publishing.

Van Couvering, J. A. & Kukla, G. (1988*b*). Pleistocene. In *Encyclopedia of Human Evolution and Prehistory*, ed. I. Tattersall, E. Delson, & J. A. Van Couvering, pp. 459–64. New York: Garland Publishing.

Van Couvering, J. A. & Kukla, G. (1988*c*). Sea-level change. In *Encyclopedia of Human Evolution and Prehistory*, ed. I. Tattersall, E. Delson, & J. A. Van Couvering, pp. 505–10. New York: Garland Publishing.

Van Schaik, C. P. (1983). Why are diurnal primates living in groups? *Behaviour*, **87**, 120–44.

Van Schaik, C. P. & Van Hooff, J.A.R.A.M. (1983). On the ultimate causes of primate social systems. *Behaviour*, **85**, 91–117.

Van Soest, P. J. (1963). Use of detergents in the analysis of fibrous seeds. II. A rapid method for the determination of fibre and lignin. *Journal of the Association of Official Analytical Chemists*, **46**, 829–35.

Van Soest, P. J. (1977). Plant fiber and its role in herbivore nutrition. *The Cornell Veterinarian*, **67**, 307–26.

Van Soest, P. J. (1982). *Nutritional Ecology of the Ruminant.* Corvallis, OR: O & B Books.

Verheyen, W. N. (1962). Contribution à la craniologie comparée des Primates: les genres *Colobus* Illiger 1811 et *Cercopithecus* Linné 1758. Annales de la Musée Royal de L'Afrique Centrale, *Sciences Zoologiques*, **105**, 1–255.

Vincent, J. F. V. (1990). The fracture properties of plants. *Advances in Botanical Research*, **17**, 235–87.

Vincent, J. F. V. (1991). Texture of plants and fruits. In *Feeding and the Texture of Food*, ed. J. F. V. Vincent & P. J. Lillford, pp. 19–33. Cambridge: Cambridge University Press.

Vincent, J. F. V., Jeronimidis, G., Khan, A. A. & Luyten, H. (1991) The wedge fracture test a new method for measurement of food texture. *Journal of Texture Studies*, **22**, 45–57.

Vogel, C. (1966). Morphologische Studien am besichtsschodel Catarrhiner Primaten. *Biblioteca Primatologia*, **4**, 1–226.

Vogel, C. (1971). Behavioral differences of *Presbytis entellus* in two different habitats. In *Proceedings of the Third International Congress of Primatology*, Zurich 1970, ed. H. Kummer. Basel: S. Karger.

Vogel, C. (1973). Acoustical communication among free-ranging common Indian langurs (*Presbytis entellus*) in two different habitats of north India. *American Journal of Physical Anthropology*, **38**, 469–80.

Vogel, C. (1984). Patterns of infant-transfer within 2 troops of common langurs (*Presbytis entellus*) near Jodhpur: testing hypotheses concerning the benefits and risks. In *Current Primate Researches*, ed. M. L. Roonwal, S. M. Mohnot & N. Rathore, pp. 361–79. Jodhpur: Department of Zoology, University of Jodhpur.

Vogel, C. & Loch, H. (1984). Reproductive parameters, adult male replacements and infanticide among free-ranging langurs (*Presbytis entellus*) at Jodhpur (Rajasthan), India. In: *Infanticide: Comparative and Evolutionary Perspectives*, ed. G. Hausfater & S. B. Hrdy, pp. 237–55. Hawthorne, NY: Aldine.

Von Orgetta, M. (1979). Neue Ergebnisse einer playnologischen Untersuchung der Lignite von Pikermi/Attica. *Annales Géologique des Pays Hellenique, Mémoires*, **1**, 909–21.

Vrba, E. S. (1980). The significance of bovid remains as indicators of environment and predation patterns. In *Fossils in the Making*, ed. A. K. Behrensmeyer & A. P. Hill, pp. 247–71. Chicago: University of Chicago Press.

Vrba, E. S. (1985). Palaeoecology of early Hominidae, with special reference to Sterkfontein, Swartkrans and Kromdraai. In *L'Environnement des Hominidés au Plio-Pléistocène*, ed. Y. Coppens, pp. 345–69. Paris: Masson.

Vrba, E. S. (1989). Late Pliocene climatic events and hominid evolution. In *Evolutionary History of the 'Robust' Australopithecines*, ed. F. E. Grine, pp. 405–26. New York: Aldine de Gruyter.

Walker, A. (1984). Mechanisms of honing in the male baboon canine. *American Journal of Physical Anthropology*, **65**, 47–60.

Walker, P. & Murray, P. (1975). An assessment of masticatory efficiency in a series of anthropoid primates with special reference to the Colobinae and Cercopithecinae. In *Primate Functional Morphology and Evolution*, ed. R. H. Tuttle, pp. 151–94. The Hague: Mouton.

Waser, P. (1975). Monthly variations in feeding and activity patterns of the mangabey, *Cercocebus albigena* (Lydekker). *East African Wildlife Journal*, **13**, 249–63.

Waser, P. (1977). Individual recognition, intragroup cohesion and intergroup spacing: evidence from sound playback to forest monkeys. *Behaviour*, **60**, 28–74.

Waser, P. & Waser, M. S. (1977). Experimental studies of primate vocalizations: specializations for long-distance propagation. *Zeitschrift für Tierpsychologie*, **43**, 239–63.

Washburn, S. L. (1942). Skeletal proportions of adult langurs and macaques. *Human Biology*, **14**, 444–72.

Wasser, S. K. & Barash, P. P. (1981). The selfish 'allomother': a comment on Scollay & DeBold (1980). *Ethology and Sociobiology*, **2**, 91–3.

Watanabe, K. (1981). Variation in group composition and population density of the two sympatric Mentawaian leaf monkeys. *Primates*, **22**, 145–60.

Waterman, P. G. (1984). Food acquisition and processing as a function of plant chemistry. In *Food Acquisition and Processing by Primates*, ed. D. J. Chivers, B. A. Wood & A. Bilsborough, pp. 177–211. New York: Plenum Press.

Waterman, P. G. (1986). A phytochemist in the African rain-forest. *Phytochemistry*, **25**, 3–17.

Waterman, P. G. (1994). Costs and benefits of secondary metabolites to the Leguminosae. In *Proceedings of the Third International Legume Conference*, ed. D. B. McKey, in press.

Waterman, P. G. & Choo, G. M. (1981). The effects of digestibility-reducing compounds in leaves on food selection by some Colobinae. *Malaysian Applied Biology*, **10**, 147–62.

Waterman, P. G., Choo, G. M., Vedder, A. L. & Watts, D. W. (1983). Digestibility, digestion-inhibitors and nutrients of herbaceous foliage and green stems from an African montane flora and comparison with other tropical flora. *Oecologia (Berlin)*, **60**, 244–9.

Waterman, P. G., Mbi, C. N., McKey, D. B. & Gartlan, J. S. (1980). African rain forest vegetation and rumen microbes: phenolic compounds as correlates of digestibility. *Oecologia (Berlin)*, **47**, 22–33.

Waterman, P. G. & McKey, D. B. (1989). Herbivory and secondary compounds in rain-forest plants. In *Tropical Rain Forest Ecosystems*, ed. H. Lieth & M. J. A. Werger, pp. 513–36. Amsterdam: Elsevier.

Waterman, P. G. & Mole, S. (1989). Extrinsic factors influencing production of secondary metabolites in plants. In *Insect–Plant Interactions*, vol. 1., ed. E. A. Bernays, pp. 107–34. Boca Raton: CRC Press.

Waterman, P. G. & Mole, S. (1994). *Methods in Ecology: Analysis of Phenolic Plant Metabolites*. Oxford: Blackwell Scientific Publications.

Waterman, P. G., Ross, J. A. M., Bennett, E. L. & Davies, A. G. (1988). A comparison of the floristics and leaf chemistry of the tree flora in two Malaysian

rain forests and the influence of leaf chemistry on populations of colobine monkeys in the Old World. *Biological Journal of the Linnaean Society*, **34**, 1–32.
Waterman, P. G., Ross, J. A. M. & McKey, D. B. (1984). Factors affecting levels of some phenolic compounds, digestibility and nitrogen content of the mature leaves of *Barteria fistulosa* (Passifloraceae). *Journal of Chemical Ecology*, **10**, 387–401.
Waterman, P. G. & Zhong, S-M. (1982). Vallesiachotamine and isovallesiachotamine from the seeds of *Strychnos tricalysoides*. *Planta Medica*, **45**, 28–30.
Watkins, B. E., Ullrey, D. E. & Whetter, P. A. (1985). Digestibility of a high-fiber biscuit-based diet by black-and-white colobus (*Colobus guereza*). *American Journal of Primatology*, **9**, 137–44.
Weitzel, V. (1983). A preliminary analysis of the dental and cranial morphology of *Presbytis* and *Trachypithecus* in relation to diet. MA thesis, Australian National University, Canberra, Australia.
Weitzel, V. & Groves, C. P. (1985). The nomenclature and taxonomy of the colobine monkeys of Java. *International Journal of Primatology*, **6**, 399–409.
Welsch, U. (1967). Die Altersveranderungen des Primatengebisses. *Gegenbaurs Morphologisches Jahrbuch*, **110**, 1–171.
Weyreter, H., Heller, R., Dellow, D., Lechner-Doll, M. & Engelhardt, W. V. (1987). Rumen fluid volume and retention time of digesta in an indigenous and a conventional breed of sheep fed a low quality, fibrous diet. *Journal of Animal Physiology and Animal Nutrition*, **58**, 89–100.
Wheatley, B. P. (1980). Feeding and ranging of East Bornean *Macaca fascicularis*. In *The Macaques*, ed. D. G. Lindburg, pp. 215–46. New York: Van Nostrand Reinhold.
White, F. (1983). *The Vegetation of Africa*. Paris: UNESCO.
White, L. J. T. (1992). Vegetation history and logging disturbance: effects on rain forest mammals in the Lopé Reserve, Gabon. PhD thesis, University of Edinburgh, Edinburgh, Scotland.
Whitesides, G. H. (1991). Patterns of foraging, ranging, and interspecific associations of Diana monkeys (*Cercopithecus diana*) in Sierra Leone, West Africa. PhD thesis, University of Miami, Coral Gables, FL.
Whitmore, T. C. (1984*a*). A vegetation map of Malesia. *Journal of Biogeography*, **11**, 461–71.
Whitmore, T. C. (1984*b*). *Tropical Rain Forests of the Far East*, 2nd ed. Oxford: Oxford University Press.
Whitten, A. J. (1980). The Kloss gibbon in Siberut rain forest. PhD thesis, University of Cambridge, Cambridge.
Whitten, A. J. (1982). A numerical analysis of tropical rain forest, using floristic and structural data, and its application to an analysis of gibbon ranging behaviour. *Journal of Ecology*, **70**, 249–71.
Whitten, A. J. & Sadar, Z. (1981). Master plan for a tropical paradise. *New Scientist*, **91**, 230–5.
Wiens, J. A. (1977). On competition and variable environments. *American Scientist*, **65**, 590–7.
Wilkie, D. S., Sidle, J. G. & Boundzanga, G. C. (1992). Mechanized logging, market hunting, and a bank loan in Congo. *Conservation Biology*, **6**, 570–80.
Williamson, P. G. (1985). Evidence for an early Pliocene rainforest expansion in east Africa. *Nature*, **315**, 467–89.
Wilson, C. C. & Wilson, W. L. (1977). Behavioral and morphological variations

among primate population in Sumatra. *Yearbook of Physical Anthropology*, **20**, 207–33.

Wilson, W. L. & Wilson, C. C. (1975). Species-specific vocalizations and the determination of phylogenetic affinities of the *Presbytis aygula-melalophos* group in Sumatra. In *Contemporary Primatology*, ed. S. Kondo, M. Kawai & A. Ehara, pp. 459–63. Basel: S. Karger.

Wing, L. D. & Buss, I. O. (1970). Elephants and forests. *Wildlife Monographs*, **19**, 1–92.

Winkler, P. (1988). Troop history, female reproductive strategies and timing of male change in hanuman langurs, (*Presbytis entellus*). *Human Evolution*, **3**, 227–37.

Winkler, P., Loch, H. & Vogel, C. (1984). Life history of hanuman langurs (*Presbytis entellus*). *Folia Primatologica*, **43**, 1–23.

Wittenberger, J. F. (1980). Group size and polygamy in social mammals. *American Naturalist*, **115**, 197–222.

Wolf, K. (1978). Preliminary report on the completion of the field phase of a study of the social behaviour of the silvered leaf-monkeys (*Presbytis cristata*) at Kuala Selangor, Peninsular Malaysia. Report, Kuala Lumpur: Department of Wildlife and National Parks.

Wolf, K. (1980). Social change and male reproductive strategy in silvered leaf-monkeys (*Presbytis cristata*) in Kuala Selangor, Peninsular Malaysia. *American Journal of Physical Anthropology*, **52**, 294.

Wolf, K. & Fleagle, J. G. (1977). Adult male replacement in a group of silvered leaf-monkeys (*Presbytis cristata*) at Kuala Selangor, Malaysia. *Primates*, **18**, 949–55.

Wolfheim, J. H. (1983). *Primates of the World: Distribution, Abundance, and Conservation*. Seattle, WA: University of Washington Press.

Wolter, R. (1980). Alimentation et colliques chez le cheval. *Pratique vétérinaire équine*, **12**, 25–31.

Wolter, R. (1982). Alimentation et pathologie chez le cheval. *Pratique vétérinaire équine*, **14**, 12–20.

Wooldridge, F. L. (1971). *Colobus guereza*: birth and infant development in captivity. *Animal Behaviour*, **19**, 481–85.

Wrangham, R. W. (1980). An ecological model of female bonded primate groups. *Behaviour*, **75**, 262–300.

Wrangham, R. W. (1987). Evolution of social systems. In *Primate Societies*, ed. B. B. Smuts, D. L. Cheney, R. M. Seyfarth, R. W. Wrangham & T. T. Struhsaker, pp. 282–96. Chicago: University of Chicago Press.

Wrangham, R. W. & Van Zinnicq Bergmann Riss, E. (1990). Rates of predation on mammals by Gombe chimpanzees, 1972–75. *Primates*, **31**, 157–70.

Wrangham, R. W. & Waterman, P. G. (1983). Condensed tannins in fruits eaten by chimpanzees. *Biotropica*, **15**, 217–22.

Wright, W. (1992). The fracture properties of grasses and their relevance to feeding in herbivores. PhD dissertation, University of Reading, England.

Wu, B.-Q. (1993). Patterns of spatial dispersion, locomotion and foraging behaviour in three groups of Yunnan snub-nosed langur (*Rhinopithecus bieti*). *Folia Primatologica*, **60**, 63–71.

Yeager, C. P. (1989). Feeding ecology of the proboscis monkey (*Nasalis larvatus*). *International Journal of Primatology*, **10**, 497–530.

Yeager, C. P. (1991). Proboscis monkey (*Nasalis larvatus*) social organization: intergroup patterns of association. *American Journal of Primatology*, **23**, 73–86.

Zapfe, H. (1991). *Mesopithecus pentelicus* Wagner aus dem Turolien von Pikermi bei Athen, Odontologie und Osteologie. *Neue Denk-Schriften Naturhistorisches Museum Wien*, **5**, 1–203.

Zhang, Y. Z., Wang, S. & Quan, G. Q. (1981). On the geographical distribution of primates in China. *Journal of Human Evolution*, **10**, 215–26.

Zhao, Q. (1988). Status of the Yunnan snub-nosed monkey. *Primate Conservation*, **9**, 131–4.

Zingeser, M. R. (1969). Cercopithecoid canine tooth honing mechanisms. *American Journal of Physical Anthropology*, **31**, 205–14.

Zucker, W. V. (1983). Tannins: does structure determine function? An ecological perspective. *American Naturalist*, **121**, 335–65.

Index

201f refers to a figure on that page, 107t refers to a table, **202–4** refers to a primary ecological description on those pages

Abies, 141
Abu, *see* Mt Abu
Abuko, The Gambia, 107, 108t, 109, 110t, 113t, 114t, 115, 116, 118, 303
Acacia robusta, 298, 299f
Acer, 166
acidosis, 243, 277
activity patterns, 248
  black-and-white colobus, 100–1
  red colobus, 111–12, 113t
adaptive strategies of Asian colobines, 168–70
Adina cordifolia, 300, 301t
African moist forest habitat, 51, 75–8, 77f, 79, 91, 93, 107, 118, 352
age of maturation, 318t
Aglaia bourdilloni, 147
agonistic behaviour, rates of, 7
Agrostistachys longifolia, 147
Alangium salvifolium, 300
Albizia gummifera, 298, 299f
alkaloids, 144, 246–7, 252, 258–60, 259f, 260, 282, 348
  assays for, 261
  distribution of, 261
  interactions with tannins, 260, 266
all-male bands, 140, 141, 144, 146t, 150t, 156, 160t, 163t, 165t, 335, 337
allomothering, *see* infant handling
Alouatta, 2, 198, 208, 218, 239, 248, 292, 296–7
angiosperm evolution, 251–2
Angolan black-and-white colobus, *see* *Colobus angolensis*
Aningeria altissima, 87
anthropoid faunas of colobus study sites, 82t
apparency theory, 252–3, 266
arboreality, 3, 5
Asian forest habitats, 78, 129–36, 130f, 132f, 138, 141, 143, 145–7, 149, 152, 154, 157, 159, 164–6, 352
Aspilia, 261
Aucoumea klaineana, 86
Australopithecus, 41
Avicennia, 159

baboons, *see* *Papio*
bacterial microflora of forestomach, *see* forestomach microbes
Bacteroides, 233
banded leaf-monkey, *see* *Presbytis melalophos*
Barro Colorado Island, Panama, 266, 292, 296–7
Barteria fistulosa, 269
Betula, 164
Bhimtal, Nepal, 130f, 142t, 320f, 321f, 331f
bilophodonty, 15, 180, 191, 194, 198–9
biomass
  of colobines, 7, 75, 156, 159, 160, 285–91, 286f, 288–90t, 291, 305–7, 306f, 349
  of primates, 85, 88
birth interval, 104, 116, 318t, 319–22, 322f
  and infanticide, 328, 330–2
birth seasons, *see* seasonality in breeding
black-and-white colobus, *see* *Colobus*
black colobus, *see* *Colobus satanas*
bloat, 270
body size, 1, 41, 47t, 72, 118, 248, 280–1, 285

and foregut fermentation 224–5, 236, 237t, 348
Bole, Ethiopia, 92t, 94t, 97, 285, 286f, 290t, 291, 295, 349, 350, 351, 352
Bouvier's red colobus, *see Procolobus badius bouvieri*
brain size, 2
Brugiera, 159
bushmeat, *see* hunting

C_3–C_4 plants, *see* carbon isotopes
Caesaria tomentosa, 140
Calophyllum inophyllum, 188, 275t
Camellia, 166
canine tooth morphology, 175–7, 175f, 176f, 177f
capped langur, *see Trachypithecus pileatus*
captive breeding, difficulties with, 335
carbon isotopes, 30–1, 33, 34, 42
carbon/nutrient balance, 268–9
Caryedes brasiliensis, 258
cattle, 229, 236, 248
ceboids, 205, 208, 221f, 222f
Cebus, 199
caeco-colic fermentation, *see* fermentation
cellulolysis, *see* cellulose digestion
cellulose, 253, 254f, 255f
cellulose digestion, 232–5, 234t, 245–6, 253
cell-wall composition, 187, 232, 253–5
Celtis, 88f
Celtis africana, 87
Celtis durandii, 294, 303
Cercocebus, 82t, 202f
Cercocebus albigena, 243, 335
Cercocebus galeritus, 89–90, 350
Cercopithecidae, 1
  early, 12, 15–16, 42
cercopithecids, *see* Cercopithecidae
Cercopithecinae
  compared with Colobinae, *see* colobines compared to cercopithecines
  competition with colobines, 125, 295
oldest fossils, 17
cercopithecine teeth, 175, 179–80, 193, 194–5
cercopithecoid, ancestral, 199
Cercopithecoides, 13f, 31, 38–9, 41, 42
Cercopithecoides kimeui, 38, 39
Cercopithecoides williamsi, 24f, 38, 201f
Cercopithecus, 3, 82t, 118, 199, 355
Cercopithecus aethiops, 40, 82t, 327
Cercopithecus albogularis, 82t, 90, 353
Cercopithecus ascanius, 82t, 243
Cercopithecus diana, 82t, 112, 119, 121, 122, 243
Cercopithecus neglectus, 350
Cercopithecus nictitans, 82t, 83
cheek pouches, 2, 4f, 8
chemistry, *see* flower chemistry, foliage chemistry, fruit chemistry, plant chemistry, seed chemistry
chimpanzee, *see Pan troglodytes*
climate
  Africa, 76, 77f
  Asia, 131f, 132
Colobina, 45
Colobinae, general description, 1–9
colobine communities, species richness of, 350–1
colobine phylogeny, 14f
colobines compared to cercopithecines, xi, 1–9, 4f, 40–2, 136, 173–80, 188, 193, 194–200, 196f, 197f, 208, 209, 221f, 222f, 241–4, 247, 312–3, 319, 324, 342, 347, 349, 353–4, 355
Colobus, 8, 13f, 48, 71, 75, **91–106**, 126–8
  biomass, 286f
  coat patterns, 49–50
  diet, 98–9, 282
  feeding sites, 100, 127
  geographic distribution, 48–9f
  habitat range, 91
  natural history summarized, 104–6, 127
  social organization, 94t, 106
  taxonomy, 49–51
  vocalizations, 50, 103, 128
Colobus angolensis, 48–51, 75–6, 91, 93, 94t, 96, 99, 106, 109, 350
Colobus flandrini, 31
Colobus guereza, 50, 82t, 88, 91, 106, 107, 354, 355
  activity pattern, 100, 248
  biomass, 290t, 291, 302f, 303
  body weight, 47t, 280, 290t
  body weight and diet, 281f
  dental microwear, 197f
  diet, 96–7, 98t, 105, 219t, 220f, 243, 245, 280, 341
  digestive physiology, 233, 234t, 236, 239, 240, 244, 246
  dispersal, 337
  feeding competition, 294
  feeding height, 100
  field studies summarized, 92f
  food plant chemistry, 248, 272t, 273, 276, 282, 309
  geographical distribution, 49f, 51, 91, 106
  group fission, 95, 338
  gut contents, weight, 237t
  gut morphology and dimensions, 210t, 212, 213t, 214t, 222f, 223f, 230f
  habitats, 6, 50, 51, 91, 105, 106, 350
  male loud call, 50, 103

Colobus guereza – (*cont.*)
population density, 285, 287, 290t, 295, 296f, 302, 354
predation on, 292
ranging behaviour, 101, 102t, 112, 115
reproduction, 104, 316t, 318t
salivary glands, 232
social behaviour and organization, 76, 93–5, 94t, 95, 287t, 312f, 314t, 316t, 323, 325, 337, 340, 341f, 343, 345f
teeth, 177
territoriality, 103, 295, 316t
Colobus polykomos, 24f, 50, 82t, 83, 91, 93, 271, 294, 351
activity pattern, 101, 248
biomass, 290t
body weight, 47t, 290t
body weight and diet, 281f
diet, 98t, 99, 105, 219t, 242
feeding height, 100, 121
food plant chemistry, 272t, 273, 278t, 282, 309
geographical distribution, 48f
male loud call, 103
population density, 290t, 355
ranging behaviour, 101, 102t
reproduction, 104, 316t, 318t, 319
social behaviour and organization, 94t, 95, 287t, 312f, 314t, 316t, 325, 345f
Colobus satanas, 50, 82t, 85, 86, 86f, 91, 93, 271, 312f
biomass, 290t, 291
body weight, 47t, 290t
diet, 97, 98t, 99, 105, 224, 225, 242, 258, 301
food plant chemistry, 272t, 273, 274f, 278t, 280, 309
geographical distribution, 48f, 51, 91, 106
male loud call, 50, 103
ranging behaviour, 101–3, 115
reproduction, 316t
social behaviour and organization, 94t, 96, 287t, 290t, 312f, 314t, 316t, 345
Colobus vellerosus, 6, 48–50, 91, 93, 94t, 95–6, 99
colonizing trees, leaf chemistry of vs climax species, 266–7, 309
Commelinum benghalense, 240
competition between species, *see* interspecific competition
competition within groups, 7–8, 349
concentrate feeders, 239, 241, 246
Congo (Zaire) Basin, 76, 78, 93
conservation action plans, 354, 356–7t, 358
conservation and community development, 358
coprophagy, 239
copulation
black-and-white colobus, 313, 316t
Hanuman langur, 313, 316t
olive colobus, 120, 122, 313, 316t
red colobus, 116, 313, 316t
copulation calls in red colobus, 116, 118
critical stress intensity factor, 192–3
crowding, 295
crowned hawk-eagle, *see* predation
Cullenia exarillata, 147
cutin, 255f, 258
Cyclobalanopsis, 166
cyclone effects in Polonnaruwa, 299–301, 301t
Cynometra alexandri, 76, 87

deer, 224, 229, 233
Dendrohyrax, 125, 205
dental biomechanics, 189–94
dental formula, 173
dental microwear, 194–5, 196f, 197f, 199–200
dental morphology, 2, 21, 64, 66, 173–80
detoxification processes, 208, 252
in forestomach, 232–3, 246–7, 252, 260, 282, 283, 348
Dharwar, India, 286f, 288t, 291, 320f, 321f, 328, 329t, 331f
Diana guenon, *see Cercopithecus diana*
diet, 221f
and gut morphology, 217–27
animal matter, in, 138, 218, 221f, 241–2
of Asian colobines summarized, 168, 169f
of colobines compared with ceboids, 218
of *Colobus angolensis*, 99
of *Colobus guereza*, 96–7, 98t, 219t
of *Colobus polykomos*, 98t, 99, 219t
of *Colobus satanas*, 97–9, 98t
of *Nasalis larvatus*, 159–60, 219t
of *Presbytis melalophos*, 155–6, 155t, 219t
of *Presbytis rubicunda*, 157–8, 158f, 219t
of *Procolobus badius*, 109–11, 110t
of *Procolobus verus*, 121
of *Pygathrix*, 164–6
of *Semnopithecus entellus*, 138–40, 139f, 143–4, 144f, 219t
of *Trachypithecus auratus*, 152–4, 153t
of *Trachypithecus cristatus*, 153t
of *Trachypithecus johnii*, 146t, 147, 148f
of *Trachypithecus obscurus*, 155f, 219
of *Trachypithecus pileatus*, 150t
of *Trachypithecus vetulus*, 143–4, 144f, 219t
seasonality of, *see* seasonality in diet
digestibility of plant material, 124, 240, 269t, 272t

digestion, 205–6, 229–49, 251–3
in forestomach, 205, 229, 231, 348
in large intestine, 205–6
in tubus gastricus and small intestine, 238
digestion inhibitors, 205, 252, 253–8, 267
dikdik, see *Madoqua kirkii*
Diospyros, 139
Diospyros mespiliformis, 90
Diospyros thomasii, 270
Dipterocarpaceae, 132f, 133, 136, 154, 157, 159, 162, 168, 262, 280, 291, 294, 303, 305, 306f, 307, 351
disease, 292, 297
distribution of living colobines, *see* geographical distribution of colobines
dog, see predation
Dolichopithecus, 40, 42
Dolichopithecus eohanuman, 25, 26f
Dolichopithecus ruscinensis, 13f, 23, 24f, 25–7, 347
dominance hierarchies, 323–5
Douala-Edéa Forest Reserve, Cameroon, 75, 77, 79, 80t, 81t, **84–6**, 85f, 91, 94t, 96, 97, 98t, 99, 101, 102t, 104, 105, 247, 263t, 268, 269t, 272t, 278, 279f, 286f, 290t, 291, 307, 308f, 309
douc, *see Pygathrix nemaeus*
Drypetes, 147
Drypetes sepiara, 133, 144, 300, 301t
duiker, 224, 248
dusky langur, *see Trachypithecus obscurus*

earliest colobines
adaptations of, 42, 347–8
fossils, 347
Echium plantagineum, 260
ecological crunch, *see* resource bottleneck
edentates, 208
Elaeodendron, 144
elastic modulus, 192–3
elephant, 89, 205, 208
energetic economy of black-and-white colobus, 100–1, 105, 127, 248
ensalivation, *see* salivary glands
Eocene, 11
equid, 208
Erythrophleum, 282
Erythrophleum ivorense, 247
Euonymus porphyreus, 166
Eurasian fossil localities, 19
Eusideroxylon zwageri, 157
evolution of social systems, 128
evolution of teeth, 198–200
evolutionary relationships, *see* phylogeny
facial morphology, 2, 4f, 42, 64
Fagus, 166
Fathala, Senegal, 107, 108t, 109, 110t, 116
feeding competition, 325
feeding ecology, 6–7, 96–100
and social organization, 7–8
female-biased transfer
in olive colobus, 120, 127
in red colobus, 109, 117, 127, 337, 342
female relationships, 117, 127–8, 324–6, 345–6
fermentation
caeco-colic, 199, 205–6, 209, 218, 225, 230f, 239–40, 248
forestomach, 6, 199, 205–6, 209, 218, 224–5, 230f, 232–6, 234t, 245, 248–9
fibre, 127, 253–5, 262, 266–7, 269t
acid detergent, 240, 244t, 253, 254, 255f, 263t, 272t, 277t, 278t
neutral detergent, 240, 253, 255f
fibre digestion, 105, 168, 232, 235, 238, 239, 240, 241, 348
fibre–protein ratio, *see* protein–fibre ratio
Ficus, 87, 143, 144, 152, 300
Ficus sycomorus, 90, 298, 299f
Flamingia semialata, 139
flower chemistry, 264, 276
foliage chemistry, 262–4, 266–7, 268, 269t, 271–6, 272t, 274f, 275f
folivory, 2, 96–7, 104, 168, 244–6
and gut complexity, 208
vs frugivory, 198–9, 208, 217, 335, 347–8
food dispersion and social organization, 8, 106, 117, 124, 340–1
food, physical properties of, 180, 183
food-processing adaptations, 231, 349, 351
food procurement techniques, 6, 349
food selection, 6, 97, 99, 109, 124, 231
and plant chemistry, 270–80
foot, morphology of, 3, 347
foraging strategy, 335
in colobines vs cercopithecines, 349
in *Colobus*, 104–5
forestomach capacity and food retention, 235–8
forestomach fermentation, *see* fermentation
forestomach microbes, 1, 6, 229, 232–3, 238, 241, 247, 253, 260, 270, 282
forestomach pH, 230f, 233, 234f, 235, 243, 246, 247
Fort Ternan, 16
fossil
bovid, 30, 34–5, 36
cercopithecine, 17, 41, 200
colobine, 3, 13f, 17–42, 200, 347
hominoid, 16
fracture
mechanics, 184–5, 184f, 186f
properties of food, 183, 186–9, 187t, 191–3, 192f, 193f

fracture – (*cont.*)
strength, 183, 187t
toughness, 183, 187t, 188–9, 191–3
frugivory vs folivory, 224–5
fruit chemistry, 265–6, 276–8
Funtumia latifolia, 282
fusion–fission society, 316t, 338

gallery forest habitat, 50f, 91, 93, 105, 298–9, 303
Gambian red colobus, *see Procolobus badius temminckii*
gastrointestinal tract, *see* gut
gelada, *see Theropithecus*
geographical distribution of colobines, 1, 3–6, 49–72, 352–3
geophagy, 86f, 231, 247–8, 260
gestation period, 317, 318t
gibbons, *see Hylobates*
Gilbertiodendron dewevrei, 76–6
giraffe, 231
Gnetum microcarpum, 188
Gola Forest, Sierra Leone, 79, 133, 293, 355
Gombe, Tanzania, 107, 108t, 109, 110t, 111, 113t, 125, 293, 342
Gomphandra coriacea, 147
Gorilla gorilla, 82t, 208, 239, 248, 267
grey-cheeked mangabey, *see Cercocebus albigena*
grey langur, *see Semnopithecus entellus*
grooming, 95, 109, 120, 323, 325, 345f
group fission, 95, 158, 338, 341
group size and structure, *see* social behaviour and social organization
group takeovers, 140, 141-2, 144, 145, 152, 158, 337–8
growth phase forest, 303
Grukna, Indonesia, 163f
guenons, *see Cercopithecus*
guereza, *see Colobus guereza*
gut, 205–27, 230f
compartment, comparative dimensions, 209–16, 213t, 214t
compartment quotients, 217, 218f, 221, 222f
dimensions in wild vs captive colobines, 211t, 212
in relation to body size, 209–10, 217, 224–5
in relation to diet, 217–27, 219t, 220f, 221f
in relation to sex, 210–12

habitat modification by humans, 42–3, 79, 85, 89, 90, 138, 162, 166, 354
and infanticide, 330, 333
Hadar, Ethiopia, 33, 34, 36, 38, 40
hallux, 3
hamadryas baboon, *see Papio hamadryas*
hand, 3, 231
Hanuman (or grey) langur, *see Semnopithecus entellus*
harassment of copulation, 313
Himalayas, 71, 141–2, 129, 132, 313, 319, 323, 349
hippopotamus, 205, 208
home range, *see* range size *and* ranging behaviour
Hominidae, 78
hominoid, 15, 16, 208, 222f
Homo, 41
Homo erectus, 27
homosexual mounting, 325
horse, 205, 239, 248
howler monkey, *see Alouatta*
hunting, see predation
Hylobates, 136, 195, 221f, 244, 307, 351
Hylobates lar, 242, 243
Hylobates syndactylus, 136, 208
Hymenaea, 265

Ilex, 166
incisor morphology, 173–5, 174f, 175f
Indri indri, 248
infant
coat colour, *see* neonatal coats
handling, 8, 49, 104, 116–17, 123, 128, 153t, 316t, 326–8, 345f, 346
kidnapping, 327
killing, *see* infanticide
infanticidal strategy models, 328, 330
infanticide, 8–9, 312, 316t, 323, 328–34
in *Procolobus badius*, 329t
in *Semnopithecus entellus*, 140, 321, 328–34, 329t, 337, 338
in *Trachypithecus cristatus*, 152, 329t
in *Trachypithecus vetulus*, 145, 329t
infanticide counter-strategies, 334
insectivory, 241–2
interaction between groups, 101, 103, 109, 115, 117, 334–5
intermale competition, *see* male–male competition
interspecific competition
among colobines, 167, 294–5, 351
between colobines and other animals, 125, 295
Intsia palembanica, 134, 188
Ituri Forest, Zaire, 51, 93, 94t, 106, 350

jackal, *see* predation
Jodhpur, India, 286f, 288t, 315t, 320f, 321f,

322f, 324, 328, 329t, 330, 331f, 332, 337
Jozani Forest, Zanzibar, 108t, 110t, 263t, 271, 276
Junbesi, Nepal, 130f, **141–2**, 142t, 315t, 320f, 321f, 324, 328, 329t, 331f

Kakachi, India, 69f, 130f, 146t, **147**, 148f, 169f, 263t, 269t, 271, 272t, 286t, 288t, 291, 308f
Kamajong, Java, 159
Kanha, India, 21, 71f, 130f, 132, 133, 134f, 135f, **138–40**, 169f, 286f, 288t, 291, 292, 315t, 320f, 321f, 322f, 329t, 330, 331f, 332, 340
Kanyawara, *see* Kibale Forest Reserve
Kasi, 69
Ketambe, Sumatra, 286f, 289t, 306f
Kibale Forest Reserve, Uganda, 51, 75, 77, 79, 80t, 81t, **87–9**, 88f, 92t, 94t, 97, 98t, 100, 101, 102t, 103, 105, 107, 108t, 109, 110t, 111, 112, 113t, 115, 116, 126, 127, 134f, 242, 243, 245, 247, 263t, 268, 269t, 271, 272t, 276, 277t, 291, 292, 294, 297, 302f, 307, 308f, 309, 310, 314t, 319, 329t, 331, 342, 354
kin selection, 342
Klainedoxa gabonensis, 85
Kleiber's Law, 217
Koobi Fora, Kenya, 34, 35, 36, 38, 39, 40
Kuala Lompat, Malaysia, 130f, 131f, 132, 133, 134f, 135f, 150t, **154–7**, 159, 169f, 241, 243, 263t, 268, 269t, 271, 272t, 276, 277t, 286f, 289t, 291, 293, 294, 297, 305, 306f, 308f, 309, 315t
Kuala Selangor, Malaysia, 130f, 152–4, 153t, 329t

Laetoli, Tanzania, 33, 37, 38, 40, 42
lagomorph, 208
Lake Turkana Basin, 34, 35
langur, *see Trachypithecus*
larynx in colobus, 53
leaf chemistry, *see* foliage chemistry
mechanical properties, 188–9
petiole feeding and chemistry, 264, 275–6
leaf-monkeys, *see Presbytis*
leaping, 3, 64, 65f
Leguminosae, 76, 83, 87, 99, 154, 157, 162, 187, 265, 270, 280, 282, 305–7, 306f
lemur, 239
leopard, *see Panthera pardus, and* predation
lianes in diet, 99, 104, 156, 157
Libypithecus, 13f, 39, 40, 41
Libypithecus markgrafi, 24f, 39
lignin, 240, 252, 253, 254f, 255f, 258, 262
limb proportions, 3, 4f, 58
lipid, 277
Litsea, 149
logging, 358
at Lopé, 87
at Tekam, 303–5, 304f, 310
effects on colobines, 89, 354–5
in Kibale Forest, 51, 89, 302–3, 302f, 310, 354–5
Lopé Reserve, Gabon, 75, 77, 79, 80t, 81t, **86–7**, 86f, 94t, 98t, 99, 100, 101, 102t, 103, 104, 105, 115, 268, 271, 286f, 290t, 291, 309
Lophira alata, 85
loud calls, 8, 48, 103, 147, 316t, 324, 335, 345f
lysozyme, 238

Maboko Island, Kenya, 12, 13f, 15, 16
Macaca, 3, 25, 41, 42, 61, 136, 350
Macaca anderssoni (=*M. robusta*), 27
Macaca fascicularis, 136, 197f, 241
Macaca mulatta, 136
Macaca nemestrina, 136, 242
Macaca radiata, 136
Macaca silenus, 136
Macaca sinica, 136, 243, 247
Macaca sylvanus, 23, 208
macaque, *see Macaca*
macropod marsupial, 205, 208, 217
Madhupur, Bangladesh, 130f, **149–52**, 150t, 151f, 169f, 286f, 288t, 292
Madoqua kirkii, 231, 239, 280
Mahale Mountains, Tanzania, 293
male-biased transfer
in black-and-white colobus, 127
in *Semnopithecus entellus*, 323
male influxes, 338
male–male competition, 103, 116, 117, 118, 128, 332, 339
male relationships, 106, 117, 127, 323–4, 343–5
male tenure, 332–3
Mallotus, 149, 300
Manas, Assam, 149, 150t
mandrill, *see Papio (Mandrillus)*
mangabey, *see Cercocebus*
Markhamia platycalyx, 7, 264, 275, 294
mating harassment, *see* harassment of copulation
matrilineal society, 311, 312f, 316t, 323, 325, 332, 335, 339, 342, 345f
Mchelelo, *see* Tana River
Megamacaca lantianensis, see Pygathrix (R.) lantianensis

Melemchi, Nepal, 130f, 142t, 286f, 288t, 315t, 320f, 321f, 328, 329t, 331f
menstrual cycle length, 317
Mentawai Islands, Indonesia, 3, 57, 58, 61, 64, **162–4**, 163t, 288t, 306f, 312, 314t, 315t
Mesopithecus, 13f, 22
Mesopithecus delsoni, 17
Mesopithecus monspessulanus, 23
Mesopithecus pentelicus, 17, 18f, 20, 21, 24f, 25
Methanobacterium ruminantium, 233
Mezzetia leptopoda, 156, 188
Mezzetia parviflora, 187t
Microcolobus, 13f, 41
Microcolobus tugenensis, 31
migration between groups, 95, 96, 109, 117, 140, 144, 151, 162, 337
Millettia atropurpurea, 187t, 188, 190, 191f
Mimusops bagshawei, 87
mineral needs and food selection, 247–8, 275–6
Miocene, 1, 3, 12, 13f, 15, 16, 17–22, 30–1, 32f, 40–2, 199, 347, 352
Miopithecus talapoin, 8, 31
molar morphology, 2, 4f, 15, 64, 66, 180, 181f, 182f, 198–9
  and biomechanics, 189–94, 190f
  and food processing, 105, 180, 348
Monocarpia marginalis, 156
monodominant forests, 76–7
monogamy, 7, 61, 159, 162, 312f, 324, 339, 342
monopolization hypothesis, 106, 324, 343–5
moose, 231
morphology, cranial, 2, 4f, 50, 58
mousedeer, *see Tragulus*
Mt Abu, India, 242, 315t, 320f, 321f, 326, 328, 329t, 331f
Mt Sontra, Vietnam, 164, 165t
Mundanthurai, India, 146t, 147, 329t, 330, 331f
Myristica, 147

Nasalinae, 45
Nasalis larvatus, 5, 23, 24f, 38, 39, 42, 45, 63f, 129, **159–62**, 168, 328, 349, 351
  biomass, 286f, 288t, 291
  body weight, 1, 47t, 61, 288t
  body weight and diet, 281
  dental microwear, 197f
  diet, 159–60, 220f, 219t, 224, 225
  dispersal, 337
  geographical distribution, 59f
  gut dimensions, 210t, 213t, 214t, 216, 222f, 223f
  habitat, 61, 159, 161
  nose, 61
  population density, 288t
  ranging behaviour, 160–1, 335
  reproduction, 316t
  social organization, 61, 161–2, 163t, 312f, 314t, 316t, 337, 338
  taxonomy, 28, 57
  territoriality, 316t
  vocalizations, 8, 316t
neonatal coats, 8, 42, 51, 52, 64, 65, 66–7, 70, 73, 104, 116, 312, 316t, 327–8, 345f, 346
Neotragus moschatus, 239
Newtonia buchanani, 87
Nilgiri langur, *see Trachypithecus johnii*
nitrogen metabolism, 240–1
non-protein amino acids, 258, 259f, 265
Nypa fruticans, 159

Olduvai, Tanzania, 35, 39
Old World monkeys, *see* Cercopithecidae
Olea welwitschii, 87
Oligocene, 11
olive colobus, *see Procolobus verus*
Omo River, 34, 35, 36, 38, 39, 40
Ootacamund, India, 145, 146t
optimal diet, 251
Orcha, India, 286t, 288t, 315t, 320f, 321f, 328, 329t, 331f

Pachystela msolo, 90
Paitan, *see* Mentawai Islands
Paleocene, 11
paleoenvironment, 11–12, 16, 20–2, 30–1, 33–6, 38–9, 42–3
palynology, *see* pollen
Pangandaran Nature Reserve, Java, 130f, **152–4**, 153t, 168, 169f, 269t, 271, 272t
Panthera pardus, 89, 126
Pan troglodytes, 8, 82t, 118, 343, 261
  hunting of red colobus by, 125–6, 293, 342
Papio, 3, 41, 42, 82t, 199, 200, 323
Papio cynocephalus, 90, 197f
Papio hamadryas, 40, 338
Papio (Mandrillus), 25
Papionini, 173
Paracolobus, 13f, 37, 38, 39, 41
Paracolobus chemeroni, 24f, 36–7, 37f
Paracolobus mutiwa, 37
Paradolichopithecus, 25, 42
Parapapio, 41, 200, 201f
Parapresbytis, 26
Parinari excelsa, 87, 302
Pasoh, Malaysia, 286f, 289t
passage rate, 236, 244, 246

patrilineal society, 312f, 316t, 323, 324, 325, 332, 335, 339, 342, 346
peccary, 205, 224
perineal swellings, *see* sexual swellings
Periyar, India, 130f, 145, 146t
phalangeroid marsupials, 198
phenology
  African sites, 81t
  Asian sites, 133–6, 135f, 143, 154
  Barro Colorado, 296–7
phylogeny, 12–17, 14f, 40–3
pig, 205
Pikermi, Greece, 17, 20, 21, 42
Piliocolobus, see Procolobus badius
Piptadeniastrum africanum, 83
plant chemistry
  and colobine biomass, 307–9, 308f
  and environment, 85, 268–70, 309
  and food selection, 251–84
plant habit, effect on secondary metabolites, 266–8
plant secondary compounds, *see* secondary metabolites
Pleistocene, 13f, 19f, 27–30, 32f, 35, 36, 38, 40, 41, 199
  forest refuge, 79
  glacial cycle and forest reduction, 12, 41, 76, 352
Pliocene, 3, 12, 13f, 19f, 22–7, 31–42, 199
pollen, 20, 33, 34
Polonnaruwa, Sri Lanka, 130f, 131f, 132, 133, 134f, **143–5**, 152, 169f, 243, 286f, 288t, 291, 292, 294, 299–301, 310, 315t, 319, 320f, 321f, 329t, 331f
polyphenols, *see* tannins
Pongo pygmaeus, 41, 136
population density, 285–310
  Colobus guereza, 285, 287, 290t, 295
  Colobus polykomos, 290t
  Colobus satanas, 290t
  Nasalis larvatus, 289t
  Presbytis melalophos, 160t, 289t, 291
  Presbytis rubicunda, 160t, 285, 289t
  Procolobus badius, 285, 290t, 293, 298
  Procolobus verus, 290t
  Semnopithecus entellus, 145, 288t, 291, 336f
  Trachypithecus johnii, 288t
  Trachypithecus obscurus, 105t, 288t, 291
  Trachypithecus phayrei, 150t
  Trachypithecus pileatus, 150t, 288t
  Trachypithecus vetulus, 145, 288t, 291
predation, 292–4
  and group living, 339–43
  by chimpanzees, 125–6, 293, 342
  by crowned hawk-eagles, 89, 125, 126, 292
  by dogs, 145, 292
  by humans, 41, 75, 78, 83, 85–6, 89, 137, 166, 293–4, 353, 355, 358
  by jackals, 292
  by leopards, 126
  in Africa, 125–6
  in Asia, 136–7
premolar morphology, 177–80, 178f, 179f
Presbytina, *see* Semnopithecina
Presbytis, 13f, 129, 168, 170, 351, 352
  biomass, 286f
  diet, 169f
  geographical distribution, 6, 57, 64, 66f
  habitat range, 6, 57, 64
  male loud calls, 8, 64–5
  morphology, 64, 168
  taxonomy, 28, 57, 64–6, 71
Presbytis comata, 66f, 159, 160t, 175f, 315t, 345f, 357t
Presbytis cruciger, 65
Presbytis eohanuman, see Dolichopithecus eohanuman
Presbytis femoralis, 65
Presbytis frontata, 67f, 288t
Presbytis hosei, 66f, 158–9, 160t, 276, 277, 278t, 288t, 294–5, 315t, 351
Presbytis melalophos, 65f, **154–7**, 160t, 167, 271
  biomass, 288t, 291
  body weight, 47t, 288t, 291
  body weight and diet, 246, 281f
  diet, 155–6, 155f, 187–8, 219t, 220f, 242–3, 282
  dispersal, 337
  effects of logging on population, 303–4, 304f
  food plant chemistry, 272t, 276–7, 278t
  geographical distribution, 66f
  gut dimensions, 210t, 211t, 212, 213t, 214t, 215, 222f, 223f
  interspecific feeding competition, 294
  population density, 288t, 291
  ranging behaviour, 156
  social behaviour and organization, 156–7, 287t, 312f, 315t, 316t, 325, 337, 345f
  taxonomy, 65
  territoriality, 156, 316t
  vocalizations, 65, 156, 316t
Presbytis potenziani, 7, 160t, **162**, 357t
  body weight, 47t, 288t
  population density, 288t
  social organization, 161t, 162, 312f, 315t, 316t, 345f
  taxonomy, 65
  vocalizations, 162, 316t, 324
Presbytis rubicunda, **157–9**, 160t, 271
  biomass, 288t

Presbytis rubicunda – (*cont.*)
body weight, 47t, 288t
body weight and diet, 246, 281f
diet, 157–8, 158f, 160t, 187, 199, 219t, 242, 243, 247
dispersal, 337
food plant chemistry, 272t, 273, 274, 276–7, 278t
foraging strategy, 349
geographical distribution, 67f
gut dimensions, 210t, 212, 213t, 214t, 216, 222
interspecific feeding competition, 294–5
population density, 281, 285, 288t
ranging behaviour, 158
social behaviour and organization, 7, 157–8, 160t, 287t, 312f, 315t, 316t, 337, 345f
sympatry with other *Presbytis* species, 65, 158–9, 294–5, 351
teeth, 175, 180
territoriality, 158, 316t
Presbytis siamensis, 65
?Presbytis sivalensis, 21
Presbytis thomasi, 66f, 159, 160t, 288t, 312f, 315f, 338, 345f
Presbytiscus, see Pygathrix (Rhinopithecus) avunculus
presaccus, evolution of, 225–7, 226f
proboscis monkey, *see Nasalis larvatus*
Procolobus, 13f, 48, 71
Procolobus badius, 51–6, 75, 82t, 83, 89, **107–18**, 126–7, 313, 351
activity patterns, 111–12, 113t
biomass, 281, 286f, 287, 290t, 302f
body weight, 39, 47t, 290t
body weight and diet, 281f
coat patterns, 52–3
diet, 109–11, 110t, 127, 242, 245, 264, 341
dispersal, 337
effects of logging on populations, 302–3
feeding sites, 111, 121, 127
food plant chemistry, 272t, 274, 277, 282
forestomach fermentation, 234t
geographical distribution, 52f, 54, 352–3
gut content weights, 237t
habitat, 6, 107, 350
infanticide, 328, 329t, 331
interspecific feeding competition, 294–5
natural history summarized, 117–18
population density, 285, 293, 298–9, 302, 354, 355
predators, 125–6, 292–4, 342
ranging behaviour, 112, 114t, 115, 335
reproduction, 116, 118, 298, 316t, 318t, 319
social behaviour and organization, 107–9, 108t, 115, 117–8, 120, 287t, 290t, 312f, 314t, 316t, 325, 327, 337, 339, 341, 342–3, 344f, 345f
taxonomy, 53–6, 55t
teeth, 175
transfers between groups, 76, 120
vocalizations, 53, 56, 89, 107, 115, 116, 118, 316t
Procolobus badius badius, 52, 53, 55t, 56, 108t, 114t, 116, 122, 271
Procolobus badius bouvieri, 54, 55t, 352, 353, 356t
Procolobus badius gordonorum, 54, 55t, 108t, 356t
Procolobus badius kirkii, 54, 55t, 107, 108t, 271, 355, 356t
Procolobus badius oustaleti, 54, 55t, 108t, 352, 353
Procolobus badius pennantii, 54, 55t, 356t
Procolobus badius preussi, 54, 55t, 56, 108t, 116, 352, 353, 356t
Procolobus badius rufomitratus, 54, 55t, 89, 90, 107, 108t, 114t, 298–9, 354
Procolobus badius temminckii, 53, 54f, 55t, 56, 107, 108t, 114t, 116, 117, 312
Procolobus badius tephrosceles, 54, 55t, 87–8, 107, 108t, 114t, 116, 271, 274, 277, 281, 302, 312
Procolobus verus, 3, 5f, 51, 56, 75, 82t, 83, **118–25**, 123f, 126, 271, 295, 351
activity patterns, 121, 124
anatomy, 56, 125
association with other species, 118, 119, 124
biomass, 286f, 290t
body weight, 1, 47t, 56, 118, 124, 280, 290t, 293
body weight and diet, 124, 246, 281
coat patterns, 56, 124
diet, 121, 124
digestive physiology, 233, 234t
feeding sites, 6, 121, 124
geographical distribution, 56, 118
gut content weights, 237t
habitat, 56, 118
natural history summarized, 124–5
oral transport of infants, 56, 76, 118, 123–4, 327
population density, 290t
ranging behaviour, 121–2
reproduction, 122–3, 316t, 318t
social organization, 76, 119–20, 124, 287t, 312f, 314t, 316t, 345f
vocalizations, 122, 316t
Prohylobates, 12, 13f
Propithecus, 198

protected forests, 358
protein content
  of foliage, 262–4, 272–3
  of fruits, 277
  of seeds, 278t
protein–fibre ratio, 154, 264t, 272–8, 272t, 274f, 275f, 281f, 308f
Protoxerus stangeri, 125
Pterocarpus marsupium, 139
Pterygota alata, 156
purple-faced langur, *see Trachypithecus vetulus*
Pygathrix, 13f, 129, 168, 170
  geographical distribution, 59f, 164
  taxonomy, 57–61
Pygathrix nemaeus, 46t, 58, 60f, **164**, 165t, 314t, 357t
  gut dimensions, 210t, 213t, 214t, 216, 222f, 223f
Pygathrix (Rhinopithecus), 5, 13f, 27, 36, 164–7, 57–61, 338, 349, 353
Pygathrix (Rhinopithecus) bieti, 5, 7, 46t, 47t, 59f, 61, **164–6**, 165t, 314t, 357t
Pygathrix (Rhinopithecus) brelichi, 5, 7, 27, 29f, 46t, 59f, 61, 62f, **164–6**, 165t, 314t, 357t
Pygathrix (Rhinopithecus) lantianensis, 27, 29f, 41
Pygathrix (Rhinopithecus) roxellana, 5, 7, 27, 46t, 59f, 61, 165t, **166–7**, 220f, 222f, 223f, 314t, 353, 357t
Pygathrix (Rhinopithecus) roxellana tingianus, 27, 28f

qualitative plant defences, *see* toxins
quantitative plant defences, *see* digestion inhibitors

rabbit, 239
rainfall
  at African study sites, 77f
  at Asian study sites, 131f
range size and ranging behaviour, 101–3, 170, 334–5
  Asian colobines compared, 170
  Colobus guereza, 101, 102t, 112, 115
  Colobus polykomos, 101, 102t
  Colobus satanas, 101, 102t, 115
  Nasalis larvatus, 160–1, 335
  Presbytis melalophos, 156
  Presbytis rubicunda, 158
  Procolobus badius, 112, 114t, 115, 335
  Procolobus verus, 121–2
  Pygathrix (Rhinopithecus) bieti, 165t, 166
  Pygathrix (Rhinopithecus) roxellana, 165t, 166–7, 335
  Semnopithecus entellus, 138t, 140, 142, 335, 336t
  Trachypithecus auratus, 153t, 154
  Trachypithecus cristatus, 153t, 154
  Trachypithecus geei, 150t
  Trachypithecus johnii, 146t, 147
  Tracypithecus obscurus, 150t
  Trachypithecus phayrei, 150t
  Trachypithecus pileatus, 150t, 335
  Trachypithecus vetulus, 145, 335
Ranthambhore, India, 242, 320f, 321f, 323, 329t
Raphia, 79
Rauvolfia vomitoria, 260, 282
red colobus, *see Procolobus badius*
red leaf-monkey, *see Presbytis rubicunda*
reproduction, 103–4, 116–17, 122–4, 313, 316t, 318t, 319–22
reproductive synchrony in females, 343–5
resouce bottlenecks 97, 104, 282, 294, 296–8, 301, 305, 351
resource defence vs predator risk hypotheses of group living, 339–40
Rhinocolobus, 13f, 24f, 36, 38, 39, 41, 42, 201f
Rhinopithecus, see Pygathrix (Rhinopithecus)
Rhizophora, 159
ruminants, 205, 206, 217, 224, 229, 232, 236, 238, 240, 248, 251, 260, 261, 280, 282, 348

saccus gastricus, 207f, 208, 225f, 230f, 233–8, 234t, 237t
Sacoglottis gabonensis, 85
Sal, *see Shorea robusta*
salivary glands, 2, 230f, 231–2
salivary proteins, 2, 257, 283
Samunsam, Sarawak, 130f, **159–62**, 160t, 163t, 286f, 288t, 291
Schleichera oleosa, 144, 301t
seasonality
  in breeding, 104, 116, 118, 122–3, 140, 141, 165t, 318t, 319, 320t, 324, 333–4, 333f, 344–5
  in diet, 97, 99, 109, 121, 170, 194, 296–8
secondary metabolites, 9, 208, 241, 251–3
  distribution within plants, 261–70
  seed chemistry, 264–5, 278–80, 278t, 279f
  dispersal, 265–6
  mechanical propeties, 187–8, 191–2, 191f, 193f, 198, 265, 280
seed-eating, 75, 97, 98t, 99, 104–5, 121, 127, 149, 168, 187–9, 190–2, 193f, 195, 198–9, 208, 224–7, 242–4, 244t, 278, 280, 282–3, 348
Semnopithecina, 45

Semnopithecus, 13f, 129, 168
taxonomic position, 57, 64, 68
Semnopithecus entellus, 70–3, 71f, 21, 38, 136, **137–45**, 167, 170, 313, 349, 354, 355
biomass, 286f, 288t, 291
body weight, 47t, 72, 136, 288t
coat coloration, 72–3
cranial morphology, 18f, 70
diet, 71, 138–40, 141, 143–4, 144f, 169f, 219t, 220f, 224, 225, 242, 243, 299–301, 305
digestive physiology, 233, 234t
dispersal, 337
food plant chemistry, 273, 282
geographical distribution, 6, 71, 72f, 136
gut dimensions, 210t, 213t, 214t, 215, 222f, 223f
habitat, 6, 71, 72, 136, 231
infant handling, 326–8
infanticidal behaviour, 8, 316t, 321, 328–34, 331f
interspecific feeding competition, 294
population density, 145, 288t, 291, 330, 331f, 336f
predators, 292
reproduction, 316t, 317, 318t, 319–22
salivary glands, 232
social behaviour and organization, 138t, 140, 141–2, 143–4, 287t, 312f, 315t, 321f, 323, 325, 337, 338, 339–40, 340f, 341, 343, 344f, 345f
taxonomy, 70–3
territoriality, 145, 316t
vocalizations, 8, 316t
Semnopithecus entellus ajax, 136, 141–2
Semnopithecus entellus entellus, 138–40
Semnopithecus entellus hypoleucos, 72
Semnopithecus entellus schistacea, 136, 141–2
Semnopithecus entellus thersites, 136, 143–5
Semnopithecus palaeindicus, 25
Sepilok Virgin Jungle Reserve, Sabah, 130f, 131f, 132, 133, 134f, 135f, **157–9**, 158f, 160t, 169f, 242, 244t, 263t, 268, 269t, 271, 272t, 276, 277t, 286f, 288t, 291, 305, 306f, 308f, 309
sexual dimorphism, 58, 61, 64, 120, 177, 334
sexual selection hypothesis for infanticide, 328, 329–32
sexual swellings, 8, 48, 317, 319, 345f, 346
olive colobus, 8, 103–4, 122–3, 123f, 124, 127, 316t, 317
red colobus, 8, 103–4, 116, 118, 122, 124, 127, 316t, 317
Simias concolor, 8, 64, 316t, 317
sheep, 236, 239
Shorea robusta, 132f, 133, 139, 149, 262
Siberut, Indonesia, *see* Mentawai Islands
silica in foliage, 258
silvered langur, *see Trachypithecus cristatus*
simakobu, *see Simias concolor*
Simias concolor, 3, 5, 13f, 28, 45, 57, 59f, 61, 63f, 64, **162–4**, 163t, 168, 288t, 316t, 345f, 357t
Simla, India, 129, 130f, 131f, **141**, 142t, 169f, 320f, 321f, 331f
Singur, India, 321f
Sivapithecus, 41
Siwaliks, 21–2, 26
sloth, 205, 217, 229, 248
snub-nosed monkey, *see Pygathrix*
social behaviour and social organization, 7–8, 93–6, 311–46, 349
Colobus guereza, 94t, 95, 312f, 314t, 316t, 323, 325, 337, 340, 341f, 343
Colobus polykomos, 94t, 95, 312f, 314t, 316t, 325
Colobus satanas, 95t, 96, 312f, 314t, 316t
Nasalis larvatus, 161–2, 163t, 312f, 314t, 316t, 335, 337
Presbytis comata, 312f, 315t
Presbytis melalophos, 156–7, 312f, 315t, 316t, 325, 337
Presbytis rubicunda, 157–8, 312f, 315t, 316t, 337
Procolobus badius, 107–9, 108t, 117, 312f, 314t, 316t, 325, 335, 337, 342, 343, 334f
Procolobus verus, 119–20, 124, 312f, 314t, 316t
Pygathrix nemaeus, 164, 165t, 314t
Pygathrix (Rhinopithecus) bieti, 165t, 166, 314t
Pygathrix (Rhinopithecus) brelichi, 165t, 166, 314t
Pygathrix (Rhinopithecus) roxellana, 165t, 166, 314t
Semnopithecus entellus, 138, 138t, 140, 141–2, 144, 312f, 315t, 316t, 324, 325, 335, 337, 338, 340f, 343, 344f
Simias concolor, 61, 64, 162–4, 314t, 316t
Trachypithecus auratus, 152, 153t, 315t
Trachypithecus cristatus, 152, 153t, 315t
Trachypithecus geei, 150t, 315t
Trachypithecus johnii, 145t, 147–8, 312f, 315t, 316t, 325, 337
Trachypithecus obscurus, 150t, 315t
Trachypithecus phayrei, 150t, 315t
Trachypithecus pileatus, 150t, 312f, 315t, 316t, 335, 337
Trachypithecus vetulus, 145, 312f, 315t, 316t, 337

social behaviour and social organization in relation to ecology, 168, 170, 334–46
social pathology hypothesis for infanticide, 330
sodium, 86f, 231, 276
soil characteristics of colobine study sites, 79, 80, 84, 85f, 86, 87, 157
soil feeling, *see* geophagy
soliciation of copulation, 313
Sonneratia, 159
Sorbus yunnanensis, 166
South African cave sites, 35–6, 38, 40
species-richness
  of African forests, 78
  of primate communities, 350
sperm competition, 118, 124, 322, 326
spider monkey, 244
Stephanoaetus coronatus, see predation by crowned hawk-eagles
stomach
  compound, 206, 208, 229, 348
  functions of, in digestion, 48, 205–6, 208–9, 224–7, 230f, 232–8
  morphology of, 1, 4f, 48, 206, 207f, 208, 224–7, 230f, 351
  simple, 206
  size of, in relation to body size, 346, 351
Strombosia scheffleri, 87
Strychnos, 247
Strychnos potatorum, 282
Strychnos tricalysoides, 280
sunbathing, 100, 105
Sungai Tekam, Malaysia, 130f, 156–7, 286f, 289t, 303, 304f, 310
suni, *see Neotragus moschatus*
sympatry of colobine species, 78, 105, 106, 350–1, 352

Tai forest, Côte d'Ivoire, 94t, 108t, 119, 120, 125, 133, 134f, 286f, 290t
tail length, 3, 61, 68
Tana River, Kenya, 77, 79, 81t, **89–90**, 90f, 107, 108t, 109, 110t, 112, 113t, 115, 117, 127, 134f, 286f, 287, 290t, 298–9, 303, 314t, 329t, 342
Tana River red colobus, *see Procolobus badius rufomitratus*
Tanjung Puting, Indonesia, 63f, 158, 163t
tannin–protein complexes, 255, 257
tannins, 2, 233, 252, 255–7, 263t, 266–7, 268, 269t, 283
  and food selection, 270, 273
  assays for, 257
  condensed, 252, 255, 256f, 269t, 272t, 273, 277t
  hydrolysable, 252, 255, 256f
  in foliage, 262, 263t, 272t, 273
  in fruits, 265–6
  in seeds, 265, 278t
tapir, 199
taxonomy of living colobines, 45–73
Taxus, 141
Teclea nobilis, 277
teeth, *see* dental morphology
Tekam, *see* Sungai Tekam
tenure length of males and infanticide, 332–3
terpenes, 154, 247, 280
terrestriality, 3, 5, 6, 16, 21, 23, 38–9, 40–2, 61, 64, 71f, 100, 104, 105, 117, 140, 328, 347, 350, 354
territoriality, 101, 122, 156, 170, 295, 316t, 335, 341
Theropithecus, 41, 42
Theropithecus brumpti, 200
Theropithecus gelada, 200, 338
Theropithecus oswaldi, 200, 202f
Thomas' leaf-monkey, *see Presbytis thomasi*
threatened species, 353–8, 356–7t
thumb reduction, 3, 5f, 6, 41–2, 45, 125
timber extraction, see logging
Tiwai Island, Sierra Leone, 77, **79–84**, 80t, 81t, 82t, 83f, 84f, 94t, 95, 98t, 99, 100, 101, 102t, 103, 104, 105, 107, 108t, 109, 110t, 111, 112, 113t, 114t, 116, 118, 119–23, 243, 244t, 247, 269t, 270, 271, 272t, 286f, 287, 290t, 307, 308f, 309, 355
Tonkin snub-nosed monkey, *see Pygathrix (Rhinopithecus) avunculus*
tooth wear, *see* dental microwear
toxins, 103, 205, 232–3, 246–7, 252, 258–61, 265, 266, 267, 278, 279f, 280
Trachypithecus, 6, 13f, 129, 168, 170, 351
  biomass, 286f
  diet, 169f
  geographical distribution, 57–8, 68
  habitat range, 58, 67
  morphology, 66–7
  taxonomy, 28, 57, 64, 68–70, 71
  vocalizations, 8
Trachypithecus auratus, 68, **152–4**, 168, 219t, 270, 271, 272t, 282
Trachypithecus barbei, 69
Trachypithecus cristatus, **152–4**
  body weight, 47t
  diet, 152–4, 153t
  digestive physiology, 233, 234t, 236
  geographical distribution, 68f, 152
  gut content weights, 237t
  gut dimensions, 210t, 213t, 214t, 215, 222f, 223f
  incisor morphology, 174f, 175
  infanticide, 328, 329t, 331

Trachypithecus cristatus – (*cont.*)
ranging behaviour, 153t
social organization, 152, 153t, 312f, 315t, 316t, 345f
teeth, 180
territoriality, 316t
Trachypithecus delacouri, see Trachypithecus francoisi
Trachypithecus francoisi, 68–9, 68f, 152, 357t
Trachypithecus geei, 68f, 69, **149**, 150t, 312f, 315t
Trachypithecus johnii, 69f, **145–8**, 271
biomass, 288t
body weight, 47t, 288t
body weight and diet, 281f
diet, 146t, 147, 148f, 281
dispersal, 337
food plant chemistry, 272t, 282
geographical distribution, 68f, 136, 145
infant handling, 326
population density, 288t
ranging behaviour, 146t, 148
social behaviour and organization, 146t, 147–8, 312f, 315t, 316t, 325, 337, 345f
territoriality, 316t
Trachypithecus leucocephalus, see Trachypithecus francoisi
Trachypithecus obscurus, 70f, 152, **154–6**, 167
biomass, 288t, 291
body weight, 47t, 288t
diet, 105t, 154–5, 155f, 219t, 220f
geographical distribution, 68f
gut dimensions, 210t, 211t, 212, 213t, 214t, 215, 222f, 223f
interspecific feeding competition, 294
locomotion, 155
population density, 150t, 288t, 291
ranging behaviour, 150t
social organization, 150t, 287t, 315t, 316t, 345f
territoriality, 150t, 316t
Trachypithecus phayrei, 68, 150t, 152
Trachypithecus pileatus, **149–52**
activity pattern, 248
biomass, 288t
body weight, 47t, 288t
diet, 149–51, 151f, 282
geographical distribution, 68f, 149
population density, 150t, 288t
predators, 292
ranging behaviour, 150t, 151, 335
social behaviour and organization, 7, 149–52, 150t, 287t, 312f, 315t, 316t, 337, 338, 345f
terrestriality, 6
territoriality, 316t
Trachypithecus senex, see Trachypithecus vetulus
Trachypithecus vetulus, **143–5**, 167
activity pattern, 248
biomass, 288t
body weight, 47t, 288t
diet, 143–4, 144f, 219t, 243, 247, 299–301
dispersal, 337
food plant chemistry, 273
geographical distribution, 68f
gut dimensions, 210t, 213t, 214t, 215
interspecific feeding competition, 294
population density, 145, 288t
predators, 292
ranging behaviour, 145
reproduction, 316t, 318t, 320–1
social organization, 145, 287t, 312f, 315t, 316f, 337, 341, 345f
territoriality, 145, 316t
Trachypithecus vetulus philbrickii, 143
Tragulus, 231, 240, 248
treefall foraging, 97, 104
tree families and colobine populations, 305–7, 306f
tree hyrax, see *Dendrohyrax*
tree kangaroo, 205, 229
troop structure, *see* social behaviour and organization
tubus gastricus, 207f, 208, 230f, 233, 234t, 238
Tylopoda, 205

underbite, 175
urea, 241
ursine black-and-white colobus, *see Colobus polykomos*

vegetation of colobus study sites, 81t
vervet, *see Cercopithecus aethiops*
Victoriapithecus, 12, 13f, 15, 16, 199
vitamin B_{12}, 232
volatile fatty acids, 208, 229, 232–5, 234t, 239, 243

Walsura piscidia, 144, 300, 301t
watercourse species, 350
Wolong, China, 165t
Wuling Mts, China, 62f, 165t

Xylopia quintasii, 280

yield strength, 183, 193
Yunling Mts, China, 165t

Zanzibar red colobus, *see Procolobus badius kirkii*
Zanzibar, Tanzania, 107, 109, 353
Zea mays, 188, 244